Foundations of Modern Physics

In addition to his ground-breaking research, Nobel Laureate Steven Weinberg is known for a series of highly praised texts on various aspects of physics, combining exceptional physical insight with a gift for clear exposition. Describing the foundations of modern physics in their historical context and with some new derivations, Weinberg introduces topics ranging from early applications of atomic theory through thermodynamics, statistical mechanics, transport theory, special relativity, quantum mechanics, nuclear physics, and quantum field theory. This volume provides the basis for advanced undergraduate and graduate physics courses as well as being a handy introduction to aspects of modern physics for working scientists.

STEVEN WEINBERG is a member of the Physics and Astronomy Departments at the University of Texas at Austin. He has been honored with numerous awards, including the Nobel Prize in Physics, the National Medal of Science, the Heinemann Prize in Mathematical Physics, and most recently a Special Breakthrough Prize in Fundamental Physics. He is a member of the US National Academy of Sciences, the UK's Royal Society, and other academies in the US and internationally. The American Philosophical Society awarded him the Benjamin Franklin medal, with a citation that said he is "considered by many to be the preeminent theoretical physicist alive in the world today." He has written several highly regarded books, including *Gravitation and Cosmology*, the three-volume work *The Quantum Theory of Fields*, *Cosmology*, *Lectures on Quantum Mechanics*, and *Lectures on Astrophysics*.

Foundations of Modern Physics

Steven Weinberg
University of Texas, Austin

CAMBRIDGE
UNIVERSITY PRESS

University Printing House, Cambridge CB2 8BS, United Kingdom

One Liberty Plaza, 20th Floor, New York, NY 10006, USA

477 Williamstown Road, Port Melbourne, VIC 3207, Australia

314–321, 3rd Floor, Plot 3, Splendor Forum, Jasola District Centre, New Delhi – 110025, India

79 Anson Road, #06–04/06, Singapore 079906

Cambridge University Press is part of the University of Cambridge.

It furthers the University's mission by disseminating knowledge in the pursuit of education, learning, and research at the highest international levels of excellence.

www.cambridge.org
Information on this title: www.cambridge.org/9781108841764
DOI: 10.1017/9781108894845

First published 2021

Printed in the United Kingdom by TJ Books Limited, Padstow Cornwall

A catalogue record for this publication is available from the British Library.

Library of Congress Cataloging-in-Publication Data
Names: Weinberg, Steven, 1933– author.
Title: Foundations of modern physics / Steven Weinberg, The University of Texas at Austin.
Description: New York : Cambridge University Press, 2021. | Includes bibliographical references and indexes.
Identifiers: LCCN 2020055431 (print) | LCCN 2020055432 (ebook) | ISBN 9781108841764 (hardback) | ISBN 9781108894845 (epub)
Subjects: LCSH: Physics.
Classification: LCC QC21.3 .W345 2021 (print) | LCC QC21.3 (ebook) | DDC 530–dc23
LC record available at https://lccn.loc.gov/2020055431
LC ebook record available at https://lccn.loc.gov/2020055432

ISBN 978-1-108-84176-4 Hardback

For Louise, Elizabeth, and Gabrielle

Contents

Preface

This book grew out of the notes for a course I gave for undergraduate physics students at the University of Texas. In this book I think I go farther forward than is usual in undergraduate courses, giving readers a taste of nuclear physics and quantum field theory. I also go farther back than is usual, starting with the struggle in the nineteenth century to establish the existence and properties of atoms, including the development of thermodynamics that both aided in this struggle and offered an alternative program.

I fear that some readers may want to skim through this early part and hurry on to what they regard as the good stuff, quantum mechanics and relativity. That would be a pity. In my experience physics students who aim at a career in atomic or nuclear or elementary particle physics often manage to get through their formal education without ever becoming familiar with entropy, or equipartition, or viscosity, or diffusion. That was true in my own case. This book, or a course based on it, may provide some students with their last chance to learn about these and other matters needed to understand the macroscopic world.

Readers may find this book unusual also in its strong emphasis on history. I make a point of saying a little about the welter of theoretical guesswork and ill-understood experiments out of which modern physics emerged in the twentieth century. This, it seems to me, is a help in understanding what otherwise may seem an arbitrary set of postulates for relativity and quantum mechanics. It is also a matter of personal taste. Research in physics seems to me to lose some of its excitement if we do not see it as part of a great historical progression. Some valuable historical works are listed in a bibliography, along with collections of original articles that I have found most helpful.

But this is not a work of history. Historians aim at uncovering how the scientists of the past thought about their own problems – for instance, how Einstein in 1905 thought about the measurement of space and time separations in developing the special theory of relativity. For this aim of historical writing it is necessary to go deeply into personal accounts, institutional development, and false starts, and to put aside our knowledge of subsequent progress. I try to be accurate in describing the state of physics in past times, but the aim of this book in discussing the problems of the past is different: it is to make clear how physicists think about these things today.

This book is intended chiefly for physics students who are well into their time as undergraduates, and for working scientists who want a brief introduction to some area of modern physics. I have therefore not hesitated to use calculus and matrix algebra, though not in advanced versions. As required by the subject matter, the mathematical level here slopes upwards through the book. Where possible I have chosen concrete rather than abstract formulations of physical theories. For instance, in Chapter 5, on quantum mechanics, I mostly represent physical states as wave functions, only coming at the end of the chapter to their representation as vectors in Hilbert space. In some sections detailed material that can be skipped without losing the thread of the theory is put into appendices. Two of these appendices present what in my unbiased opinion are improved derivations of important results: the appendix to Section 2.6 gives a revised version of Einstein's derivation of his formula for the diffusion constant in Brownian motion, and the appendix to Section 6.4 presents a revision of Fermi's calculation of the rate of alpha decay.

In my experience, with some judicious pruning, the material of the book up to about the middle of Chapter 5 can be covered in a one-term undergraduate course. But I think that to go over the whole book would take a full two-term academic year.

This book treats such a broad range of topics that it is impossible to go very far into any of them. Certainly its treatment of quantum mechanics, statistical mechanics, transport theory, nuclear physics, and quantum field theory is no substitute for graduate-level courses on these topics, any one of which would occupy at least a whole year. This book presents what I think, in an ideal world, the ambitious physics student would already know when he or she enters graduate school. At least, it is what I wish that I had known when I entered graduate school.

In any case, I hope that the student or reader may be sufficiently interested in what I do discuss that they will want to go into these topics in greater detail in more specialized books or courses, and that they will find in this book a good preparation for such further studies.

I am grateful to many students and colleagues for pointing out errors in the lecture notes on which this book is based and for the expert and friendly assistance I have received from Simon Capelin and Vince Higgs, the editors at Cambridge University Press who guided the publication of this book.

STEVEN WEINBERG

1
Early Atomic Theory

It is an old idea that matter consists of atoms, tiny indivisible particles moving in empty space. This theory can be traced to Democritus, working in the Greek city of Abdera, on the north shore of the Aegean sea. In the late 400s BC Democritus proclaimed that "atoms and void alone exist in reality." He offered neither evidence for this hypothesis nor calculations on which to base predictions that could confirm it. Nevertheless, this idea was tremendously influential, if only as an example of how it might be possible to account for natural phenomena without invoking the gods. Atoms were brought into the materialistic philosophy of Epicurus of Samos, who a little after 300 BC founded one of the four great schools of Athens, the Garden. In turn, the idea of atoms and the philosophy of Epicurus were invoked in the poem *On the Nature of Things* by the Roman Lucretius. After this poem was rediscovered in 1417 it influenced Machiavelli, More, Shakespeare, Montaigne, and Newton, among others. Newton in his *Opticks* speculated that the properties of matter arise from the clustering of atoms into larger particles, which themselves cluster into larger particles, and so on. As we will see, Newton made a stab at an atomic theory of air pressure, but without significant success.

The serious scientific application of the atomic theory began in the eighteenth century, with calculations of the properties of gases, which had been studied experimentally since the century before. This is the topic with which we begin this chapter. Applications to chemistry and electrolysis followed in the nineteenth century and will be considered in subsequent sections. The final section of this chapter describes how the nature of atoms began to be clarified with the discovery of the electron. In the following chapter we will see how it became possible to estimate the atoms' masses and sizes.[1]

[1] Further historical details about some of these matters can be found in Weinberg, *The Discovery of Subatomic Particles*, listed in the bibliography.

1.1 Gas Properties

Experimental Relations

The upsurge of enthusiasm for experiment in the seventeenth century was largely concentrated on the properties of air. The execution and reports of these experiments did not depend on hypotheses regarding atoms, but we need to recall them here because their results provided the background for later theories of gas properties that did rely on assumptions about atoms.

It had been thought by Aristotle and his followers that the suction observed in pumps and bellows arises from nature's abhorrence of a vacuum. This notion was challenged in the 1640s by the invention of the barometer by the Florentine polymath Evangelista Torricelli (1608–1647). If nature abhors a vacuum, then when a long glass tube with one end closed is filled with mercury and set upright with the closed end on top, why does the mercury flow out of the bottom until the column is only 760 mm high, with empty space appearing above the mercury? Is there a limit to how much nature abhors a vacuum? Torricelli argued that the mercury is held up instead by the pressure of the air acting on the open end of the glass tube (or on the surface of a bath of mercury in which the open end of the tube is immersed), which is just sufficient to support a column of mercury 760 mm high. If so, then it should be possible to measure variations in air pressure using a column of mercury in a vertical glass tube, a device that we know as a barometer. Such measurements were made from 1648 to 1651 by Blaise Pascal (1623–1662), who found that the height of mercury in a barometer is decreased by moving to the top of a mountain, where less air extends above the barometer.

The quantitative properties of air pressure soon began to be studied experimentally, before there was any correct theoretical understanding of gas properties. In 1662, in the second edition of his book *New Experiments Physico-Mechanical Concerning the Spring of the Air and its Effects*, the Anglo-Irish aristocrat Robert Boyle (1627–1691) described experiments relating the pressure (the "spring of the air") and volume of a fixed mass of air. He studied a sample of air enclosed at the end of a glass tube by a column of mercury in the tube. The air was compressed at constant temperature by pushing on the mercury's surface, revealing what came to be known as *Boyle's law*, that for constant temperature the volume of a gas of fixed mass and composition is inversely proportional to the pressure, now defined by Boyle as the force per area exerted on the gas.

Temperature Scales

A word must be said about the phrase "at constant temperature." Boyle lived before the establishment of our modern Fahrenheit and Celsius scales, whose forerunners go back respectively to 1724 and 1742. But, although in Boyle's time no meaningful numerical value could be given to the temperature of any given body, it was nevertheless possible to speak with precision of two bodies being at the *same* temperature: they are at the same temperature if when put in contact neither body is felt to grow appreciably hotter or colder. Boyle's glass tube could be kept at constant temperature by immersing it in a large bath, say of water from melting ice. Later the Fahrenheit temperature scale was established by defining the temperature of melting ice as 32 °F and the temperature of boiling water at mean atmospheric pressure as 212 °F, and defining a 1 °F increase of temperature by etching 212 − 32 equal divisions between 32 and 212 on the glass tube of a mercury thermometer. Likewise, in the Celsius scale, the temperatures of melting ice and boiling water are 0 °C and 100 °C, and 1 °C is the temperature difference required to increase the volume of mercury in a thermometer by 1% of the volume change in heating from melting ice to boiling water. As we will see in the next chapter, there is a more sophisticated universal definition of temperature, to which scales based on mercury thermometers provide only a good approximation.

After the temperature scale was established it became possible to carry out a quantitative study of the relation between volume and temperature, with pressure and mass kept fixed by enclosing the air in a vessel with flexible walls, which expand or contract to keep the pressure inside equal to the air pressure outside. This relation was announced in an 1802 lecture by Joseph Louis Gay-Lussac (1775–1850), who attributed it to unpublished work in the 1780s by Jacques Charles (1746–1823). The relation, subsequently known as *Charles' Law*, is that at constant pressure and mass the volume of gas is proportional to $\mathcal{T} - \mathcal{T}_0$, where $\mathcal{T}$ is the temperature measured for instance with a mercury thermometer and $\mathcal{T}_0$ is a constant whose numerical value naturally depends on the units used for temperature: $\mathcal{T}_0 = -459.67$ °F $= -273.15$ °C. Thus $\mathcal{T}_0$ is absolute zero, the minimum possible temperature, at which the gas volume vanishes. Using Celsius units for temperature differences, the absolute temperature $T \equiv \mathcal{T} - \mathcal{T}_0$ is known today as the temperature in degrees Kelvin, denoted K.

Theoretical Explanations

In Proposition 23 of his great book, the *Principia*, Isaac Newton (1643–1727) made an attempt to account for Boyle's law by considering air to consist of particles repelling each other at a distance. Using little more than dimensional analysis, he showed that the pressure p of a fixed mass of air is inversely proportional to the volume V if the repulsive force between particles separated by a distance r falls off as $1/r$. But as he pointed out, if the repulsive force goes as $1/r^2$, then $p \propto V^{-4/3}$. He did not claim to offer any reason why the repulsive force should go as $1/r$ and, as we shall see, it is not forces that go as $1/r$ but rather forces of very short range that act only in collisions that mostly account for the properties of gases.

It was the Swiss mathematical physicist Daniel Bernoulli (1700–1782) who made the first attempt to understand the properties of gases theoretically, on the assumption that a gas consists of many tiny particles moving freely except in very brief collisions. In 1738, in the chapter, "On the Properties and Motions of Elastic Fluids, Especially Air" of his book *Hydrodynamics*, he argued that in a gas (then called an "elastic fluid") with n particles per unit volume moving with a velocity v that is the same (because of collisions) in all directions, the pressure is proportional to n and to v^2, because the number of particles that hit any given area of the wall in a given time is proportional to the number in any given volume, to the rate at which they hit the wall, which is proportional to v, and to the force that each particle exerts on the wall, which is also proportional to v. For a fixed mass of gas n is inversely proportional to the volume V, so pV is proportional to v^2. If (as Bernoulli thought) v^2 depends only on the temperature, this explains Boyle's law. If v^2 is proportional to the absolute temperature, it also gives Charles' law.

Bernoulli did not give much in the way of mathematical details, and did not try to say to what else the pressure might be proportional besides nv^2, a matter crucial for the history of chemistry. These details were provided by Rudolf Clausius (1822–1888) in 1857, in an article entitled "The Nature of the Motion which We Call Heat." Below is a more-or-less faithful description of Clausius' derivation, in a somewhat different notation.

Suppose a particle hits the wall of a vessel and remains in contact with it for a small time t, during which it exerts a force with component F along the inward normal to the wall. Its momentum in the direction of the inward normal to the wall will decrease by an amount Ft, so if the component of the velocity of the particle before it strikes the wall is $v_\perp > 0$, and it bounces back elastically with normal velocity component $-v_\perp$, the change in the inward normal component of momentum is $-2mv_\perp$, where m is the particle mass, so

$$F = 2mv_\perp/t \ .$$

Now, suppose that this goes on with many particles hitting the wall over a time interval $T \gg t$, all particles with the same velocity vector $\mathbf{v}$. The number $\mathcal{N}$ of particles that will hit an area A of the wall in this time is the number of particles in a cylinder with base A and height $v_\perp T$, or

$$\mathcal{N} = nAv_\perp T \ ,$$

where n is the number density, the number of particles per volume. Each of these particles is in contact with the wall for a fraction t/T of the time T, so the total force exerted on the wall is

$$F\mathcal{N}(t/T) = 2mv_\perp/t \times nAv_\perp T \times (t/T) = 2nmv_\perp^2 A \ .$$

We see that all dependence on the times t and T cancels. The pressure p is defined as the force per area, so this gives the relation

$$p = 2nmv_\perp^2 \ . \tag{1.1.1}$$

This is for the unphysical case in which every particle has the same value of $v_\perp$, positive in the sense that the particles are assumed to be going toward the wall. In the real world, different particles will be moving with different speeds in different directions, and Eq. (1.1.1) should be replaced with

$$p = 2nm \times \frac{1}{2}\langle v_\perp^2 \rangle = nm\langle v_\perp^2 \rangle \ , \tag{1.1.2}$$

the brackets indicating an average over all gas particles, with the factor $1/2$ inserted in the first expression because only 50% of these particles will be going toward any given wall area.

To express $\langle v_\perp^2 \rangle$ in terms of the root mean square velocity, Clausius assumed without proof that "on the average each direction [of the particle velocities] is equally represented." In this case, the average square of each component of velocity equals $\langle v_\perp^2 \rangle$, and the average of the squared velocity vector is then

$$\langle \mathbf{v}^2 \rangle = \langle v_1^2 \rangle + \langle v_2^2 \rangle + \langle v_3^2 \rangle = 3\langle v_\perp^2 \rangle$$

and therefore Eq. (1.1.2) reads

$$p = nm\langle \mathbf{v}^2 \rangle /3 \ . \tag{1.1.3}$$

This is essentially the result $p \propto n\langle \mathbf{v}^2 \rangle$ of Bernoulli, except that, with the factor $m/3$, Eq. (1.1.3) is now an equality, not just a statement of proportionality. For a fixed mass M of gas occupying a volume V, the number density is $n = M/mV$, so Clausius could use Boyle's law (which he called Mariotte's law), which states that pV is constant for fixed temperature, to conclude that for a given gas $\langle \mathbf{v}^2 \rangle$ depends only on the temperature. Further, as Clausius remarked, Eq. (1.1.3) together with Charles' law (which Clausius called the law of Gay-Lussac) indicates that $\langle \mathbf{v}^2 \rangle$ is proportional to the absolute temperature T. If we like, we can adopt a modern notation and write the constant of proportionality as $3k/m$, so that

$$m\langle \mathbf{v}^2 \rangle /3 = kT \ , \tag{1.1.4}$$

and therefore Eq. (1.1.3) reads

$$p = nkT \ , \tag{1.1.5}$$

where k is a constant, in the sense of being independent of p, n, and T. But the choice of notation does not tell us whether k varies from one type of gas to another or whether it depends on the molecular mass m. Clausius could not answer this question, and did not offer any theoretical justification for Boyle's law or Charles' law. Clausius deserves to be called the founder of thermodynamics, discussed in Sections 2.2 and 2.3, but these are not questions that can be answered by thermodynamics alone. As we will see in the following section, experiments in the chemistry of gases indicated that k is the same for all gases, a universal constant now known as Boltzmann's constant, but the theoretical explanation for this and for Boyle's law and Charles' law had to wait for the development of kinetic theory and statistical mechanics, the subject of Section 2.4.

As indicated by the title of his article, "The Nature of the Motion which We Call Heat," Clausius was concerned to show that, at least in gases, the phenomenon of heat is explained by the motion of the particles of which gases are composed. He defended this view by using his theory to calculate the specific heat of gases, a topic to be considered in the next chapter.

1.2 Chemistry

Elements

The idea that all matter is composed of a limited number of elements goes back to the earliest speculations about the nature of matter. At first, in the century before Socrates, it was supposed that there is just one element: water (Thales) or air (Anaximenes) or fire (Heraclitus) or earth (perhaps Xenophanes). The idea of four elements was proposed around 450 BC by Empedocles of Acragas (modern Agrigento). In *On Nature* he identified the elements as "fire and water and earth and the endless height of air." Classical Chinese sources list five elements: water, fire, earth, wood, and metal.

Like the theory of atoms, these early proposals of elements did not come accompanied with any evidence that these really are elements, or any suggestion how such evidence might be gained. Plato in *Timaeus* even doubled down and stated that the difference between one element and another arises from the shapes of the atoms of which the elements are composed: earth atoms are tiny cubes, while the atoms of fire, air, and water are other regular polyhedra – solids bounded respectively by 4, 8, or 20 identical regular polygons, with every edge and every vertex of each solid the same as every other edge or vertex of that solid.

By the end of the middle ages this list of elements had come to seem implausible. It is difficult to identify any particular sample of dirt as the element earth, and fire seems more like a process than a substance. Alchemists narrowed the list of elements to just three: mercury, sulfur, and salt.

Modern chemistry began around the end of the eighteenth century, with careful experiments by Joseph Priestley (1733–1804), Henry Cavendish (1743–1810), Antoine Lavoisier (1743–1794), and others. By 1787 Lavoisier had worked out a list of 55 elements. In place of air there were several gases: hydrogen, oxygen, and nitrogen; air was identified as a mixture of nitrogen and oxygen. There were other non-metals on the list of elements: sulfur, carbon, and phosphorus, and a number of common metals: iron, copper, tin, lead, silver, gold, mercury. Lavoisier also listed as elements some chemicals that we now know are tightly bound compounds: lime, soda, and potash. And the list also included heat and light, which of course are not substances at all.

Law of Combining Weights

Chemistry was first used to provide quantitative information about atoms by John Dalton (1766–1844), the son of a poor weaver. His laboratory notebooks from 1802 to 1804 describe careful measurements of the weights of elements combining in compounds. He discovered that these weights are always in fixed ratios. For instance, he found that when hydrogen burns in oxygen, 1 gram of hydrogen combines with 5.5 grams of oxygen, giving 6.5 grams of water, with nothing left over. Under the assumption that one particle of water consists of one atom of hydrogen and one atom of oxygen, one oxygen atom must weigh 5.5 times as much as one hydrogen atom.

As we will see, water was soon discovered to be H_2O: two atoms of hydrogen to each atom of oxygen. If Dalton had known this, he would have concluded that an oxygen atom weighs 5.5 times as much as *two* hydrogen atoms, i.e., 11 times the weight of one hydrogen atom. Of course, more accurate measurements later revealed that 1 gram of hydrogen combines with about 8 grams of oxygen, so one oxygen atom weighs eight times the weight of two hydrogen atoms, or 16 times as much as one hydrogen atom. Atomic weights soon became defined as the weights of atoms relative to the weight of one hydrogen atom, so the atomic weight of oxygen is 16. (This is only approximate. Today the atomic weight of the atoms of the most common isotope of carbon is defined to be precisely 12; with this definition, the atomic weights of the most common isotopes of hydrogen and oxygen are measured to be 1.007825 and 15.99491.)

The following table compares Dalton's assumed formulas for a few common compounds with the correct formulas:

Compound	*Dalton formula*	*True formula*
Water	HO	H_2O
Carbon dioxide	CO_2	CO_2
Ammonia	NH	NH_3
Sulfuric acid	SO_2	H_2SO_4

Here is a list of the approximate true atomic weights for a few elements, the weights deduced by Dalton, and (in the column marked with an asterisk) the weights Dalton would have calculated if he had known the true chemical formulas.

Element	*True*	*Dalton*	*Dalton**
H	1	1	1
C	12	4.3	8.6
N	14	4.2	12.6
O	16	5.5	11
S	32	14.4	57.6

To make progress in measuring atomic weights, it was evidently necessary to find some way of working out the correct formulas for various chemical compounds. This was provided by the study of chemical reactions in gases.

Law of Combining Volumes

On December 31, 1808, Gay-Lussac read a paper to the Societe Philomathique in Paris, in which he announced his observation that gases at the same temperature and pressure always combine in definite proportions of *volumes*. For instance, two liters of hydrogen combine with one liter of oxygen to give water vapor, with no hydrogen or oxygen left over. Likewise, one liter of nitrogen combines with three liters of hydrogen to give ammonia gas, with nothing left over. And so on.

The correct interpretation of this experimental result was given in 1811 by Count Amadeo Avogadro (1776–1856) in Turin. Avogadro's principle states that equal volumes of gases at the same temperature and pressure always contain equal numbers of the gas particles, which Avogadro called "molecules," particles that may consist of single atoms or of several atoms of the same or different elements joined together. The observation that water vapor is formed from a volume of oxygen combined with a volume of hydrogen twice as large shows, according to Avogadro's principle, that molecules of water are formed from twice as many molecules of hydrogen as molecules of oxygen, which is not what Dalton had assumed.

There was a further surprise in the data. Two liters of hydrogen combined with one liter of oxygen give not one but *two* liters of water vapor. This is not what one would expect if oxygen and hydrogen molecules consist of single atoms and water molecules consist of two atoms of hydrogen and one atom of oxygen. In that case two liters of hydrogen plus one liter of oxygen would produce *one* liter of water vapor. Avogadro could conclude that if, as seemed plausible, molecules of water contain two atoms of hydrogen and one atom of oxygen, the molecules of oxygen and hydrogen must each contain two atoms. That is, taking water molecules as H_2O, the reaction for producing molecules of water is

$$2H_2 + O_2 \rightarrow 2H_2O \ .$$

The use of Avogadro's principle rapidly provided the correct formulas for gases such as CO_2, NH_3, NO, and so on. Knowing these formulas and measuring the *weights* of gases participating in various reactions, it was possible to correct Dalton's atomic weights and calculate more reliable values for the atomic weights of the atoms in gas molecules, relative to any one of them. Taking the atomic weight of hydrogen as unity, this gave atomic weights close to 12 for carbon, 14 for nitrogen, 16 for oxygen, 32 for sulfur, and so on. Then, knowing these atomic weights, it became possible to find atomic weights for many other elements, not just those commonly found in gases, by measuring the weights of elements combining in various chemical reactions.

The Gas Constant

As we saw in the previous section, in 1857 Clausius had shown that in a gas consisting of n particles of mass m per volume with mean square velocity $\langle \mathbf{v}^2 \rangle$, the pressure is $p = nm\langle \mathbf{v}^2 \rangle/3$. Using Charles' law, he concluded that $\langle \mathbf{v}^2 \rangle$ is proportional to absolute temperature. Writing this relation as $m\langle \mathbf{v}^2 \rangle/3 = kT$ with k some constant gives Eq. (1.1.5), $p = nkT$. But this in itself does not tell us how k varies from one gas to another. This is answered by Avogadro's principle. With N particles in a volume V, the number density is $n = N/V$, so Eq. (1.1.5) can be written

$$pV = NkT \ . \tag{1.2.1}$$

If as stated by Avogadro the number of molecules in a gas with a given pressure, volume, and temperature is the same for any gas, then $k = pV/NT$ must be the same for any gas. Clausius did not draw this conclusion, perhaps because there was then no known theoretical basis for Avogadro's principle. The universality of the constant k, and hence Avogadro's principle, were explained later by kinetic theory, to be covered in the next chapter. The constant k came to be called *Boltzmann's constant*, after Ludwig Boltzmann, who as we shall see was one of the chief founders of kinetic theory.

The *molecular weight* μ of any compound is defined as the sum of the atomic weights of the atoms in a single molecule. The actual mass m of a molecule is its molecular weight times the mass m_1 of a hypothetical atom with atomic weight unity:

$$m = \mu m_1 \ . \tag{1.2.2}$$

In the modern system of atomic weights, with the atomic weight of the most common isotope of carbon defined as precisely 12, $m_1 = 1.660539 \times 10^{-24}$ g, which of course was not known in Avogadro's time. A mass M contains $N = M/m = M/m_1\mu$ molecules, so the ideal gas law (1.2.1) can be written

$$pV = MkT/m_1\mu = (M/\mu)RT \tag{1.2.3}$$

where R is the gas constant

$$R = k/m_1 \, . \tag{1.2.4}$$

Physicists in the early nineteenth century could use Eq. (1.2.3) to measure R, and they found a value close to the modern value $R = 8.314$ J/K. This would have allowed a determination of m_1 and hence of the masses of all atoms of known atomic weight if k were known, but k did not become known until the developments described in Section 2.6.

Avogadro's Number

Incidentally, a mole of any element or compound of molecular weight μ is defined as μ grams, so in Eq. (1.2.3) the ratio M/μ expressed in grams equals the number of moles of gas. Since $N = M/m_1\mu$, one mole contains a number of molecules equal to $1/m_1$ with m_1 given in grams. This is known as *Avogadro's number*. But of course Avogadro did not know Avogadro's number. It is now known to be 6.02214×10^{23} molecules per mole, corresponding to unit molecular weight $m_1 = 1.66054 \times 10^{-24}$ grams. The measurement of Avogadro's number was widely recognized in the late nineteenth century as one of the great challenges facing physics.

1.3 Electrolysis

Early Electricity

Electricity was known in the ancient world, as what we now call static electricity. Amber rubbed with fur was seen to attract or repel small bits of light material. Plato in *Timaeus* mentions "marvels concerning the attraction of amber." (This is where the word electricity comes from; the Greek word for amber is "elektron.")

Electricity began to be studied scientifically in the eighteenth century. Two kinds of electricity were distinguished: resinous electricity is left on an amber rod when rubbed with fur, while vitreous electricity is left on a glass rod when rubbed with silk. Unlike charges were found to attract each other, while like charges repel each other. Benjamin Franklin (1706–1790) gave our modern terms positive and negative to vitreous and resinous electricity, respectively.

In 1785 Charles-Augustin de Coulomb (1736–1806) reported that the force F between two bodies carrying charges q_1 and q_2 separated by a distance r is

$$F = \frac{k_e q_1 q_2}{r^2} \tag{1.3.1}$$

where k_e is a universal constant. For like and unlike charges the product $q_1 q_2$ is positive or negative, respectively, indicating a repulsive or attractive force. Coulomb had no way of actually measuring these charges, but he could reduce the charge on a body by a factor 2 by touching it to an uncharged body of the same material and size, and observe that this reduces the force between it and any other charged body by the same factor 2. The introduction of our modern units of electric charge had to wait until the quantitative study of magnetism.

Early Magnetism

Magnetism too was known in the ancient world, as what we now call permanent magnetism. The Greeks knew of naturally occurring lodestones that could attract or repel small bits of iron. Plato's *Timaeus* refers to lodestones as "Heraclean stones." (Our word magnet comes from the city Magnesia in Asia Minor, near where lodestones were commonly found.)

Very early the Chinese also discovered the lodestone and used it as a magnetic compass (a "south-seeking stone") for purposes of geomancy and navigation. Each lodestone has a south-seeking pole at one end, attracted to a point near the South Pole of the Earth, and a north-seeking pole at the other end, attracted to a point near the Earth's North Pole. Magnetism was first studied scientifically by William Gilbert (1544–1603), court physician to Elizabeth I. It was observed that the south-seeking poles of different lodestones repel each other, and likewise for the north-seeking poles, while the south-seeking pole of one lodestone attracts the north-seeking pole of another lodestone. Gilbert concluded that one pole of a lodestone is pulled toward the north and the other toward the south because the Earth itself is a magnet, with what in a lodestone would be its south-seeking and north-seeking poles respectively near the Earth's North Pole and South Pole.

Electromagnetism

It began to be possible to explore the relations between electricity and magnetism quantitatively with the invention in 1809 of electric batteries by Count Alessandro Volta (1745–1827). These were stacks of disks of two different metals separated by cardboard disks soaked in salt water. Such batteries drive steady currents of electricity through wires attached to the ends of the stacks, with positive and negative terminals identified respectively as the ends of the stacks from which and towards which electric current flows.

In July 1820 Hans Christian Oersted (1777–1851) in Copenhagen noticed that turning on an electric current deflected a nearby compass needle, and concluded that electric currents exert force on magnets. Conversely, he found also that magnets exert force on wires carrying electric currents.

These discoveries were carried further in Paris a few months later by Andrè-Marie Ampère (1775–1836), who found that wires carrying electric current exert force on each other. For two parallel wires of length L carrying electric currents (charge per second) I_1 and I_2, and separated by a distance $r \ll L$, the force is

$$F = \frac{k_m I_1 I_2 L}{r}\,, \tag{1.3.2}$$

where k_m is another universal constant. The force is repulsive if the currents are in the same direction; attractive if in opposite directions. One ampere is defined so that $F = 10^{-7} \times L/r$ newtons if $I_1 = I_2 = 1$ ampere. (That is, $k_m \equiv 10^{-7}$ N/ampere^2.) The electromagnetic unit of electric charge, the coulomb, is defined as the electric charge carried in one second by a current of one ampere. A modern ammeter measures electric currents by observing the magnetic force produced by current flowing through a wire loop.

The connection between electricity and magnetism was strengthened in 1831 by Michael Faraday (1791–1867), at the Royal Institution in London. He discovered that changing magnetic fields generate electric forces that can drive currents in conducting wires. This is the principle underlying the generation of electric currents today. Electricity began soon after to have important practical applications, with the invention in 1831 of the electric telegraph by the American painter Samuel F. B. Morse (1791–1872).

Finally, in the 1870s, the great Scottish physicist James Clerk Maxwell (1831–1879) showed that the consistency of the equation for the generation of magnetic fields by electric currents required that magnetic fields are also generated by changing electric fields. In particular, while oscillating magnetic fields produce oscillating electric fields, also oscillating electric fields produce oscillating magnetic fields, so a self-sustaining oscillation in both electric and magnetic fields can propagate in apparently empty space. Maxwell calculated the speed of its propagation and found it to equal $\sqrt{2k_e/k_m}$,[2] numerically about equal to the measured speed of light, suggesting strongly that light is such a self-sustaining oscillation in electric and magnetic fields. We will see more of Maxwell's equations in subsequent chapters, especially in Chapters 4 and 5.

[2] This quantity is independent of the units used for electric charge as long as the currents appearing in Eq. (1.3.2) are defined as the rates of flow of charge in the same units as used in Eq. (1.3.1). It is obviously also independent of the units used for force, as long as the same force units are used in Eqs. (1.3.1) and (1.3.2).

Discovery of Electrolysis

Electrolysis was discovered in 1800 by the chemist William Nicholson (1753–1815) and the surgeon Anthony Carlisle (1768–1840). They found that bubbles of hydrogen and oxygen would be produced where wires attached respectively to the negative and positive terminals of a Volta-style battery were inserted in water. Sir Humphrey Davy (1778–1829), Faraday's boss at the Royal Institution, carried out extensive experiments on the electrolysis of molten salts, finding for instance that, in the electrolysis of molten table salt, sodium, a previously unknown metal, was produced at the wire attached to the negative terminal of the battery and a greenish gas, chlorine, was produced at the wire attached to the other, positive, terminal. Davy's electrolysis experiments added several metals aside from sodium to Lavoisier's list of elements, including aluminum, potassium, calcium, and magnesium.

A theory of electrolysis was worked out by Faraday. In modern terms, a small fraction (1.8×10^{-9} at room temperature) of water molecules are normally dissociated into positive hydrogen ions (H^+), which are attracted to the wire attached to the negative terminal of a battery, and negative hydroxyl ions (OH^-), which are attracted to the wire attached to the positive terminal. At the wire attached to the negative terminal, two H^+ ions combine with two units of negative charge from the battery to form a neutral H_2 molecule. At the wire attached to the positive terminal, four OH^- ions give one O_2 molecule plus two H_2O molecules plus four units of negative charge, which flow through the battery to the negative terminal.[3]

Likewise, a small fraction of molten table salt (NaCl) molecules are normally dissociated into Na^+ ions and Cl^- ions. At the wire attached to the negative terminal of a battery, one Na^+ ion plus one unit of negative charge gives one atom of metallic sodium (Na); at the wire attached to the positive terminal, two Cl^- ions give one chlorine (Cl_2) molecule and two units of negative charge, which flow through the battery to the negative terminal.

In Faraday's theory, it takes one unit of electric charge to convert a singly charged ion such as H^+ or Cl^- to a neutral atom or molecule, so since molecules of molecular weight μ have mass μm_1, it takes $M/m_1\mu$ units of electric charge to convert a mass M of singly charged ions to a mass M of neutral atoms or molecules of molecular weight μ. Experiment showed that it takes about 96 500 coulombs (e.g., one ampere for about 96 500 seconds) to convert μ grams (that is, one mole) of singly charged ions to neutral atoms or molecules. (This is called a faraday; the modern value is 96 486.3 coulombs/mole.) Hence

[3] We now know that it is negative charge, i.e., electrons, that flows through a battery. As far as Faraday knew, it was equally possible that positive charges flow through a battery, in which case at the wire attached to the negative terminal two H^+ ions would give an H_2 molecule plus two units of positive charge, which would flow though the battery to the wire attached to the positive terminal, where four OH^- ions plus four units of positive charge would give an O_2 molecule and two H_2O molecules.

Faraday knew that $e/m_1 \simeq 96\,500$ coulombs/gram, where e is the unit of electric charge, which was called an "electrine" in 1874 by the Irish physicist George Johnstone Stoney (1826–1911). Having measured the faraday, if physicists knew the value of e then they would know m_1, but they didn't have this information until later. Also, no one then knew that e is the charge of an actual particle.

1.4 The Electron

As sometimes happens, in 1858 a new path in fundamental physics was opened with the invention of a practical device, in this case an improved air pump. In his pump the Bonn craftsman Heinrich Johann Geissler (1814–1879) used a column of mercury as a piston, in this way greatly reducing the leakage of air through the piston that had troubled all previous air pumps. With his pump Geissler was able to reduce the pressure in a closed glass tube to about a ten-thousandth of the typical air pressure on the Earth's surface.

With such a near vacuum in a glass tube, electric currents could travel without wires through the tube. It was discovered that an electric current would flow from a cathode, a metal plate attached to the negative terminal of a powerful electric battery, fly through a hole in an anode, another metal plate attached to the positive pole of the battery, and light up a spot on the far wall of the tube. Adding small amounts of various gases to the interior of the tube caused these cathode rays to light up, with orange or pink or blue-green light emitted along the path of the ray, when neon, helium, or mercury vapor was added. Using Geissler's pumps, Julius Plücker (1801–1868) in 1858–1859 found that cathode rays could be deflected by magnetic fields, thus moving the spot of light where the ray hits the glass at the tube end.

In 1897 Joseph John Thomson (1856–1940), the successor to Maxwell as Cavendish Professor at Cambridge, began a series of measurements of the deflection of cathode rays. In his experiments, after the ray particles pass through the anode they feel an electric or magnetic force F exerted at a right angle to their direction of motion for a distance d along the ray. They then drift in a force-free region for a distance $D \gg d$ until they hit the end of the tube. If a ray particle has velocity v along the direction of the ray, it feels the electric or magnetic force for a time d/v and then drifts for a longer time D/v. A force F normal to the ray gives ray particles of mass m a component of velocity perpendicular to the ray that is equal to the acceleration F/m times the time d/v, so by the time they hit the end of the tube they have been displaced by an amount

$$\text{displacement} = (F/m) \times (d/v) \times (D/v) = \frac{FdD}{mv^2} .$$

The forces exerted on a charge e by an electric field E or a magnetic field B at right angles to the ray are

$$F_{\text{elec}} = eE\,, \quad F_{\text{mag}} = evB$$

so

$$\text{electric displacement} = \frac{eEdD}{mv^2}\,,$$

$$\text{magnetic displacement} = \frac{eBdD}{mv}\,.$$

Thomson wanted to measure e/m. He knew D, d, E, and B, but not v. He could eliminate v from these equations if he could measure both the electric and magnetic displacements, but the electric displacement was difficult to measure. A strong electric field tends to ionize any residual air in the tube, with positive and negative ions pulled to the negatively and positively charged plates that produce the electric fields, neutralizing their charges. Finally Thomson succeeded in measuring the electric as well as the magnetic deflection by using a cathode ray tube with very low air pressure. (Both the electric and magnetic displacements were only a few inches.) This gave results for the ratio of charge to mass ranging from 6×10^7 to 10^8 coulombs per gram.

Thomson compared this with the result that Faraday had found in measurements of electrolysis, that $e/m_1 \approx 10^5$ coulombs per gram, where e is the electric charge of a singly ionized atom or molecule (such as a sodium ion in the electrolysis of NaCl) and m_1 is the mass of a hypothetical atom of atomic weight unity, close to the mass of the hydrogen atom. He reasoned that if the particles in his cathode rays are the same as those transferred in electrolysis, then their charge must be the same as e, so their mass must be about $10^{-3}m_1$. Thomson concluded that since the cathode ray particles are so much lighter than ions or atoms, they must be the basic constituents of ions and atoms.

Thomson had still not measured e or m. He had not even shown that cathode rays are streams of particles; they might be streams of electrically charged fluid, with any volume of fluid having a ratio of charge to mass equal to his measured e/m. Nevertheless, in the following decade it became widely accepted that Thomson had indeed discovered a particle present in atoms, and the particle came to be called the electron.

2
Thermodynamics and Kinetic Theory

The successful uses of atomic theory described in the previous chapter did not settle the existence of atoms in all scientists' minds. This was in part because of the appearance in the first half of the nineteenth century of an attractive competitor, the physical theory of thermodynamics. As we shall see in the first three sections of this chapter, with thermodynamics one may derive powerful results of great generality without ever committing oneself to the existence of atoms or molecules. But thermodynamics could not do everything. Section 2.4 will describe the advent of kinetic theory, which is based on the assumption that matter consists of very large numbers of particles, and its generalization to statistical mechanics. From these thermodynamics could be derived, and together with the atomic hypothesis it yielded results far more powerful than could be obtained from thermodynamics alone. Even so, it was not until the appearance of direct evidence for the graininess of matter, described in Section 2.5, that the existence of atoms became almost universally accepted.

2.1 Heat and Energy

The first step in the development of thermodynamics was the recognition that heat is a form of energy. Though so familiar to us today, this was far from obvious to the physicists and chemists of the early nineteenth century. Until the 1840s heat was widely regarded as a fluid, named *caloric* by Lavoisier. Caloric theory was used to calculate the speed of sound by Pierre-Simon Laplace (1749–1827) in 1816, the conduction of heat by Joseph Fourier (1768–1830) in 1807 and 1822, and the efficiency of steam engines by Sadi Carnot (1796–1832) in 1824, whose work as we will see in the next section became a foundation of thermodynamics. Adding to the confusion, other scientists considered heat as some sort of wave. This reflected uncertainty regarding the nature of what is now called infrared radiation, discovered by William Herschel (1738–1822) in 1800.

Heat as Energy

In 1798 Benjamin Thompson (1753–1814), an American expatriate in England, offered evidence against the idea that heat is a fluid. (Thompson is also known as Count Rumford, a title he was given when he later served as military adviser in Austria.) It was well known that boring a cannon produces heat, which might be supposed to be due to the liberation of caloric from the iron, but Rumford observed that if the heat is carried away by immersing the cannon in running water while it is being bored there is no limit to the heat that can be produced.

The first measurement of the energy in heat was provided in the mid-1840s by James Prescott Joule (1818–1889). In his apparatus a falling weight turned paddles in a tank of water, heating the water. The gravitational force on a mass m kilograms is m times the acceleration of gravity, 9.8 meters/sec^2 or 9.8 newtons per kilogram. Work is force times distance, so dropping one kilogram a distance of one meter gave it an energy equal to 9.8 newton meters, now also known as 9.8 joules. Joule found that the paddles driven by this dropping weight would raise the temperature of 100 grams of water by 0.023 °C, so the paddles produced heat equal to 0.023×100 calories, the calorie being defined as the heat required to raise the temperature of one gram of water by one degree Celsius. Hence Joule could conclude that 9.8 joules is equivalent to 2.3 calories, so one calorie is equivalent to $9.8/2.3 = 4.3$ joules. The modern value is 4.184 joules.

In 1847 the Prussian physician and physicist Hermann von Helmholtz (1821–1894) put forward the idea of the universal conservation of energy, whether in the form of kinetic or potential or chemical energy or heat. But what sort of energy is heat? For some nineteenth century physicists the question was irrelevant. They developed the science of heat known as thermodynamics, which did not depend on any detailed model of heat energy. But there was one context in which the nature of heat energy seemed evident. In his great 1857 paper, *The Nature of the Motion which We Call Heat*, Clausius found that at least part of the heat energy of gases is the kinetic energy of their molecules.

Kinetic Energy

The concept of kinetic energy was long familiar. If a steady force F is exerted on a particle of mass m, it produces an acceleration F/m, so after a time t the velocity of the body is $v = Ft/m$. The distance traveled in this time is t times the average velocity $v/2$, and the work done on the particle is the force times this distance:

$$F \times t \times Ft/2m = F^2t^2/2m = mv^2/2 .$$

Instead of this work going into heating a tub of water, as in the experiment of Joule, it goes into giving the particle an energy $mv^2/2$.

This energy has the special property of being conserved when bodies come into contact in collisions. Consider a collision between two rigid balls A and B with initial vector velocities $\mathbf{v}_A$ and $\mathbf{v}_B$. For the moment suppose that the time interval t over which this force acts is sufficiently brief that the forces acting on the balls do not change appreciably during this time. The force that A exerts on B is equal and opposite to the force $\mathbf{F}$ that B exerts on A, so Newton's second law tells us that the final velocities of A and B are $\mathbf{v}'_A = \mathbf{v}_A + \mathbf{F}t/m_A$ and $\mathbf{v}'_B = \mathbf{v}_B - \mathbf{F}t/m_B$. Hence, as Newton showed, momentum is conserved:

$$m_A\mathbf{v}'_A + m_B\mathbf{v}'_B = m_A\mathbf{v}_A + m_B\mathbf{v}_B\ . \tag{2.1.1}$$

Neglecting changes in acceleration during the brief time t, the vector displacements traveled by A and B equal t times the average velocities, $[\mathbf{v}_A + \mathbf{v}'_A]/2$ and $[\mathbf{v}_B + \mathbf{v}'_B]/2$, respectively. If the balls remain in contact during this time interval, then these displacements must be the same, so

$$\mathbf{v}_A + \mathbf{v}'_A = \mathbf{v}_B + \mathbf{v}'_B\ . \tag{2.1.2}$$

To derive a second conservation law, rewrite Eq. (2.1.2) as $\mathbf{v}'_B - \mathbf{v}'_A = \mathbf{v}_A - \mathbf{v}_B$ and square this, giving

$$\mathbf{v}'^2_B - 2\mathbf{v}'_B \cdot \mathbf{v}'_A + \mathbf{v}'^2_A = \mathbf{v}^2_B - 2\mathbf{v}_B \cdot \mathbf{v}_A + \mathbf{v}^2_A\ .$$

Multiply this with m_Am_B and add the square of Eq. (2.1.1), so that the scalar products cancel. Dividing by $2(m_A+m_B)$, the result is another conservation law,

$$\frac{m_A}{2}\mathbf{v}'^2_A + \frac{m_B}{2}\mathbf{v}'^2_B = \frac{m_A}{2}\mathbf{v}^2_A + \frac{m_B}{2}\mathbf{v}^2_B\ . \tag{2.1.3}$$

Equations (2.1.1) and (2.1.3) have been derived here only for the case in which the particles are in contact only for a brief time interval during which the force acting between the bodies is constant, but this is not an essential requirement for we can break up any time interval into a large number of brief intervals in each of which the change in the force is negligible, Then, since $m_A\mathbf{v}_A + m_B\mathbf{v}_B$ and $m_A\mathbf{v}^2_A/2 + m_B\mathbf{v}^2_B/2$ do not change in each interval, they do not change at all, as long as the bodies exert forces on each other only when they are in contact.

In 1669 Christiaan Huygens (1629–1695) reported in *Journal des Sçavans* that he had confirmed the conservation of the total of $m\mathbf{v}^2/2$, probably by observing collisions of pendulum bobs, for which initial and final velocities could be precisely determined. Newton in the *Principia* called the conserved quantity $m\mathbf{v}$ the *quantity of motion*, while Huygens gave the name *vis viva* ("living force") to the conserved quantity $m\mathbf{v}^2/2$. These two quantities have since become known as *momentum* and *kinetic energy*.

On the other hand, it was essential in deriving the conservation of kinetic energy that we assumed that particles interact only when in contact. This is generally a good approximation in gases, but it is not valid in the presence of long range forces, such as electromagnetic or gravitational forces. In such cases kinetic energy is *not* conserved – it is only the sum of kinetic energy plus some sort of potential energy that does not change.

Specific Heat

The total kinetic energy of N molecules of gas of mass m and mean square velocity $\langle \mathbf{v}^2 \rangle$ is $Nm\langle \mathbf{v}^2 \rangle/2$. Clausius had found the relation (1.1.4) between mean square velocity and absolute temperature, according to which $m\langle \mathbf{v}^2 \rangle/2 = 3kT/2$, where k is some constant (later identified as a universal constant of nature), so the total kinetic energy is $3NkT/2$. A mass M of gas of molecular weight μ contains $N = M/\mu m_1$ molecules, so the total kinetic energy is $3MRT/2\mu$, where $R = k/m_1$ is the gas constant (1.2.4). Clausius concluded that to raise the temperature of a mass M of gas of molecular weight μ by an amount dT at constant volume, so that the gas does no work on its container, requires an energy $dE = 3MRdT/2\mu$. The ratio dE/MdT is known as the specific heat, so Clausius found that the specific heat of a gas at constant volume is

$$C_v = 3R/2\mu \; . \tag{2.1.4}$$

This result must be distinguished from the value for a different sort of specific heat, measured at constant pressure, such as when the gas is in a container with an expandable wall, for which the volume V can change to keep the pressure p equal to the pressure of the surrounding air or other medium. When pressure pushes a surface of area A a small distance dL, the work done is the force pA times dL, which equals pdV where $dV = AdL$ is the change in volume. According to the ideal gas law (1.2.3), $pV = RTM/\mu$, so if the temperature is increased by an amount dT, then at constant pressure the gas does work $pdV = RMdT/\mu$, and this temperature increase therefore requires an energy $3MRdT/2\mu + MRdT/\mu = 5MRdT/3$. In other words, the specific heat at constant pressure is

$$C_p = 5R/2\mu \; . \tag{2.1.5}$$

This result is often expressed in terms of the ratio of specific heats,

$$\gamma \equiv C_p/C_v \; . \tag{2.1.6}$$

So Clausius found that if all the heat of a gas is contained in the kinetic energy of its molecules then $\gamma = 5/3$.

This did not agree with measurements of the specific heats of common diatomic gases, such as oxygen or hydrogen, which Clausius cited as giving $\gamma = 1.421$. Later, it was found that γ does indeed equal 5/3 for a monatomic gas like mercury vapor, but this left the question, in what form is the energy in ordinary gases that are not monatomic?

To deal with this issue, Clausius suggested that the internal energy of a gas is larger than the kinetic energy of the molecules, say by a factor $1 + f$, with f some positive number. Then instead of Eq. (2.1.4) we have

$$C_v = (1 + f) \times 3R/2\mu \,, \tag{2.1.7}$$

and in place of Eq. (2.1.5),

$$C_p = (1 + f)3R/2\mu + R/\mu \,. \tag{2.1.8}$$

The specific heat ratio is then

$$\gamma = 1 + \frac{2}{3(1 + f)} \,. \tag{2.1.9}$$

This is often expressed (especially in astrophysics) as a formula for the internal energy density $\mathcal{E}$ in terms of the pressure and γ:

$$\mathcal{E} = 3RTM(1 + f)/2\mu V = 3(1 + f)p/2 = p/(\gamma - 1) \,. \tag{2.1.10}$$

The observation that $\gamma \simeq 1.4$ for diatomic gases like O_2 and H_2 indicated that the internal energy of these gases is larger than the kinetic energy of its molecules by a factor $1 + f \simeq 5/3$. Measurements gave values of γ for more complicated molecules like H_2O or CO_2 even closer to unity, indicating that f is even larger for these molecules. The reason for these values for f and γ did not become clear until the formulation of the equipartition of energy, to be discussed in Section 2.4.

Adiabatic Changes

It often happens that work is done adiabatically, that is, without the transfer of heat. In this case the conservation of energy tells us that the work done by an expanding fluid must be balanced by a decrease in its internal energy $\mathcal{E}V$:

$$0 = p\,dV + d(\mathcal{E}V) = (p + \mathcal{E})dV + V d\mathcal{E} \,. \tag{2.1.11}$$

For an ideal gas, the internal energy per unit volume $\mathcal{E}$ is given by Eq. (2.1.10), so this tells us that

$$0 = \gamma p\,dV + V\,dp$$

and so, in an adiabatic process,

$$p \propto V^{-\gamma} \propto \rho^{\gamma} \,, \tag{2.1.12}$$

or, since for a fixed mass $p \propto T/V$,

$$T \propto V^{1-\gamma} \propto \rho^{\gamma-1} \,. \tag{2.1.13}$$

This is in contrast with an isothermal process, for which T is constant and $p \propto V^{-1}$.

Equation (2.1.12) has an immediate consequence for the speed of sound. At audible frequencies the conduction of heat is typically too slow to be effective, so the expansion and compression of a fluid carrying a sound wave is adiabatic. It is a standard result of hydrodynamics, proved by Newton, that the speed of sound is

$$c_s = \sqrt{\frac{\partial p}{\partial \rho}} \ .$$

Newton thought that p would be proportional to ρ in a sound wave, which would give $c_s = \sqrt{p/\rho}$, but in fact at audible frequencies the pressure is given by the adiabatic relation (2.1.12), and c_s is larger than Newton's value by a factor $\sqrt{\gamma}$.

2.2 Absolute Temperature

We have been casually discussing temperature, but what precisely do we mean by this? It is not hard to give a precise meaning to a statement that one body has a higher temperature than another, by a generalization of common experience that is sometimes known as the second law of thermodynamics (the first law being the conservation of energy). Observation of heat flow shows that if heat can flow spontaneously from a body A to a body B, then it cannot flow spontaneously from B to A. We can say then that the temperature $t_{\rm A}$ of A is higher than the temperature $t_{\rm B}$ of B. Likewise, we say that two bodies are at the same temperature if heat cannot flow from either one to the other spontaneously, without work being done on these bodies. Temperature defined in this way is observed to be *transitive*: If heat can flow spontaneously from a body A to a body B, and from B to a body C, then it can flow spontaneously from A to C. This is a property shared with real numbers – if a number a is larger than number b, and if b is larger than c, then a is larger than c – and is a necessary condition for temperatures to be represented by real numbers.

But this does not give a precise meaning to any particular numerical value of temperature, or even to numerical ratios of temperatures. If, for some definition of temperature t, a comparison of values of t tells us the direction of heat flow then the same would be true of any monotonic function $\mathcal{T}(t)$. Conventionally temperatures are defined by thermometers. With a column of some liquid such as mercury or alcohol in a glass tube, we mark off the heights of the column when the tube is placed in freezing or boiling water, and for a Celsius temperature scale etch on the tube marks that divide the distance from freezing to boiling to a hundred equal parts. The trouble is that different liquids expand differently with increasing temperature, and the temperatures measured in this way with a mercury or alcohol thermometer will not be precisely equal. We can try instead to give significance to numerical values of temperature by using a gas thermometer, relying on the ideal gas law $pV = MRT/\mu$, but this law is approximate, holding precisely only for molecules of negligible size that interact only in contact in collisions. How can we give precise meaning to numerical values of temperature without relying on approximate relations?

Surprisingly, as shown by Rudolf Clausius in his 1850 paper[1] "On the Moving Force of Heat," it is possible by find a definition of temperature T with absolute significance by the study of thermodynamic engines known as Carnot cycles.

Sadi Carnot (1796–1832) was a French military engineer, the son of Lazare Carnot, organizer of military victory in the French Revolution, and uncle of a later president of the Third Republic. In 1824 Carnot in *Reflections on the Motive Power of Fire* set out to study the efficiency of steam engines, explaining that "Already the steam engine works our mines, impels our ships, excavates our ports and our rivers, forges iron, fashions wood, grinds grains, spins and weaves our clothes, transports our heaviest burdens, etc." (A few years later he might also have mentioned the beginning of steam-propelled locomotives, with the opening of the Liverpool–Manchester railroad in 1830.) Carnot invented an idealized engine, known as a *Carnot cycle*, which as we shall see is maximally efficient and provides a natural definition of absolute temperature.

In the Carnot cycle, a working fluid (such as steam in a cylinder fitted with a piston) goes through four frictionless steps:

1. Isothermal: The working fluid does work on its environment, for instance by pushing a piston against external pressure, but keeping a constant temperature by absorbing heat Q_2 from a hot reservoir at temperature t_2. (We will continue to use lower case t to indicate temperature defined in any way that indicates the direction of heat flow, without specifying any physical significance to its particular numerical values.)
2. Adiabatic: The working fluid, perfectly insulated from its environment and with no internal friction, does more work, with its temperature dropping to the temperature t_1 of a cold reservoir but with no heat flowing in or out.
3. Isothermal: Work is done on the fluid, for instance by pushing in the piston, with its temperature kept constant by its giving up heat Q_1 to the cold reservoir.
4. Adiabatic: With the working fluid again completely insulated from its environment, work is done on it, bringing its volume back down to its original value and its temperature back up to the temperature t_2 of the hot reservoir.

[1] This paper is reprinted in Brush, *The Kinetic Theory of Gases – An Anthology of Classic Papers with Historical Commentary*, listed in the bibliography.

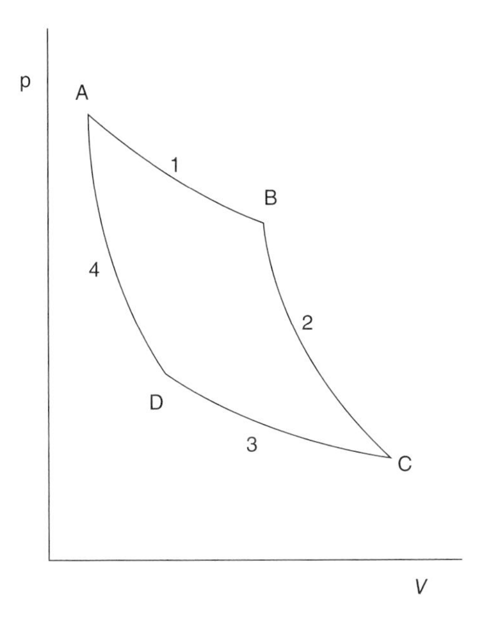

Figure 2.1 A Carnot cycle (not drawn to scale).

A graph of the pressure versus the volume of the working fluid in this cycle is a closed curve, with the net work W done on the environment equal to $\oint p dV$ – that is, to the area enclosed by the curve. (See Figure 2.1.) As long as steps 2 and 4 are truly adiabatic, the conservation of energy tells us that this work is

$$W = Q_2 - Q_1 \tag{2.2.1}$$

and the efficiency of this cycle is

$$W/Q_2 = \frac{Q_2 - Q_1}{Q_2} < 1 \,. \tag{2.2.2}$$

(We call this the efficiency, having in mind that, as for a steam engine, we have to pay for the heat Q_2 taken up at the higher temperature t_2, while the heat Q_1 given up at the lower temperature t_1 is wasted.)

Any Carnot cycle is *reversible*, because any frictionless adiabatic or isothermal process follows the same track, depending only on its endpoints, whichever direction the process takes. But not all thermodynamic cycles, which take a working fluid through a series of steps back to the original temperature and volume, are reversible even though of course they all conserve energy. For reversibility it is not enough that all steps be either isothermal or adiabatic – there also should be no friction, which if present would provide an internal source of heat that is not available to do work.

The importance of Carnot cycles in thermodynamics rests on the following theorem:
I. The efficiency of the Carnot cycle C described above is at least as great as that for any general thermodynamical cycle C′, not necessarily reversible, which begins with the working fluid absorbing a heat Q'_2 from a reservoir at the same high temperature t_2, then emitting heat at the same lower temperature t_1, and then returning to its original temperature and volume, in the process doing net work W'. That is,

$$W/Q_2 \geq W'/Q'_2 \,. \tag{2.2.3}$$

II. All Carnot cycles that take heat from a reservoir at the same temperature t_2, using it to do work, and giving up waste heat to a reservoir at the same lower temperature t_1, have the same efficiency, which depends only on t_2 and t_1.

Proof:[2] Like any positive real number, the ratio of the work done in the Carnot cycle C and in a general cycle C′ can be approximated to an arbitrary accuracy by a ratio of positive real integers N and N':

$$W/W' = N'/N \,. \tag{2.2.4}$$

Since any Carnot cycle by definition is reversible, the cycle C has an inverse C^{-1}. This is a refrigeration cycle, following the same steps as for C but in the opposite order, so that by doing work W on the fluid an amount of heat Q_1 is taken from the reservoir at temperature t_1 and heat $Q_2 > Q_1$ is delivered to the reservoir at temperature $t_2 > t_1$. Suppose we perform a compound cycle C^*, consisting of N repetitions of C^{-1} and N' repetitions of C′. According to Eq. (2.2.4), the net work done by the working fluid is

$$W^* = N'W' - NW = 0 \,.$$

Also, the net heat taken from the hot reservoir at temperature t_2 is

$$Q_2^* = N'Q'_2 - NQ_2 \,.$$

Now, since no work is done in the compound cycle, according to the fundamental property of temperature t, it is not possible for positive-definite net heat to be transferred *to* a reservoir at temperature t_2 *from* a lower temperature t_1, so the net heat Q_2^* taken *from* the hot reservoir in the cycle C^* must be positive-definite or zero. Hence, using Eq. (2.2.4),

$$0 \leq \frac{Q_2^*}{NW} = \frac{N'Q'_2 - NQ_2}{NW} = \left(\frac{Q'_2}{W'} - \frac{Q_2}{W}\right) \,,$$

[2] This treatment and that of the following section is based on that given by Fermi, *Thermodynamics*, listed in the bibliography.

and therefore

$$\frac{W}{Q_2} \geq \frac{W'}{Q'_2} \tag{2.2.5}$$

as was to be proved in the first part of the theorem.

As to the second part of the theorem, note that if C′ is also a Carnot cycle then, by the same reasoning,

$$\frac{W'}{Q'_2} \geq \frac{W}{Q_2} ,$$

so the efficiencies are equal:

$$\frac{W}{Q_2} = \frac{W'}{Q'_2} . \tag{2.2.6}$$

This has now been proved for any pair of Carnot cycles, operating between the same temperatures t_2 and t_1, whatever the values of the heat taken from the reservoir at temperature t_2 and given up to the reservoir at temperature t_1, so the common efficiency can only depend on t_2 and t_1, as was to be proved.

We shall write this relation in terms of the *inefficiency*:

$$1 - \frac{W}{Q_2} = \frac{Q_1}{Q_2} \equiv F(t_1, t_2) \tag{2.2.7}$$

with F the same function for all Carnot cycles. We next prove that the function $F(t_1, t_2)$ takes the form

$$F(t_1, t_2) = T(t_1)/T(t_2) \tag{2.2.8}$$

for some function $T(t)$. For this purpose we consider a compound cycle consisting of a Carnot cycle operating between the temperatures t_2 and $t_0 \leq t_2$ followed by a Carnot cycle operating between the temperatures t_0 and $t_1 \leq t_0$, with all the waste heat that is given to the reservoir at temperature t_0 in the first cycle taken up from this reservoir in the second cycle. Since $(Q_0/Q_2)(Q_1/Q_0) = Q_1/Q_2$, the inefficiency (2.2.7) of the compound cycle is the product of the inefficiencies of the individual cycles, so

$$F(t_1, t_2) = F(t_1, t_0)F(t_0, t_2) . \tag{2.2.9}$$

From Eq. (2.2.7) it is evident that $F(t_2, t_0)F(t_0, t_2) = 1$, so Eq. (2.2.9) may be written

$$F(t_1, t_2) = \frac{F(t_1, t_0)}{F(t_2, t_0)} . \tag{2.2.10}$$

This holds for any t_0 with $t_2 \geq t_0 \geq t_1$, so we can define $T(t) \equiv F(t, t_0)$ with an arbitrary choice of t_0 in this range, and then Eq. (2.2.10) is the desired result (2.2.8).

Now, efficiencies are never greater than 100%, so the ratio $F(t_1, t_2) = T(t_1)/T(t_2)$ in Eq. (2.2.7) must be positive, and so $T(t)$ has the same sign for all temperatures. Since only the ratios of the Ts appear in the efficiency, we are free to choose this sign to be positive, so that $T(t) \geq 0$ for all t. Also, inefficiencies are never greater than 100%, so Eq. (2.2.8) shows that $T(t_1) \leq T(t_2)$ for any t_1 and t_2 with $t_1 \leq t_2$. That is, $T(t)$ is a monotonically increasing function of t and can therefore be used to judge the direction of spontaneous heat flow as well as t itself.

We can therefore define the absolute temperature T by just using $T(t)$ as the temperature in place of t. That is, using Eqs. (2.2.7) and (2.2.8), we define absolute temperature T by the statement that a Carnot cycle running between any two temperatures T_2 and T_1 has

$$\frac{Q_1}{Q_2} = \frac{T_1}{T_2} . \tag{2.2.11}$$

A Carnot cycle running between an upper temperature T_2 and a lower temperature T_1 has an efficiency

$$\frac{W}{Q_2} = \frac{Q_2 - Q_1}{Q_2} = \frac{T_2 - T_1}{T_2} . \tag{2.2.12}$$

Of course, this only defines T up to a constant factor, leaving us free to use what units we like for temperature. But we are not free to shift $T(t)$ by adding a constant term. Indeed, since in this Carnot cycle heat flows from a reservoir at temperature T_2 to one at temperature T_1, we must have $T_2 > T_1$, and therefore in order for the efficiency (2.2.11) to be a positive quantity, the lower temperature must have $T_1 > 0$. Because any heat reservoir must have T positive-definite, we see that T is the absolute temperature, in the same sense as was found for gases by Charles.

The temperature defined by Carnot cycles is identical (up to a choice of units) to the temperature given by a gas thermometer, which for the moment we will call T^g, in the approximation that the gas is ideal. To see this, let us label the states of the gas as A at the start of the isothermal expansion 1 (and at the end of the adiabatic compression 4); as B at the start of the adiabatic expansion 2 (and the end of the isothermal expansion 1); as C at the start of the isothermal compression 3 (and the end of the adiabatic expansion 2); and as D at the start of the adiabatic compression 4 (and the end of the isothermal compression 3). Since the expansion from A to B is isothermal, during this phase the internal energy of the gas, which is given by Eqs. (1.2.3) and (2.1.10) as $\mathcal{E}V = RTM/(\gamma - 1)\mu$, does not change, and so the heat drawn from the hot reservoir is the work done:

$$Q_2 = \int_A^B p dV = \frac{MRT_2^g}{\mu} \int_A^B \frac{dV}{V} = \frac{MRT_2^g}{\mu} \ln\left(\frac{V_B}{V_A}\right) .$$

Likewise, the heat given up to the cold reservoir in the isothermal compression from C to D is

$$Q_1 = \frac{MRT_1^g}{\mu} \ln\left(\frac{V_C}{V_D}\right) .$$

Further, since the expansion from B to C and the contraction from D to A are adiabatic, Eq. (2.1.13) gives $V \propto (T^g)^{-1/(\gamma-1)}$, and so during these parts of the cycle

$$V_C/V_B = V_D/V_A = \left(\frac{T_2^g}{T_1^g}\right)^{1/(\gamma-1)} ,$$

and therefore $V_B/V_A = V_C/V_D$, and the logarithmic factors in Q_2 and Q_1 are equal. The efficiency is then

$$\frac{Q_2 - Q_1}{Q_2} = \frac{T_2^g - T_1^g}{T_2^g}$$

in agreement with Eq. (2.2.12) if $T = T^g$, up to a possible constant factor.

2.3 Entropy

In macroscopic classical thermodynamics we characterize the state of a system by a set of variables that can be specified independently. For instance, for a fluid of fixed mass and chemical composition, in a vessel with adjustable volume (say, with a movable piston) the state is specified by giving the values of any two of the thermodynamic variables – pressure, volume, temperature, energy, etc. – the remaining variables being determined in equilibrium or in adiabatic variations by these two values and some equation of state, such as the ideal gas law (1.2.3). Many of the consequences of macroscopic classical thermodynamics can be deduced from the existence of another thermodynamic variable, known as the *entropy*, introduced in 1854 by Rudolf Clausius, that like other thermodynamic variables depends only on the state of the system, although its definition seems to indicate that it also depends on the way that the system is prepared.

Suppose a system is prepared in a given state 1 by starting with it in a standard state (labeled 0 below) and then taken to 1 on a path P through the space of independent variables used to define thermodynamic states, in which by a series of small reversible changes at varying absolute temperatures T it picks up small net amounts dQ of heat energy from the environment. (The heat energy increment dQ is taken as positive if the system takes heat from the environment and negative if it gives up heat.) Then the entropy of this state is defined by

$$S_1 = S_0 + \int_P \frac{dQ}{T} , \tag{2.3.1}$$

where the integral is taken over any reversible path from state 0 to state 1, and S_0 is whatever entropy we choose to ascribe to the standard state 0. The remarkable thing is that the integral here is independent of the particular reversible path chosen, so that this really defines the entropy S_1 up to a common constant term S_0 as a function of the state of the system, not of how it is prepared, provided that T is the absolute temperature defined, as in the previous section, by the efficiency of Carnot cycles. Furthermore, with the entropy defined in this way, for any path P' from state 0 to state 1 that may or may not be reversible, we have

$$\int_{P'} \frac{dQ}{T} \leq S_1 - S_0 . \tag{2.3.2}$$

Proof: The first step in proving these results is to prove the following lemma: for an arbitrary cycle, reversible or irreversible, that takes a system Σ from any state back to the same state, taking in and giving up heat at various temperatures, we have

$$\oint \frac{dQ}{T} \leq 0 . \tag{2.3.3}$$

After establishing this lemma, the rest of the proof will be straightforward.

To prove this lemma we can approximate the cycle by a sequence of brief isothermal steps, in each of which the system Σ takes in heat (if dQ is positive) or gives heat up (if dQ is negative) at a momentary temperature T. We can imagine that, at each step, the heat taken in or given up is given up or taken in by another system, which undergoes a Carnot cycle between the momentary temperature T and a fixed temperature T_0. In this Carnot cycle, the ratio of the heat dQ given up by the Carnot cycle to Σ and the heat dQ_0 taken by the Carnot cycle from the reservoir at temperature T_0 is given by Eq. (2.2.11):

$$\frac{dQ}{dQ_0} = \frac{T}{T_0} ,$$

or in other words

$$dQ_0 = T_0 \left(\frac{dQ}{T} \right) .$$

Hence in the complete cycle the Carnot cycles take in a total net heat $T_0 \oint dQ/T$ from the reservoir at temperature T_0. Since the system Σ and each of the Carnot cycles return to their original states, if this heat taken in at temperature T_0 were positive-definite then it would have to go into work, which is impossible since work cannot be done by taking heat from a reservoir at a fixed temperature with no changes elsewhere. (If it could, then this work by producing friction could be used to transfer some heat to any body, even one at a temperature higher than T_0.) So we conclude that the integral $\oint dQ/T$ is at most zero, as was to be shown.

The rest is easy. Note that if two paths P and P' are both reversible paths that go from state 0 to state 1, then PP'^{-1} is a closed cycle, where P'^{-1} is path P' taken in reverse, from state 1 to state 0. It follows then from the inequality (2.3.3) that

$$0 \geq \oint_{PP'^{-1}} \frac{dQ}{T} = \int_P \frac{dQ}{T} - \int_{P'} \frac{dQ}{T} .$$

But $P'P^{-1}$ is also a closed cycle, so

$$0 \geq \oint_{P'P^{-1}} \frac{dQ}{T} = \int_{P'} \frac{dQ}{T} - \int_P \frac{dQ}{T} .$$

These two results are consistent only if for reversible paths both cyclic integrals vanish, in which case

$$\int_{P'} \frac{dQ}{T} = \int_P \frac{dQ}{T} . \tag{2.3.4}$$

We can therefore define the entropy up to an additive constant as in Eq. (2.3.1), where P is any reversible path.

Finally, if P' is a general path from state 0 to state 1, reversible or irreversible, while P is a reversible path from state 0 to state 1, then $P'P^{-1}$ (but not necessarily PP'^{-1} !) is a closed cycle, so the inequality (2.3.3) gives

$$0 \geq \oint_{P'P^{-1}} \frac{dQ}{T} = \int_{P'} \frac{dQ}{T} - \int_P \frac{dQ}{T} ,$$

and therefore, using Eq. (2.3.1),

$$\int_{P'} \frac{dQ}{T} \leq S_1 - S_0$$

as was to be shown.

In the special case of a completely isolated system Σ, no heat can be taken into Σ or given up by Σ, so the integrand in the integral on the left-hand side of Eq. (2.3.2) must vanish and therefore $S_1 \geq S_0$. *In isolated systems the entropy can only increase.* On the other hand, if an isolated system is undergoing only reversible changes, then according to Eq. (2.3.1) the entropy is constant.

There is another definition of entropy, used in information theory as well as in physics. If a system can be in any one of a number of states characterized by a continuous (generally multidimensional) parameter α, with a probability $\mathcal{P}(\alpha)\, d\alpha$ of being in states with this parameter in a narrow range $d\alpha$ around α, then the entropy is

$$S = -k \int \mathcal{P}(\alpha) \ln \mathcal{P}(\alpha)\, d\alpha \,, \tag{2.3.5}$$

where k is the universal constant, known as Boltzmann's constant, appearing in Eq. (1.2.1). As we shall see in the next section, according to kinetic theory, with a suitable choice of S_0 the thermodynamic entropy (2.3.1) equals the information-theoretic entropy (2.3.5).

The mere fact that the entropy S defined by (2.3.1) depends only on the thermodynamic state has far-reaching consequences. Consider a fixed mass of fluid in a vessel with variable volume. The independent thermodynamic variables here can be taken as the volume V and the temperature T, with pressure p, internal energy E, and entropy S all functions of V and T. The work done by the fluid pressure p in increasing the fluid volume by a small amount dV is pdV, so the heat required to change the temperature by an infinitesimal amount dT and the volume by an infinitesimal amount dV is

$$dQ = dE + pdV \,,$$

so according to Eq. (2.3.1), the change in the entropy is given by

$$TdS = dE + pdV \,. \tag{2.3.6}$$

In other words

$$\frac{\partial S(V,T)}{\partial T} = \frac{1}{T}\frac{\partial E(V,T)}{\partial T} \tag{2.3.7}$$

$$\frac{\partial S(V,T)}{\partial V} = \frac{1}{T}\frac{\partial E(V,T)}{\partial V} + \frac{p(V,T)}{T} \,. \tag{2.3.8}$$

To squeeze information about pressure and internal energy from these formulas, we use the fact that partial derivatives commute. From Eq. (2.3.7) we have

$$\frac{\partial}{\partial V}\left(\frac{\partial S(V,T)}{\partial T}\right) = \frac{1}{T}\frac{\partial^2 E(V,T)}{\partial T \partial V}$$

while, from Eq. (2.3.8),

$$\frac{\partial}{\partial T}\left(\frac{\partial S(V,T)}{\partial V}\right) = \frac{1}{T}\frac{\partial^2 E(V,T)}{\partial T \partial V} - \frac{1}{T^2}\frac{\partial E(V,T)}{\partial V} + \frac{\partial (p(V,T)/T)}{\partial T} \,.$$

Setting these equal gives a relation between the derivatives of E and p:

$$0 = -\frac{1}{T^2}\frac{\partial E(V,T)}{\partial V} + \frac{\partial \big(p(V,T)/T\big)}{\partial T} \,. \tag{2.3.9}$$

This is for a fixed mass. Since $E(V,T)$ is an extensive variable, it must be proportional to this mass but does not otherwise have to depend on volume. In fact, it is frequently a good approximation to suppose that, apart from its proportionality to mass, $E(V,T)$ is independent of volume. This is the case if the fluid consists of infinitesimal particles that interact only in contact in collisions; since there is nothing with the dimensions of length that can enter in the calculation of the energy, $E(V, T)$ cannot here depend on volume. In this case Eq. (2.3.9) yields Charles' law, that for fixed volume V the pressure $p(V, T)$ is proportional to T. This shows again that the absolute temperature T in the ideal gas law (1.2.3) is the same up to a constant factor as the temperature T defined by the efficiency (2.2.11) of Carnot cycles.

Although this result was obtained without having a formula for the entropy, for some purposes it is useful actually to know what the entropy is. In a homogeneous medium, the entropy S of any mass M of matter may conveniently be written as $S = Ms$, where s is the entropy per unit mass, a function of temperature and various densities known as the specific entropy. Dividing Eq. (2.3.6) by M, we have then

$$T ds = d(\mathcal{E}/\rho) + pd(1/\rho) , \tag{2.3.10}$$

where as before $\mathcal{E} \equiv E/V$ is the internal energy density and $\rho \equiv M/V$ is the mass density. We consider an ideal gas, for which $T = p\mu/R\rho$ while $\mathcal{E}$ and p are related by Eq. (2.1.10): $\mathcal{E} = p/(\gamma - 1)$. Then Eq. (2.3.10) gives

$$\frac{p\mu}{R\rho} ds = \frac{1}{\gamma - 1}\left(\frac{dp}{\rho} + \gamma pd\left(\frac{1}{\rho}\right)\right) = \frac{\rho^{\gamma-1}}{\gamma - 1} d\left(\frac{p}{\rho^\gamma}\right) ,$$

so

$$ds = \frac{R/\mu}{\gamma - 1}\frac{\rho^\gamma}{p} d\left(\frac{p}{\rho^\gamma}\right) .$$

The solution is

$$s = \frac{R/\mu}{\gamma - 1} \ln\left(\frac{p}{\rho^\gamma}\right) + \text{constant} . \tag{2.3.11}$$

We see that the result of Section 2.2 that $p \propto \rho^\gamma$ for adiabatic processes is just the statement that s is constant in these processes, which of course it must be since in an adiabatic process the heat input dQ vanishes.

In many stars there are regions in which convection effectively mixes matter from various depths. Since heat conduction is usually ineffective in stars, little heat flows into or out of a bit of matter as it rises or falls, and so it keeps the same specific entropy. These regions therefore have a uniform specific entropy, and therefore a uniform value for the ratio p/ρ^γ. For instance, this is the case in the Sun for distances from the center greater than about 65% of the Sun's radius out to a thin surface layer.

Neutral Matter

We have been mostly concerned with matter in which in each mass there is a non-vanishing conserved quantity, the number of particles. There is a different context, with no similar conserved numbers, in which thermodynamics yields more detailed information about pressure and energy. In the early universe, at temperatures above about 10^{10} K, there is so much energy in radiation and electron–antielectron pairs that the contribution to the energy of the excess of matter over antimatter may be neglected. Here there is no number density on which the pressure and energy density $\mathcal{E} \equiv E/V$ can significantly depend, so here $E(V,T) = V\mathcal{E}(T)$ and $p(V,T) = p(T)$; thus here Eq. (2.3.9) is an ordinary differential equation for $p(T)$:

$$0 = -\frac{\mathcal{E}(T)}{T^2} + \frac{d}{dT}\left(\frac{p(T)}{T}\right)$$

or, in other words,

$$p'(T) = \frac{\mathcal{E}(T) + p(T)}{T} \,. \tag{2.3.12}$$

Thermodynamics alone does not fix any relation between $\mathcal{E}(T)$ and $p(T)$, but given such a relation this result gives both as functions of temperature. For instance, as an example of the power of thermodynamics, it was known in the nineteenth century as a consequence of Maxwell's theory of electromagnetism that the pressure of electromagnetic radiation is one-third of its energy density. Setting $p = \mathcal{E}/3$ in Eq. (2.3.12) gives $\mathcal{E}'(T) = 4\mathcal{E}(T)$, so

$$\mathcal{E}(T) = 3p(T) = aT^4 \,, \tag{2.3.13}$$

where a is a constant, known as the radiation energy constant. But, as we shall see in Section 3.1, it was not possible to understand the value of a until the advent of quantum mechanics in the early twentieth century.

The Laws of Thermodynamics

It is common to summarize the content of classical thermodynamics in three laws. As already mentioned, the first law is just the conservation of energy, discussed in the context of heat energy in Section 2.1, and the second, usually attributed to Clausius, on which the discussion of thermodynamic efficiency in Section 2.2 is based, can be stated as the principle that without doing work it is not possible to transfer heat from a cold reservoir to one at higher temperature. We have seen that this leads to the existence of a quantity, the entropy, which depends only on the thermodynamic state and satisfies Eq. (2.3.1) when reversible changes are made in this state. This can instead be taken as the second law of thermodynamics.

There are several formulations of the third law, some given by Walther Nernst (1864–1941) in 1906–1912. The most fruitful, it seems, is that it is possible to assign a common value to the entropy (conventionally taken as zero) for all systems at absolute zero temperature, so that at absolute zero the integral in Eq. (2.3.1) must converge. This has the consequence, in particular, that the specific heat dQ/dT must vanish for $T \to 0$. This seems to contradict the results of Section 2.1 for ideal gases, which give a temperature-independent specific heat whether for fixed volume or fixed pressure. The contradiction is avoided in practice because no substance remains close to an ideal gas as the temperature approaches absolute zero. We will see when we come to quantum mechanics that if an otherwise free particle is confined in any fixed volume, then it cannot have precisely zero momentum, as required for a classical ideal gas at absolute zero temperature. On the other hand, solids can exist at absolute zero temperature, and in that limit their specific heats do approach zero.

2.4 Kinetic Theory and Statistical Mechanics

We saw in the previous chapter how by the mid nineteenth century the ideal gas law had been established through the work especially of Bernoulli and Clausius. But, though derived by considering the motions of individual gas molecules, in its conclusions it dealt only with bulk gas properties such as pressure, temperature, mass density, and energy density. For many purposes, including the calculation of chemical or transport processes, it was necessary to go further and work out the detailed probability distribution of the motion of individual gas particles. This was done in the kinetic theory of James Clerk Maxwell and Ludwig Boltzmann (1844–1906). Kinetic theory was later generalized to the formalism known as statistical mechanics, especially by the American theorist Josiah Willard Gibbs (1839–1906). As it turned out, these methods went a long way toward not only establishing a correspondence with thermodynamics but also explaining the principles of thermodynamics on the assumption that macroscopic matter is composed of very many particles, and thereby helping to establish the reality of atoms.

The Maxwell–Boltzmann Distribution

Maxwell in 1860 considered the form of the probability distribution function $P(v_x, v_y, v_z)$ for the x, y, and z components of the velocity of any molecule in a gas in equilibrium.[3] The probability distribution function is defined so that the probability that these components are respectively between v_x and $v_x + dv_x$, between v_y and $v_y + dv_y$, and between v_z and $v_z + dv_z$, is of the form

$$P(v_x, v_y, v_z)dv_x dv_y dv_z \ .$$

[3] J. C. Maxwell, Phil. Mag. **19**, 19; **20**, 21 (1860). This article is included in Brush, *The Kinetic Theory of Gases – An Anthology of Classic Papers with Historical Commentary*, listed in the bibliography.

He assumed (without offering a real justification) that the probability that any component of velocity of a particle is in a particular range is not correlated with the other components of the velocity. Then $P(v_x, v_y, v_z)$ must be proportional to a function of v_x alone, with a coefficient that depends only on v_y and v_z, and likewise for v_y and v_z, so $P(v_x, v_y, v_z)$ must take the form of a product:

$$P(v_x, v_y, v_z) = f(v_x)g(v_y)h(v_z)\ .$$

Rotational symmetry requires further that P can depend only on the magnitude of the velocity, not on its direction, and hence only on $v_x^2 + v_y^2 + v_z^2$. The only function of $v_x^2 + v_y^2 + v_z^2$ that takes the form $f(v_x)g(v_y)h(v_z)$ is proportional to an exponential:

$$P(v_x, v_y, v_z) \propto \exp\big(- C(v_x^2 + v_y^2 + v_z^2)\big)\ .$$

The constant C must be positive in order that P should not blow up for large velocity, which would make it impossible to set the total probability equal to unity, as it must be. Taking C to be positive, and setting the total probability (the integral of P over all velocities) for each particle equal to one, gives the factor of proportionality:

$$P(v_x, v_y, v_z) = \left(\frac{C}{\pi}\right)^{3/2} \exp\big(- C(v_x^2 + v_y^2 + v_z^2)\big)\ .$$

We can use this to calculate the mean square velocity components:

$$\overline{v_x^2} = \overline{v_y^2} = \overline{v_z^2} = \frac{1}{2C}.$$

Clausius had introduced an absolute temperature T by setting $m\overline{v_\perp^2} = kT$, where k is a constant to be determined experimentally and $v_\perp$ is the component of the velocity in a direction normal to the container wall, which for an isotropic velocity distribution can be taken as any direction, so the constant C must be given by $C = m/2kT$ and the Maxwell distribution takes the form

$$P(v_x, v_y, v_z) = \left(\frac{m}{2\pi kT}\right)^{3/2} \exp\big(- m(v_x^2 + v_y^2 + v_z^2)/2kT\big)\ . \tag{2.4.1}$$

As we saw at the end of Section 2.2, the requirement that in an ideal gas $m\overline{v_\perp}^2 = kT$, which led here to $C = m/2kT$, also ensures that, up to an arbitrary constant factor, T is the absolute temperature defined by the efficiency of Carnot cycles.

The formula for the probability distribution P was derived in 1868 in a more convincing way by Boltzmann.[4] He defined a quantity

[4] L. Boltzmann, Sitz. Ber. Akad. Wiss. (Vienna), part II, **66**, 875 (1872). A translation into English of this article is included in Brush, *The Kinetic Theory of Gases – An Anthology of Classic Papers with Historical Commentary*, listed in the bibliography.

$$H \equiv \overline{\ln P} = \int_{-\infty}^{+\infty} dv_x \int_{-\infty}^{+\infty} dv_y \int_{-\infty}^{+\infty} dv_z \; P(\mathbf{v}) \ln P(\mathbf{v}) \,,$$

and showed that collisions of gas particles always lead to a decrease in H until a minimum is reached, at which $P(\mathbf{v})$ is the Maxwell–Boltzmann distribution function. A generalization of this H-theorem was given in 1901 by Gibbs.[5] The generalization and proof are given below, along with the application to gases.

The General H-Theorem

Consider a large system with many degrees of freedom, such as a gas with many molecules (but not necessarily a gas). The states of the system are parameterized by many variables, which we summarize with a symbol α. (For instance, for a monatomic gas α stands for the set of positions $\mathbf{x}_1$, $\mathbf{x}_2$, etc. and momenta $\mathbf{p}_1$, $\mathbf{p}_2$, etc. of atoms $1, 2, \ldots$ For a gas of multi-atom molecules, α would also include the orientations and their rates of change for each molecule.) We denote an infinitesimal range of these parameters by $d\alpha$. (For instance, for a monatomic gas $d\alpha$ stands for the product $d^3\mathbf{x}_1 d^3\mathbf{p}_1 d^3\mathbf{x}_2 d^3\mathbf{p}_2 \ldots$, known as the phase space volume.) We define $\mathcal{P}(\alpha)$ so that the probability that the parameters of the system are in an infinitesimal range $d\alpha$ around α is $\mathcal{P}(\alpha)d\alpha$, with $\mathcal{P}$ normalized so that $\int d\alpha \mathcal{P}(\alpha) = 1$. Define

$$H \equiv \overline{\ln \mathcal{P}} = \int \mathcal{P}(\alpha)\, d\alpha \, \ln \mathcal{P}(\alpha) \,. \tag{2.4.2}$$

Gibbs showed that H always decreases until it reaches a minimum value, at which $\mathcal{P}(\alpha)$ is proportional to the exponential of a linear combination of conserved quantities, such as the total energy.

Proof: Define a differential rate $\Gamma(\alpha \to \beta)$ such that the rate at which a system in state α makes a transition to a state within a range $d\beta$ around state β is $\Gamma(\alpha \to \beta)\, d\beta$. The probability $\mathcal{P}(\alpha)d\alpha$ can either increase because the system in a range $d\beta$ of states around β makes a transition to the range $d\alpha$ of states around α, or decrease because the system in the range of states $d\alpha$ around α makes a transition to some other state in a range $d\beta$ around β, so

$$\frac{d\mathcal{P}(\alpha)d\alpha}{dt} = \int d\beta \left[\mathcal{P}(\beta)\, \Gamma(\beta \to \alpha) d\alpha - \mathcal{P}(\alpha)\, d\alpha \Gamma(\alpha \to \beta)\right] \,,$$

or, cancelling the differentials $d\alpha$,

$$\frac{d\mathcal{P}(\alpha)}{dt} = \int d\beta \left[\mathcal{P}(\beta)\, \Gamma(\beta \to \alpha) - \mathcal{P}(\alpha)\Gamma(\alpha \to \beta)\right] \,. \tag{2.4.3}$$

[5] J. W. Gibbs, *Elementary Principles of Statistical Mechanics, Developed with Especial Reference to The Rational Foundation of Thermodynamics* (Scribner, New York, 1902).

(Note that this makes $\int d\alpha \mathcal{P}(\alpha)$ time-independent, as it must be. Cancelling $d\alpha$ is justified because phase space volumes such as $d\alpha$ do not change with time,) Now use $(d/dt)y \ln y = (\ln y + 1)(dy/dt)$, which gives here

$$\frac{dH}{dt} = \int \int d\alpha d\beta \, \big(\ln \mathcal{P}(\alpha) + 1\big)[\mathcal{P}(\beta)\, \Gamma(\beta \to \alpha) - \mathcal{P}(\alpha)\, \Gamma(\alpha \to \beta)] \, .$$

Interchange α and β in the second double integral arising from the second term in square brackets:

$$\frac{dH}{dt} = \int \int d\alpha d\beta \; \mathcal{P}(\beta) \, \ln\left(\frac{\mathcal{P}(\alpha)}{\mathcal{P}(\beta)}\right) \Gamma(\beta \to \alpha) \, . \tag{2.4.4}$$

Now use the inequality that $y \ln(x/y) \leq x - y$ for any positive numbers x and y. (To prove this, note that $y \ln(x/y) - x + y$ vanishes for $x = y$, while its derivative with respect to x is $-(x - y)/x$, so it monotonically approaches zero from below for $x < y$ and then decreases monotonically for $x > y$.) From this inequality, we have

$$\frac{dH}{dt} \leq \int \int d\alpha d\beta \; [\mathcal{P}(\alpha) - \mathcal{P}(\beta)] \, \Gamma(\beta \to \alpha) \, . \tag{2.4.5}$$

Again interchange α and β, now in the first double integral:

$$\frac{dH}{dt} \leq \int \int d\alpha d\beta \; \mathcal{P}(\beta) \, [\Gamma(\alpha \to \beta) - \Gamma(\beta \to \alpha)] \, . \tag{2.4.6}$$

In the original proof it was assumed that the laws of physics are invariant under reversal of the direction of time's flow, and therefore $\Gamma(\beta \to \alpha) = \Gamma(\alpha \to \beta)$, so that Eq. (2.4.6) says that H decreases with time, in accord with the H-theorem. In studies of the decay of neutral K-mesons in 1964–1970 it was found that time-reversal invariance is not exact.[6] Fortunately, the H-theorem survives, because on very general grounds in quantum mechanics it can be shown without using time-reversal invariance that[7]

$$\int d\alpha \, [\Gamma(\alpha \to \beta) - \Gamma(\beta \to \alpha)] = 0 \, . \tag{2.4.7}$$

With Eq. (2.4.6), this is enough to require that $dH/dt \leq 0$, as was to be shown.

Let us pause for a moment to reflect how remarkable is this result. The decrease of H with time indicates a fundamental difference between past and future, even though this result would hold even if the underlying microscopic laws of physics were entirely symmetric under the direction of time's flow, and indeed as we have seen it was first derived under the assumption of

[6] K. R. Shubert *et al.*, Phys. Lett. **13**, 138 (1964). This had been strongly suggested by an earlier experiment of J. H. Christensen, J. W. Cronin, V. L. Fitch, and R. Turlay, Phys. Rev. Lett. **13**, 138 (1964).

[7] For a very general proof, with references to earlier work by others, see S. Weinberg, *The Quantum Theory of Fields*, Vol. I, pp. 150–151 (Cambridge University Press, Cambridge, UK, 1995, 2005).

time-reversal invariance. This distinction between past and future is obvious in everyday life: A glass tumbler that falls on the floor will shatter, giving up its kinetic energy to heat in the floor, but glass fragments lying on the floor will not draw energy from the floor and leap up to reassemble as a tumbler. But from where does this distinction come? It is the introduction of the concept of probability into physics that creates an asymmetry between past and future. We can try rewriting the fundamental equation (2.4.3) for the rate of change of probability by replacing t with $-t$,

$$\frac{d\mathcal{P}(\alpha)}{-dt} = \int d\beta[\mathcal{P}(\beta)\,[-\Gamma(\beta \to \alpha)] - \mathcal{P}(\alpha)\,[-\Gamma(\alpha \to \beta)]]\,,$$

but then we would have to replace the rates Γ with $-\Gamma$, which makes no sense because these rates have to be positive. It is the condition that $\Gamma \geq 0$ together with Eq. (2.4.3) that fixes the direction of time's flow. We see this also in the derivation of Eq. (2.4.5), which follows from Eq. (2.4.4) only if we assume that $\Gamma(\beta \to \alpha)$ is positive.

Canonical and Grand Canonical Ensembles

Let's now return to the H-theorem. The decrease in H will stop when H reaches a minimum value, at which it is stationary for any physically possible infinitesimal change in $\mathcal{P}(\alpha)$. For an arbitrary infinitesimal change $\delta\mathcal{P}(\alpha)$, we have

$$\delta H = \int \delta\mathcal{P}(\alpha)\, d\alpha\, [\ln \mathcal{P}(\alpha) + 1]\,.$$

Now, $\delta\mathcal{P}(\alpha)$ is not entirely arbitrary but is constrained by the condition that variations in $\mathcal{P}$ cannot change either the total probability $\int \mathcal{P}(\alpha)d\alpha = 1$ or the mean value of any conserved quantity such as the total energy $E(\alpha)$. In order that δH should vanish for *any* variation in $\mathcal{P}(\alpha)$ that preserves $\int d\alpha\mathcal{P}(\alpha)$ and the mean values of all conserved quantities, it is necessary and sufficient that $\ln \mathcal{P}(\alpha)$ should be a linear combination of a constant and any conserved quantities. For instance, if the total energy $E(\alpha)$ is the only conserved quantity (as it is for radiation) then if we denote the coefficient of $E(\alpha)$ in this linear combination as $-1/\Theta$ we have

$$\mathcal{P}(\alpha) = \exp\left[\mathcal{C} - \frac{E(\alpha)}{\Theta}\right]\,,$$

with the constant factor $e^{\mathcal{C}}$ fixed by the requirement that $\int \mathcal{P}(\alpha)\, d\alpha = 1$. We will show below that, with this probability distribution, the quantity $-H$ has the defining property (2.3.1) of entropy provided that Θ is proportional to the absolute temperature T, $\Theta = kT$, so the canonical ensemble is usually written

$$\mathcal{P}(\alpha) = \exp\left[\mathcal{C} - \frac{E(\alpha)}{kT}\right]\,. \qquad (2.4.8)$$

The value of k expresses how we convert units of temperature into units of energy, and since it is just a matter of our system of units it cannot depend on what sort of system is described by this distribution.

More generally, there may be some other conserved quantities $\mathcal{N}_i$. For instance, in a gas consisting of molecules of different types, even if these molecules are undergoing chemical reactions, under ordinary conditions the numbers $\mathcal{N}_i$ of atoms of type i do not change. In such cases, $\ln \mathcal{P}(\alpha)$ will in general contain a term proportional to each conserved quantity $\mathcal{N}_i(\alpha)$, with a coefficient that we will denote as μ_i/kT, where μ_i (or sometimes μ_i/kT) is a quantity known as the *chemical potential*. The probability density is then

$$\mathcal{P}(\alpha) = \exp\left[C - \frac{[E(\alpha) - \sum_i \mu_i \mathcal{N}_i(\alpha)]}{kT}\right] . \tag{2.4.9}$$

A multi-particle system with probabilities distributed in this way is said to form a *grand canonical ensemble*. For instance, in a gas of H_2, O_2, and H_2O molecules there are two chemical potentials, for hydrogen and oxygen atoms, so in equilibrium at a given temperature we can derive one set of ratios among the three molecular densities without knowing anything about the values of the chemical potentials, but we need to know these potentials to derive all the densities.

Connection with Thermodynamics

Aside from a constant factor, H is the entropy, as defined by Clausius in thermodynamics. To see this, suppose we slowly add heat dQ to our system, preserving the equilibrium form (2.4.8) or (2.4.9) of the distribution $\mathcal{P}$ and the values of all conserved quantities other than the energy, but shifting the average total energy $\overline{E}$ by δQ. Then

$$\begin{aligned}\delta H &= \int d\alpha\, \delta\mathcal{P}(\alpha)[\ln(\mathcal{P}(\alpha)) + 1] \\ &= -\frac{1}{kT}\int d\alpha\, \delta\mathcal{P}(\alpha) E(\alpha) = -\frac{1}{kT}\delta\overline{E} = -\frac{\delta Q}{kT}\end{aligned}$$

which is the defining equation $dS = dQ/T$ of the entropy if (apart from an arbitrary constant term) we define

$$S = -kH = -k\int d\alpha\; \mathcal{P}(\alpha)\; \ln \mathcal{P}(\alpha) , \tag{2.4.10}$$

thus justifying Eq. (2.3.4). The decrease in H implies the increase in entropy, thus justifying one consequence of the second law of thermodynamics. This was shocking to some physicists of the nineteenth century, who regarded thermodynamics as an independent theory, just as fundamental as Newtonian mechanics.

Compound Systems

Equation (2.4.10) makes it easy to justify a fundamental property of the entropy, that it is *extensive*. Suppose a system can be regarded as composed of two parts, whose states are described by parameters α_1 and α_2, and that the probabilities in these two parts are uncorrelated, so that the probability $\mathcal{P}(\alpha_1, \alpha_2)\, d\alpha_1\, d\alpha_2$ that the system is in a state with parameters in the infinitesimal ranges $d\alpha_1$ and $d\alpha_2$ around α_1 and α_2 is a product of probabilities for the separate parts:

$$\mathcal{P}(\alpha_1, \alpha_2)\, d\alpha_1\, d\alpha_2 = \mathcal{P}_1(\alpha_1)\, d\alpha_1 \times \mathcal{P}_2(\alpha_2)\, d\alpha_2\,, \tag{2.4.11}$$

with

$$\int d\alpha_1\, \mathcal{P}_1(\alpha_1) = \int d\alpha_2\, \mathcal{P}_2(\alpha_2) = 1\,.$$

Then Eq. (2.4.10) gives

$$S = -k \int d\alpha_1 d\alpha_2\, \mathcal{P}_1(\alpha_1)\, \mathcal{P}_2(\alpha_2) \left[\ln \mathcal{P}_1(\alpha_1) + \ln \mathcal{P}_2(\alpha_2) \right] = S_1 + S_2\,, \tag{2.4.12}$$

where S_1 and S_2 are the entropies of the two parts of the system:

$$S_1 = -k \int d\alpha_1\; \mathcal{P}_1(\alpha_1)\; \ln \mathcal{P}_1(\alpha_1)\,, \quad S_2 = -k \int d\alpha_2\; \mathcal{P}_2(\alpha_2)\; \ln \mathcal{P}_2(\alpha_2)\,.$$

More generally, the difference $S - S_1 - S_2$ is a measure of the degree to which probabilities in the two subsystems are correlated, and is known as the *entanglement entropy*.

Gases

In a gas $E(\alpha)$ is the sum of the energies E_a of the individual particles. The probability distribution (2.4.9) is then equal to a product of probability distributions for the individual particles:

$$\mathcal{P}(\alpha) = \prod_a p(E_a, N_{ia}) \tag{2.4.13}$$

where $p(E_a, N_{ia})$ are the probability distribution functions for the individual particle properties:

$$p(E_a, N_{ia}) \propto \exp\left(- \left(E_a + \sum_i N_{ia} \mu_i \right) / kT \right)\,, \tag{2.4.14}$$

in which N_{ia} is the number of atoms of type i in the ath molecule. The constant of proportionality must be chosen to make the individual total probabilities equal to unity. If all the molecules have the same chemical formula, so that $N_{ia} = N_i$ is the same for all molecules, then we can absorb the factor $\exp(-\sum_i N_i \mu_i / kT)$ into the constant of proportionality, and simply write

$$\mathcal{P}(\alpha) = \prod_a p(E_a) \quad \text{where} \quad p(E_a) \propto \exp(-E_a/kT) . \tag{2.4.15}$$

In particular, the distribution of the momentum $\mathbf{p}_a$ arises from the kinetic energy term $\mathbf{p}_a^2/2m_a$ in E_a, and Eq. (2.4.15) yields the Maxwell distribution (2.4.1) but, as we have now seen, derived by Boltzmann in a more convincing way.

Equipartition

One of the most useful results of statistical mechanics is the equipartition of energy in cases where the total energy $E(\alpha)$ can be written as the sum of individual energies proportional to squares of independent quantities ξ_n:

$$E(\xi_1, \xi_2, \ldots) = \sum_n c_n \xi_n^2 . \tag{2.4.16}$$

For instance, for a gas of N monatomic atoms of mass m, the index n runs over $3N$ values; the ξ_n are the three components of each atom's momentum and $c_n = 1/2m$. Molecules that are not monatomic can rotate as well as move. Here n runs over $6N$ values, with the ξ_n including the three components of each atom's momentum and the three components of its angular momentum. For an angular momentum $\mathbf{J}$, the rotational energy is

$$\frac{J_1^2}{2I_1} + \frac{J_2^2}{2I_2} + \frac{J_3^2}{2I_3} ,$$

where the I_i characterize the moments of inertia of the molecule. Here the extra ξ_n variables are the components of angular momentum, with the c_n the corresponding values of $1/2I$. But for a gas of diatomic molecules there is essentially no energy in rotations around the line separating the atoms, so here the ξ_n include only the components of each molecule's angular momentum $\mathbf{J}_\perp$ in the two directions normal to this line, and n runs over $5N$ values. For an ensemble of simple harmonic oscillators the ξ_n include both the displacement from rest of each oscillator and the displacement's rate of change. As we shall see in Section 3.1, the energy of a radiation field can also be expressed as in Eq. (2.4.16), with the ξ_n the Fourier transforms of each component of the electric and magnetic fields.

Whatever the nature of the ξ_n, because of the factorization of the exponential the probability of finding any one ξ_n in a range $d\xi_n$ takes the form $A_n \exp(-E_n/kT)\, d\xi_n$, with $E_n = c_n \xi_n^2$ and with proportionality constant A_n fixed by the condition that the total probability for each ξ_n is unity, so that $A_n \int \exp(-E_n/kT)\, d\xi_n = 1$. Thus the mean value of E_n is

$$\overline{E_n} \equiv \frac{\int_{-\infty}^{\infty} d\xi_n \, c_n \xi_n^2 \, \exp(-c_n \xi_n^2/kT)}{\int_{-\infty}^{\infty} d\xi_n \, \exp(-c_n \xi_n^2/kT)} = \frac{\int_0^\infty d\sqrt{E_n} \, E_n \, \exp(-E_n/kT)}{\int_0^\infty d\sqrt{E_n} \, \exp(-E_n/kT)}$$

$$= \frac{\int_0^\infty dE_n \, E_n^{1/2} \, \exp(-E_n/kT)}{\int_0^\infty dE_n \, E_n^{-1/2} \exp(-E_n/kT)} = \frac{(kT)^{3/2}\Gamma(3/2)}{(kT)^{1/2}\Gamma(1/2)} = kT/2 \, . \quad (2.4.17)$$

It is a fortunate aspect of kinetic theory that these mean energies do not depend on the coefficients c_n, or indeed on much else about the physical system aside from the distribution of the total energy among individual quadratic degrees of freedom.

In any gas the kinetic energy of the nth particle is $m_n \mathbf{p}_n^2/2$. The average of each of the three terms in this kinetic energy is $kT/2$, so the average kinetic energy of each particle is $3kT/2$. Equation (1.1.4) gives $m\overline{\mathbf{v}^2}/3 = kT$, where this k is the constant k in the gas constant (1.2.4), so we see that this k is the same as the constant k in the general probability distribution (2.4.8) or (2.4.9) of statistical mechanics.

For a generic polyatomic molecule the mean rotational energy associated with the three degrees of freedom J_i is $3kT/2$, but, as already mentioned, for a diatomic molecule meaningful rotation is only possible around the two axes perpendicular to the linear molecule, so the mean rotational energy is only $2kT/2$. That is, if we write the mean translational plus rotational energy per molecule as $3kT/2 \times (1 + f)$, as in Section 2.1, then $f = 0$ for monatomic molecules, $f = 2/3$ for diatomic molecules, while $f = 1$ for other molecules. Equation (2.1.9) gives the specific heat ratio as $\gamma = 1+2/3(1+f)$, so $\gamma = 5/3$ for monatomic gases, $\gamma = 7/5$ for diatomic molecules (which explains why experiments on gases like O_2 and H_2 gave results near $\gamma = 1.4$ in Clausius' time), and $\gamma = 4/3$ for other molecules.

Of course, molecules can also vibrate as well as rotate and move, and energy can also go into exciting the clouds of electrons that hold them together. For reasons that only became clear with the advent of quantum mechanics, these degrees of freedom can only be excited at temperatures much higher than is common in our environment.

Entropy as Disorder

The entropy can be regarded as a measure of the disorder of a system. To see this, it is easiest to approximate the parameters of a system as taking a discrete set of values α_ν instead of a continuum of values α. We can connect the continuum and discrete descriptions by dividing the continuum into tiny ranges $\alpha_\nu \leq \alpha \leq \alpha_\nu + \delta\alpha$ (for simplicity treating α here as if it were one-dimensional) and approximating $\mathcal{P}(\alpha)$ as a constant $\mathcal{P}_\nu/\delta\alpha$ in each interval, so that the probability that α is in this interval is

$$\int_{\alpha_\nu}^{\alpha_\nu+\delta\alpha} d\alpha\, \mathcal{P}(\alpha) = \delta\alpha \left(\frac{P_\nu}{\delta\alpha}\right) = P_\nu \ . \tag{2.4.18}$$

Then the entropy (2.4.10) is

$$S = -k \sum_\nu \delta\alpha \left(\frac{P_\nu}{\delta\alpha}\right) \ln \left(\frac{P_\nu}{\delta\alpha}\right) = \Sigma - k \sum_\nu P_\nu \ln P_\nu \tag{2.4.19}$$

where Σ is a constant,

$$\Sigma = k \ln \delta\alpha \ . \tag{2.4.20}$$

Since there was an arbitrary constant in Eq. (2.4.10), we can absorb Σ into the definition of that constant, and define the entropy simply as

$$S = -k \sum_\nu P_\nu \ \ln P_\nu \ . \tag{2.4.21}$$

Since $0 \leq P_\nu \leq 1$, each $\ln P_\nu$ is negative and S is positive. The entropy reaches its minimum value, zero, in the completely ordered state in which just a single P_ν equals one and all others vanish. In disordered systems with non-vanishing probabilities for different states S is positive-definite. In the completely disordered state with all P_ν equal, the entropy reaches its maximum possible value, equal to k times the logarithm of the number $1/P_\nu$ of intervals.

2.5 Transport Phenomena

So far, we have been concerned with systems in which thermodynamic variables such as temperature, pressure, density, etc. are constant in time and space, or vary very slowly. But many of the most interesting physical phenomena are associated with the transport of such quantities over time from one place to another in inhomogeneous media. As we shall see in the following section, the study of such transport phenomena gave physicists their first reliable values for the masses of individual atoms and molecules.

Conservation Laws

In many cases we have to deal with conserved quantities, such as the number of molecules or the total electric charge. By a quantity being conserved is meant that the net rate of increase of the quantity (negative if a decrease) in any volume plus the net rate at which this quantity flows out of the volume (negative if flowing in) vanishes. The current $\mathcal{J}$ of this quantity is defined so that the net rate of outward flow is $\int_A \mathbf{dA} \cdot \mathcal{J}$ where A is the surface surrounding the volume V, and $\mathbf{dA}$ is an element of area of this surface, taken as a vector pointing outward from the surface. Hence if this quantity has a density $\mathcal{N}$ and a current $\mathcal{J}$, then the conservation condition is

$$\frac{\partial}{\partial t}\int_V d^3x\,\mathcal{N}+\int_A \mathbf{dA}\cdot\mathcal{J}=0\,. \tag{2.5.1}$$

Using Gauss's theorem, we can write the second term in Eq. (2.5.1) as an integral over the volume of the divergence of the current, so Eq. (2.5.1) is equivalent to

$$\int_V d^3x\left[\frac{\partial}{\partial t}\mathcal{N}+\boldsymbol{\nabla}\cdot\mathcal{J}\right]=0\,,$$

and, since this must be true for any volume, the integrand must vanish:

$$\frac{\partial}{\partial t}\mathcal{N}+\boldsymbol{\nabla}\cdot\mathcal{J}=0\,. \tag{2.5.2}$$

For instance, if matter is carried from one place to another only by a bulk motion with velocity $\mathbf{v}$, then the mass density ρ satisfies an equation of the form (2.5.2), with the mass current given by $\rho\mathbf{v}$:

$$\frac{\partial}{\partial t}\rho+\boldsymbol{\nabla}\cdot(\rho\mathbf{v})=0\,. \tag{2.5.3}$$

Momentum Flow

Such conservation laws are ubiquitous in physics. We will be concerned now with a particular set of conserved quantities in fluids, the components of momentum. The density of the ith component (with $i = 1, 2, 3$) of momentum is ρv_i, where ρ is the mass density and v_i is the ith component of the bulk velocity. Their conservation provides the fundamental dynamical equation for fluids. The conservation equation here takes the general form

$$\frac{\partial}{\partial t}(\rho v_i)+\sum_j \frac{\partial}{\partial x_j}T_{ji}=0\,, \tag{2.5.4}$$

where T_{ji} is the jth component of the current of the ith component of momentum, and the sums here and below run over the directions 1, 2, 3.[8]

By analogy with the case of the mass current $\rho\mathbf{v}$, we might think that the jth component T_{ji} of the current of the ith component of momentum is $\rho v_i \times v_j$. This would be the case if momentum like mass were carried from place to place only by the bulk motion of the fluid. But of course fluid elements exert forces on one another, both pressure and viscous forces, with a consequent transfer of

[8] T_{ji} is the purely spatial part of a larger array, a tensor with time as well as space components that serves in the general theory of relativity as the source of the gravitational field.

momentum. So, to keep an open mind, let us write the jth component of the current of the ith component of momentum as

$$T_{ji} = \rho v_j v_i + \tau_{ji} \, , \tag{2.5.5}$$

with τ_{ji} a correction term arising from forces acting within the fluid. According to Eq. (2.5.4), the i-component of the internal force per unit volume is $-(\partial/\partial x_j)\tau_{ji}$.

So what is τ_{ji}? An answer was first given in 1822 by Claude-Louis Navier (1785–1836), of the Corps des Ponts et Chaussées, and later in his own formulation by Sir George Stokes (1819–1903). Rather than trying to reproduce their reasoning, we give a treatment below that has a more modern flavor, relying largely on principles of invariance.

First, we can learn a little about the momentum current T_{jk} by imposing the condition that angular momentum should satisfy a conservation condition. The density of the ith component of angular momentum is $\rho(\mathbf{x} \times \mathbf{v})_i$, so for instance the rate of change of its $i = 3$ component is

$$\frac{\partial}{\partial t}\big(\rho(\mathbf{x} \times \mathbf{v})_3\big) = \frac{\partial}{\partial t}\big(\rho(x_1 v_2 - x_2 v_1)\big) = -x_1 \sum_j \frac{\partial T_{j2}}{\partial x_j} + x_2 \sum_j \frac{\partial T_{j1}}{\partial x_j}$$
$$= -\sum_j \frac{\partial}{\partial x_j}\big(x_1 T_{j2} - x_2 T_{j1}\big) + T_{12} - T_{21} \, .$$

In order for this to take the form of a conservation law we must have $T_{12} = T_{21}$, and, since there is nothing special about the 1- and 2- directions, T_{ji} must be entirely symmetric,

$$T_{ij} = T_{ji} \, , \tag{2.5.6}$$

and then of course the same is also true of the term τ_{ji} in Eq. (2.5.5):

$$\tau_{ij} = \tau_{ji} \, . \tag{2.5.7}$$

Next, we assume that there are no preferred directions in the environment of the fluid, so that τ_{ji} is a spatial *tensor* – that is, it transforms under rotations in such a way that $\sum_{ij} a_i a_j \tau_{ji}$ is invariant under rotations for any vector $\mathbf{a}$, such as $\mathbf{v}$. In the absence of external fields, the tensor τ_{ji} must be constructed from rotationally invariant quantities like ρ and T and vectors like $\mathbf{v}$, together with their space and time derivatives, but no other vectors that would reflect a preferred direction in the environment.

One obvious such tensor is δ_{ji} times any function f of rotationally invariant quantities, where δ_{ji} is the diagonal matrix with all ones on the main diagonal. Here $\sum_{ij} a_i a_j \delta_{ji} f$ is the rotational invariant $\mathbf{a}^2 f$. We can separate a term of this form in τ_{ji} by defining a quantity

$$p \equiv \frac{1}{3} \sum_i \tau_{ii} \,, \tag{2.5.8}$$

and writing τ_{ji} as

$$\tau_{ji} = p\delta_{ji} + \Delta\tau_{ji} \,, \tag{2.5.9}$$

where $\Delta\tau_{ji}$ is both symmetric and traceless:

$$\Delta\tau_{ij} = \Delta\tau_{ji} \,, \qquad \sum_i \Delta\tau_{ii} = 0 \,. \tag{2.5.10}$$

The term $p\delta_{ji}$ in T_{ji} gives a force per unit volume $-\nabla p$ in Eq. (2.5.4), so p can be identified as the fluid's pressure. Of course, there is an infinite number of ways of constructing the symmetric traceless tensor $\Delta\tau_{ji}$ from the velocity and rotational invariants and their derivatives. One simple example is $[v_i v_j + v_j v_i - 2\delta_{ij}\mathbf{v}^2/3]f$, where f is any function of the rotational invariants. Fortunately, we can eliminate many of these possibilities (including this one) by using the principle of Galilean relativity.

Galilean Relativity

The principle of Galilean relativity[9] requires that the laws governing fluids should be the same for an observer $\mathcal{O}$ who uses space coordinates $\mathbf{x}$ and for an observer $\mathcal{O}'$ moving at any constant velocity $-\mathbf{u}$ with respect to $\mathcal{O}$, and who therefore uses coordinates related to those of $\mathcal{O}$ by

$$\mathbf{x}' = \mathbf{x} + \mathbf{u}t \,. \tag{2.5.11}$$

Aside from this change of coordinates, the moving observer sees a mass density ρ' that is the same as ρ:

$$\rho'(\mathbf{x}', t) = \rho(\mathbf{x}, t) \,. \tag{2.5.12}$$

But, for the observer $\mathcal{O}'$, his own velocity $-\mathbf{u}$ is subtracted from the velocity seen by observer $\mathcal{O}$:

$$\mathbf{v}'(\mathbf{x}', t) = \mathbf{v}(\mathbf{x}, t) + \mathbf{u} \,. \tag{2.5.13}$$

To check whether the equation (2.5.3) of mass conservation is left invariant by Galilean transformations, take the partial derivative of Eq. (2.5.12) with respect to time, holding $\mathbf{x}$ (but not $\mathbf{x}'$!) fixed:

$$\frac{\partial \rho'(\mathbf{x}', t)}{\partial t} + \sum_i \frac{\partial \rho'(\mathbf{x}', t)}{\partial x'_i} u_i = \frac{\partial \rho(\mathbf{x}, t)}{\partial t} \,.$$

[9] In the twentieth century this came to be called Galilean relativity to distinguish it from the Einstein special principle of relativity. Both principles state that the laws of nature are unaffected by the uniform motion of an observer; as we will see in Chapter 4, it is only the details of the transformation to a moving frame of reference that distinguishes Einsteinian from Galilean relativity.

Therefore

$$\frac{\partial \rho'(\mathbf{x}',t)}{\partial t} + \nabla' \cdot \big(\mathbf{v}'(\mathbf{x}',t)\rho'(\mathbf{x}',t)\big)$$
$$= \frac{\partial \rho(\mathbf{x},t)}{\partial t} + \nabla' \cdot \big((\mathbf{v}'(\mathbf{x}',t) - \mathbf{u})\rho'(\mathbf{x}',t)\big)$$
$$= \frac{\partial \rho(\mathbf{x},t)}{\partial t} + \nabla \cdot (\mathbf{v}(\mathbf{x},t)\rho(\mathbf{x},t)) = 0 \tag{2.5.14}$$

and so the equation (2.5.3) of mass conservation does satisfy the principle of Galilean relativity.

By following the same reasoning, we can see that the momentum conservation law (2.5.4) would be invariant under Galilean transformations if T_{ji} were simply given by the term $\rho v_j v_i$ in Eq. (2.5.5). Hence the principle of Galilean relativity requires that the term τ_{ji} in Eq. (2.5.5) be separately invariant under Galilean transformations, and according to Eqs. (2.5.8) and (2.5.9) the same must be true of p and $\Delta\tau_{ji}$. Because of the term $\mathbf{u}$ in the Galilean transformation (2.5.13), Galilean relativity rules out terms in $\Delta\tau_{ji}$ such as in the example $v_i v_j + v_j v_i - 2\delta_{ij}\mathbf{v}^2/3$ mentioned above, which involves $\mathbf{v}$ itself rather than its gradient.

Navier–Stokes Equation

There are still an infinite variety of terms that might appear in $\Delta\tau_{ji}$, containing any number of factors of gradients of any order of density and/or velocity. But in order to keep the units consistent, the more gradients are contained as factors in any term in $\Delta\tau_{ji}$, the more powers of some length that is characteristic of the microscopic properties of the fluid must appear in the coefficient of that term. If these lengths characterizing the fluid, such as the distance between molecules and the mean free path, are all much less than the scale of distances over which fluid properties such as density and velocity vary, then $\Delta\tau_{ji}$ is dominated by a term proportional to the minimum number of gradients.[10] So we should look for a possible term in $\Delta\tau_{ij}$ proportional to a single gradient.

It is not possible to construct a symmetric traceless tensor proportional to a single gradient of the density, so a tensor proportional to a single gradient must be linear in the gradient of the velocity. There is a unique symmetric traceless tensor of this sort:

$$\Delta\tau_{ji} = -\eta\left[\frac{\partial v_j}{\partial x_i} + \frac{\partial v_i}{\partial x_j} - \frac{2}{3}\delta_{ij}(\nabla \cdot \mathbf{v})\right], \tag{2.5.15}$$

[10] This sort of reasoning has become common in the quantum theory of fields, leading to what are known as effective-field theories.

where η is a coefficient (Galilean-invariant, like ρ and p) known as the *viscosity* of the fluid.[11] A minus sign is inserted in Eq. (2.5.15) in order that, with $\eta > 0$, the heat produced by viscous fluid flow should be positive.[12] Using Eqs. (2.5.5), (2.5.9), and (2.5.15), we see that the momentum conservation equation (2.5.4) takes the form

$$\frac{\partial}{\partial t}(\rho v_i) = -\frac{\partial p}{\partial x_i} + \sum_j \frac{\partial}{\partial x_j}\left[-\rho v_i v_j + \eta\left[\frac{\partial v_j}{\partial x_i} + \frac{\partial v_i}{\partial x_j} - \frac{2}{3}\delta_{ij}(\nabla \cdot \mathbf{v})\right]\right] . \tag{2.5.16}$$

This is the Navier–Stokes equation.

Viscosity

The measurement of viscosity was well within the capabilities of nineteenth century physicists. In a classic calculation using the Navier–Stokes equation, Stokes found that a uniform fluid with viscosity η exerts a drag force F on a spherical ball of radius a moving with velocity v through the fluid, given by

$$F = 6\pi \eta a v . \tag{2.5.17}$$

For instance, if a ball of mass m falls through a fluid, it accelerates until the viscous force balances the force mg of gravity (neglecting buoyancy), when it has the terminal velocity

$$v_{\text{terminal}} = \frac{mg}{6\pi \eta a} .$$

The viscosity of gases could also be measured by observing the effect of a surrounding gas on the motion of a pendulum.

It was harder to calculate η on the basis of a theory of molecules than to measure it. For some time the best that could be done theoretically was a rough estimate of this viscosity.

To make this estimate, consider a uniform fluid experiencing a shear flow. For instance, suppose $\mathbf{v}$ has only one component, v_1, which depends only on x_3. (The fluid could be enclosed between two flat plates, each in the 1–2 plane, with their separation in the 3-direction, and with one of the plates moving in the

[11] Often η is called the *shear* viscosity. The reason is that, if we were to insist on using whatever formula for the pressure p holds in the absence of fluid gradients, then p would not be precisely given by Eq. (2.5.8), and $\Delta\tau_{ji}$ as defined by Eq. (2.5.9) would not be precisely traceless, so it would have a term proportional to $\delta_{ij}(\nabla \cdot \mathbf{v})$, with a coefficient known as the *bulk viscosity*. For complicated reasons the bulk viscosity is generally much less than the shear viscosity (for instance, see S. Weinberg, Astrophys. J. **168**, 175 (1971)) and in any case would have no effect in our present calculation.

[12] For the details of this argument, see Sections 16 and 49 of Landau and Lifshitz, *Fluid Mechanics*, listed in the bibliography.

1-direction and the other at rest.) In this case, Eqs. (2.5.5), (2.5.9), and (2.5.15) give the 3-component of the current of the 1-component of momentum:

$$T_{31} = \Delta\tau_{31} = -\eta \frac{\partial v_1}{\partial x_3} . \tag{2.5.18}$$

To find η, let us use molecular theory to calculate the rate per unit area at which the 1-component of momentum crosses a plane normal to the 3-axis, which we will take as the plane $x_3 = 0$. This current arises because, in addition to being carried along in the 1-direction by the bulk velocity $\mathbf{v}$, each molecule has a fluctuating "peculiar velocity" $\Delta\mathbf{v}$. We make the far-reaching approximation that, because of rapid collisions, all directions of this peculiar velocity are equally likely. Then the number per unit volume whose peculiar velocity vector $\Delta\mathbf{v}$ makes an angle with the +3-axis between θ and $\theta + d\theta$ is the ratio of the solid angle $2\pi \sin\theta \, d\theta$ to 4π, times the total number density n, or $n \sin\theta \, d\theta/2$. As we saw in our calculation of gas pressure in Section 1.1, the number of these molecules striking an area dA in this plane in a time dt is the number in a cylinder with base dA and height $\Delta v_\perp \, dt = \cos\theta \, |\Delta\mathbf{v}| \, dt$, where $\Delta v_\perp$ is the component of the peculiar velocity normal to the plane $x_3 = 0$ and $|\Delta\mathbf{v}|$ is the magnitude of the peculiar velocity. This number of molecules is equal to

$$dA \times \cos\theta \, |\Delta\mathbf{v}| \, dt \times n \sin\theta \, d\theta/2 .$$

Since v_1 is assumed to be a function $v_1(x_3)$ only of x_3, a molecule that reaches the plane $x_3 = 0$ having traveled a distance r will have a 1-component of momentum $mv_1(-r\cos\theta)$, where m is the mass of the molecule. (In addition to the momentum carried by this bulk velocity, the peculiar momenta of molecules will also have 1-components, but under the assumption that all directions of peculiar velocity are equally likely, these 1-components cancel when we integrate over the azimuthal angle around the 3-direction.) A minus sign appears in the argument of v_1 because a molecule with a positive (or negative) 3-component $v_\perp$ of peculiar velocity, for which $\cos\theta > 0$ (or $\cos\theta < 0$), arrives at the plane $x_3 = 0$ from negative (or positive) values of x_3. The rate per unit area and per unit time at which the 1-component of momentum flows through the plane $x_3 = 0$ is then

$$T_{31} = \int_0^\pi \left(\frac{n\overline{|\Delta\mathbf{v}|}}{2} \right) \cos\theta \, \sin\theta \, d\theta \int_0^\infty mv_1(-r\cos\theta) \, P(r) \, dr ,$$

where $P(r)dr$ is the probability that a molecule that reaches the plane $x_3 = 0$ has traveled a distance between r and $r + dr$ since its last collision with another molecule, and the bar again denotes an average over molecules. As long as the mean distance between collisions is small compared with the scale of distances over which the fluid properties vary, all directions are equivalent, and $P(r)dr$ is also the probability that from a random starting position a molecule will travel a distance between r and $r + dr$ before its first collision. (Note that this formula applies for molecules with negative as well as positive values of $\cos\theta$, because molecules with negative values of $\cos\theta$ have a negative 3-component of peculiar velocity and therefore cross the plane $x_3 = 0$ traveling from positive to negative values of x_3, and thus contribute a negative amount to the flow of the 1-component of momentum through this plane.)

We again make the crucial assumption, which led to the Navier–Stokes equation, that the typical distances traveled by molecules are much smaller than the scale of distances over which the bulk properties of the fluid vary. Here this implies that $v_1(-r\cos\theta)$ changes little over the range of r for which $P(r)$ is not negligible. This allows us to use a Taylor expansion

$$v_1(-r\cos\theta) = v_1(0) - r\cos\theta\left(\frac{\partial v_1}{\partial x_3}\right)_{x_3=0} + \cdots .$$

The first term makes no contribution to the current, because the integral over θ of $\cos\theta\sin\theta$ vanishes. This leaves us with the contribution of the next term,

$$T_{31} = -\frac{nm\ell\overline{|\Delta\mathbf{v}|}}{2}\left(\frac{\partial v_1}{\partial x_3}\right)_{x_3=0}\int_0^\pi \cos^2\theta\,\sin\theta\,d\theta = -\frac{n\ell\overline{|\Delta\mathbf{v}|}}{3}\left(\frac{\partial v_1}{\partial x_3}\right)_{x_3=0},$$

where ℓ is the mean free path[13]

$$\ell \equiv \int_0^\infty r\,P(r)\,dr.$$

Comparing this with our formula (2.5.18) for T_{31}, taken from Eq. (2.5.15), we find for η the positive value

$$\eta = \frac{1}{3}m\ell\,n\,\overline{|\Delta\mathbf{v}|}\,. \tag{2.5.19}$$

Mean Free Path

Now we need to estimate the mean free path ℓ. Suppose we make the crude approximation that a molecule will collide with another molecule if its center passes within an effective cross-sectional area σ around the center of the other molecule. (For instance, if molecules were balls of radii a, then they would collide if their centers approached within a distance $2a$, so here $\sigma = \pi(2a)^2$.) The probability that a molecule that has already traveled a distance r without colliding will collide before it travels a further distance dr is the ratio of the total effective area $4\pi r^2 dr\,n\,\sigma$ of all the molecules in the shell between r and $r + dr$ to the area $4\pi r^2$ of this shell, and is therefore $n\sigma\,dr$. The probability that the collision occurs in the distance between r and $r + dr$ is then $n\sigma\,dr$

[13] The notion of a mean free path was introduced by Rudolf Clausius in "On the Mean Lengths of the Paths Described by the Separate Molecules of Gaseous Bodies," Ann. Phys. **105**, 239 (1858).

times the probability $p(r)$ that it had *not* collided before it had traveled the distance r. To calculate $p(r)$, we note that $p(r+dr)$ equals $p(r)$ times the probability $1 - n\sigma dr$ that the molecule will not collide before it travels to $r + dr$, so $p'(r) = -p(r)n\sigma$ and, since $p(0) = 1$, the probability of traveling a distance r without colliding is $p(r) = \exp(-n\sigma r)$. The probability of a collision in a distance from r to $r + dr$ is then

$$P(r)dr = n\sigma dr \times p(r) = n\sigma dr \, \exp(-n\,\sigma r) \ .$$

The average distance traveled between collisions is then

$$\ell \equiv \int_0^\infty r\,P(r)\,dr = \int_0^\infty r\,n\sigma dr\,\exp(-n\,\sigma r) = \frac{1}{n\sigma} \ . \tag{2.5.20}$$

This formula for ℓ is often used for media more complicated than a gas of hard balls, by taking σ as some sort of effective cross section.

Using the result (2.5.20) in Eq. (2.5.19) gives an estimate of the viscosity:

$$\eta \simeq \frac{m\,\overline{|\Delta\mathbf{v}|}}{3\sigma} \ .$$

The Maxwell–Boltzmann distribution (2.4.1) gives the mean value of $|\Delta\mathbf{v}|$ as

$$\overline{|\Delta\mathbf{v}|} = \sqrt{\frac{kT}{2\pi m}} = \sqrt{\frac{RT}{2\pi\mu}} \ ,$$

where $R = k/m_1$ is the gas constant and $\mu = m/m_1$ is the molecular weight. The viscosity is therefore

$$\eta \simeq \frac{m}{3\sigma}\sqrt{\frac{RT}{2\pi\mu}} \ . \tag{2.5.21}$$

Quantitatively this result correctly only gives the order of magnitude of η, but it has an important qualitative consequence, that the viscosity is independent of the gas density. This result was first found by Maxwell.[14] In a letter to Stokes,[15] he commented that "This is certainly very unexpected, that the friction should be as great in a rare gas as in a dense gas. The reason is that for the rare gas the mean path is greater, so that the frictional action extends to greater distance."

One reason for finding this result surprising is that it raises the question whether a gas that is so rare that it is practically a vacuum can have any viscosity? It was this point that had led Aristotle in his book *Physics* to argue that a vacuum is impossible. He concluded from his experience with motion under the influence of friction that the velocity imparted to a body by a given force is inversely proportional to the resistance, thus anticipating Stokes' law

[14] J. C. Maxwell, "Illustrations of the Dynamical Theory of Gases," Phil. Mag. **19**, 19; **20**, 21 (1860).

[15] Quoted on p. 27 of Brush, *The Kinetic Theory of Gases*, listed in the bibliography.

(2.5.17), and so he reasoned that in a vacuum where there can be no resistance all velocities would be infinite.

But, as we have seen, the derivation of the Navier–Stokes equation and Stokes' law rests on the assumption that the mean free path ℓ is much smaller than the scale of distances over which the fluid velocity varies. For a gas of sufficiently low density this will no longer be the case, and the concept of viscosity loses any meaning. For instance, when a spacecraft or a missile re-enters the Earth's atmosphere at very great altitude, where the mean free path is much larger than the dimensions of the re-entering body, the drag force F on the body is at first *not* proportional to its velocity as would be required by Stokes' law for spheres, but rather is $F = C_D \rho A v^2$. Here ρ is the air density; A is the vehicle's cross-sectional area; v is its velocity; and C_D is a dimensionless "drag coefficient" that depends on the shape of the body. This is the *Knudsen regime*. Only when the body reaches lower altitudes with much smaller mean free paths does the drag force become proportional to velocity and the body approach terminal velocity.

The fact that viscous drag is independent of gas density was regarded as a confirmation of the molecular theory of gas dynamics. Maxwell himself checked the validity of this result by measuring the viscosity of a gas at fixed temperature, with pressure and hence density varying by a factor 60. The observed constancy of viscosity over this large range of gas density tended to confirm the molecular theory of gases, but in itself it revealed nothing about the nature of molecules.

Diffusion

The general formulation above of the transport of momentum in a gas can be extended to the transport of other physical quantities in general fluids. One such quantity is the number density ν of particles suspended in a fluid. These can be large molecules, such as molecules of sugar dissolved in water, or the tiny bits of organic matter expelled from pollen grains noticed in 1827 by the botanist Robert Brown (1773–1858), or artificial little balls used in studies of diffusion to be discussed in the next section. The conservation of these particles requires that their number density $\nu(\mathbf{x}, t)$ satisfies an equation of the general form (2.5.2):

$$\frac{\partial}{\partial t}\nu + \nabla \cdot \big(\nu \mathbf{v} + \mathbf{j}\big) = 0\,, \tag{2.5.22}$$

where $\mathbf{v}$ is the fluid bulk velocity. As in Eq. (2.5.5) we again separate the convective term $\nu\mathbf{v}$ in the current from the diffusion term $\mathbf{j}$. Since by itself $(\partial/\partial t)\nu + \nabla \cdot (\nu\mathbf{v})$ is Galilean-invariant, the diffusion term $\mathbf{j}$ must be a Galilean-invariant vector, and if the scale over which the density ν varies is much larger than relevant mean free paths then it is dominated by a term with a single gradient, which can only be of the form

$$\mathbf{j} = -D\nabla\nu \;, \tag{2.5.23}$$

where D is a coefficient known as the diffusion constant.

For instance, if the fluid is at rest and D is independent of time and position then Eq. (2.5.22) takes the form

$$\frac{\partial}{\partial t}\nu = D\nabla^2\nu \;. \tag{2.5.24}$$

Here is one solution:

$$\nu(\mathbf{x},t) = \frac{N}{(4\pi Dt)^{3/2}}\exp\left(-\frac{\mathbf{x}^2}{4Dt}\right) \;, \tag{2.5.25}$$

where N is a constant equal to the number $\int \nu\, d^3x$ of particles suspended in the fluid. (This is one way of seeing that the coefficient D defined by Eq. (2.5.23) must be positive.) This distribution is spherically symmetric and localized within a radius of order $\sqrt{4Dt}$, which spreads with time owing to the diffusion of the suspended particles through the fluid.

A vivid description of how diffusion arises from the microscopic motion of suspended particles was given in 1905 by Albert Einstein[16] (1879–1955). Consider a time interval τ that is short compared with the times over which the distribution function changes appreciably but long enough that typical suspended particles collide many times with the molecules of the fluid. In this time the position of each suspended particle jumps by some random vector amount $\mathbf{\Delta}$. These amounts differ from one suspended particle to another, in a way that is governed by some sort of statistical distribution. Then for vanishing bulk velocity $\mathbf{v}$, the number density $\nu(\mathbf{x},t)$ is changed in this time interval to

$$\nu(\mathbf{x}, t+\tau) = \overline{\nu(\mathbf{x}+\mathbf{\Delta}, t)} \;,$$

the bar indicating an average over the suspended particles. Assuming that $\nu(\mathbf{x},t)$ is slowly varying over times of order τ and distances of order $|\mathbf{\Delta}|$, we can expand both sides as Taylor series in τ and $\mathbf{\Delta}$:

$$\tau\frac{\partial}{\partial t}\nu(\mathbf{x},t) + \cdots = \sum_i \overline{\Delta_i}\frac{\partial\nu(\mathbf{x},t)}{\partial x_i} + \frac{1}{2}\sum_{ij}\overline{\Delta_i\Delta_j}\frac{\partial^2\nu(\mathbf{x},t)}{\partial x_i\partial x_j} + \cdots \;.$$

Under the assumption that all directions of $\mathbf{\Delta}$ are equally likely, we have

$$\overline{\Delta_i} = 0, \qquad \overline{\Delta_i\Delta_j} = \frac{\delta_{ij}}{3}\overline{|\mathbf{\Delta}|^2} \;;$$

so, to leading order,

$$\tau\frac{\partial}{\partial t}\nu(\mathbf{x},t) = \frac{1}{6}\overline{|\mathbf{\Delta}|^2}\,\nabla^2\nu(\mathbf{x},t) \;.$$

[16] A. Einstein, Ann. Phys. **17**, 549 (1905).

Comparing this with the diffusion equation (2.5.24) for zero bulk velocity, we see that the mean square displacement increases as

$$\overline{|\boldsymbol{\Delta}|^2} = 6\tau D\ . \tag{2.5.26}$$

This is in accord with the particular solution (2.5.25) of the diffusion equation; calculating the integral $\overline{\mathbf{x}^2} = \int N^{-1}\nu(\mathbf{x},t)\mathbf{x}^2\, d^3x$ gives $\overline{\mathbf{x}^2} = 6Dt$.

The diffusion constant can be measured by observing this spreading out of the suspended particles with time. The calculation by Einstein of the diffusion constant D in terms of fundamental constants and its use to measure these constants are discussed in the next section.

2.6 The Atomic Scale

We have seen various ways in which observations in the nineteenth century provided physicists with the values of only the *ratios* of quantities that characterize the scale of individual atoms or molecules. The study of gases allowed measurement of the gas constant $R \equiv k/m_1$ (where k is Boltzmann's constant and m_1 is the mass an atom would have if it had atomic weight unity, related to Avogadro's number by $N_A = 1/m_1$); the study of electrolysis allowed measurement of the faraday, $F \equiv e/m_1$ (where e is the minimum electric charge that is transferred in electrolysis; and, under the assumption that the charge of the electron is the same as the unit e of electric charge transferred in electrolysis, the study of the bending of cathode rays allowed measurement of e/m_e. Furthermore, under the assumption that molecules are tightly packed in liquids and solids, knowledge of the mass density of a liquid or solid gave an approximate value for the ratio of the mass to the volume of individual molecules. A measurement of m_1 or k or e or m_e or the size of any molecule would yield results for all these quantities. No accurate measurements of any of these individual quantities were possible before the twentieth century, which is not to say that nineteenth century chemists and physicists did not try.

Nineteenth Century Estimates

According to Eq. (2.5.21), the viscosity of a gas of known temperature and molecular weight is given by known quantities times m/σ, where m is the mass of the gas molecules and σ is their effective cross-sectional area. Defining an effective radius a by $\sigma \equiv \pi a^2$ and setting the density in liquid form equal to $m/(4\pi a^3/3)$ gave a rough estimate of both a and m. In this way, in 1865 Josef Loschmidt (1821–1895) estimated that air molecules have a diameter of about 10^{-7} cm, and in effect that $m_1 \approx 2 \times 10^{-23}$ g, about ten times too large.

Using this and similar studies of gas properties, G. J. Stoney (1826–1911) in 1874 estimated in effect that $m_1 \approx 10^{-25}$g. Then, using the value of e/m_1 measured in electrolysis, he estimated that $e \approx 10^{-20}$ coulombs. He called this the *electrine*.

Soon after the discovery of the electron, efforts were made to measure its charge directly. At Thomson's Cavendish Laboratory, J. S. E. Townsend (1868–1957) studied falling clouds of water droplets that formed around electrically charged ions in gases produced in electrolysis. If the droplets have radius a and mass m then, as discussed in Section 2.5, they reach a terminal velocity $mg/6\pi\eta a$ at which viscous drag balances gravity, where η is the air viscosity. Measuring the terminal velocity and air viscosity then gave a value for m/a. A second relation between m and a was provided by the known density ρ of liquid water, which gives $m = 4\pi a^3\rho/3$, so both m and a were known. The droplets were collected, and their total mass and charge were measured. The ratio of the total mass to the known mass m of each droplet gave the number of droplets, and the ratio of the total electric charge to the number of droplets then gave the charge per droplet. This charge was reported to be always close to integer multiples of the same unit of charge, which Townsend estimated to be 1.1×10^{-19} coulombs, about 10 times the value found by Stoney. Similar results, none very accurate, were obtained by Thomson himself and by H. A. Wilson (1874–1964).

The early years of the twentieth century saw a great improvement in scientists' knowledge of atomic magnitudes. This improvement came from three chief sources:

- accurate direct measurement of the electric charges carried by oil droplets gave a value for e;
- measurements of effects due to the diffusion of small spheres suspended in a fluid gave a value for Avogadro's number[17] $N_A = 1/m_1$;
- the study of black body radiation gave a value for k.

Electronic Charge

One of the problems with the water droplets studied by Townsend *et al.* in their estimates of the charge of the electron was that the masses of the droplets did not remain fixed during the experiment, because water evaporates. To avoid this, Robert Andrews Millikan (1868–1953) in 1906 studied individual oil

[17] This is the way in which these experimental results were quoted by physicists at the time and have generally been described by historians since then, but it is misleading. The formulas used to analyze these experiments actually involved RN_A, where R is the gas constant appearing in the ideal gas law (1.2.3). Since R had already been measured, the measured value of RN_A could be used to find N_A. But since $R = k/N_A$, they were really measuring k, not N_A. I suppose that the results were cited in terms of N_A rather than k because Avogadro's number was much more familiar to physicists of the time than the Boltzmann constant of statistical mechanics.

drops that had picked up electric charge from air ionized by X-rays. Unlike the water droplets in the experiments of Townsend *et al.*, these oil drops were large enough that Millikan could study the motion of individual drops. As in the earlier experiments, Millikan could measure the mass m and radius a of individual drops from their terminal velocity in the absence of any external electric fields, using the known density of oil and viscosity of air. Then, when he turned on a strong vertical electric field E, a drop carrying electric charge q would feel an electric force qE in addition to the gravitational force mg, so the terminal velocity would be altered by an amount $qE/6\pi\eta a$. Measuring changes in the terminal velocity, and knowing m and a, it was then possible to calculate the changes in the drops' electric charges. For instance, in one run the changes in the electric charge q (in units of 10^{-19} coulombs) were

$$9.91, \quad -11.61, \quad 1.66, \quad 5.00, \quad 1.68, \quad -8.31, \quad 6.67, \quad 5.02, \quad \text{etc.},$$

all close to integer multiples of 1.66×10^{-19} coulombs. After repeated runs, Millikan concluded that the fundamental unit of electric charge is $e = (1.592 \pm 0.003) \times 10^{-19}$ coulombs. (The modern value is $e = 1.6021765 \times 10^{-19}$ coulombs.) This immediately allowed the calculation of m_1 (from the faraday e/m_1), and then k (from the ideal gas constant k/m_1), and so on. Even more importantly, the observation that droplet charges come close to integer multiples of a unit charge gave direct evidence for the discreteness of electric charge.

Brownian Motion

The diffusion of particles suspended in a fluid depends on the size and shape of the particles, as well as on the fluid properties and fundamental constants. Where the particles are molecules, such as sugar molecules dissolved in water, it is not possible to deduce relevant information about their size and shape with any precision from the properties of solids or liquids composed of these molecules. In the first decade of the twentieth century Einstein had the idea of learning about fundamental constants from observation of the diffusion of artificial particles, like little spherical balls, whose shape, size, and mass were accurately known. (This diffusion is a special case of what is termed "Brownian motion," after the botanist Robert Brown mentioned in the previous section.)

Einstein took notice of the common observation that it is possible to have a time-independent inhomogeneous equilibrium distribution of particles such as little balls suspended in a fluid, in which the effect of diffusion is cancelled by a steady external force F acting on each ball. For example, this force could be the combined force of gravity and buoyancy, so that it has magnitude $F = g(m - m_{disp})$ (where g is the gravitational acceleration, m is the ball's mass, and m_{disp} is the mass of the fluid displaced by the ball), and it acts in the $-z$ direction, where z is altitude. In equilibrium this is balanced by a kind of pressure, known as the *osmotic pressure*. With the balls in thermal equilibrium with the fluid at a uniform temperature T, and therefore with a kinetic energy given by the equipartition of energy as $3kT/2$, their random motion, which is responsible for diffusion, exerts a pressure that, according to the same arguments used to derive the ideal gas law (1.1.5), has the value $p(z) = \nu(z)kT$, where $\nu(z)$ is the number of these balls per unit volume at vertical coordinate z. In equilibrium at uniform absolute temperature T the balance of forces acting on such suspended particles in a slab of area A between altitudes z and $z + dz$ then requires that

$$[\nu(z) - \nu(z + dz)]AkT = F \times \nu(z)Adz \ ,$$

and therefore the force creates a decrease in the density of little balls with increasing altitude

$$\nu'(z)kT = -F\nu(z) \ . \tag{2.6.1}$$

Einstein pointed out that in addition to a balance of forces, in equilibrium there has to be a balance of currents. According to Stokes' law, the external force F acting on each little ball gives it a downward velocity $v = F/6\pi\eta a$, where a is the ball radius and η is the fluid viscosity. If not compensated by diffusion, this would give these balls a current $-v\nu = -F\nu/6\pi\eta a$, the minus sign indicating that this current is in the downward direction. But because ν decreases with increasing altitude, diffusion produces an upward current given by Eq. (2.5.23) as $-D\nu'(z)$, The cancellation of these two currents in equilibrium requires that

$$F\nu(z)/6\pi\eta a = -D\nu'(z) \ . \tag{2.6.2}$$

Einstein used Eq. (2.6.1) to eliminate the quantity $\nu'/F\nu$ in Eq. (2.6.2), and concluded that[18]

$$D = \frac{kT}{6\pi\eta a} \ . \tag{2.6.3}$$

In the appendix to this section a more direct derivation of this formula is given (taking account of a possible correction) for the case where there is no external force, and diffusion is actually taking place.

Unlike sugar molecules or the grains observed in Brownian motion, the artificial little balls used for this purpose could be chosen to have a known uniform radius a, so by measuring D at a given temperature T in a fluid of known viscosity η, it was possible to find the Boltzmann constant k.

[18] A. Einstein, "On the Motion of Small Particles Suspended in Liquids at Rest Required by the Molecular Theory of Heat," Ann. Phys. **17**, 549 (1905). Because F has dropped out of the final formula for D, Einstein's result is independent of the nature of the force acting on the suspended particles, though for simplicity we have assumed that this force is independent of position.

Within a few years after Einstein's 1905 paper, an experimental study of the diffusion of small bodies was carried out in Paris by Jean Perrin (1870–1942). Perrin measured k (or as he said, Avogadro's number) by observing the decrease with altitude of the density of little balls suspended in a vertical column of fluid. Equation (2.6.1) has the elementary solution

$$\nu(z) \propto \exp(-Fz/kT) \; . \tag{2.6.4}$$

(Perrin gave this solution in the form $\nu(z) \propto \exp[-N_A Fz/RT]$.) Using the known value of the combined gravitational and buoyancy force F on the little balls gave a value for $N_A/R = 1/k$, which by using the known value of the gas constant R, Perrin reported[19] as a value for Avogadro's number $N_A = R/k = 7.05 \times 10^{23}$/mole, corresponding to $m_1 = 1.42 \times 10^{24}$ g. As was usual at the time, no figure was given for the uncertainty of the measurement.

Perrin also used microscopic measurements over several minutes of the root mean square diffusion of suspended balls in the horizontal direction, in which no force is acting. He found that as expected the mean square displacement is proportional to the elapsed time. Using Eq. (2.5.26) gave a value for the diffusion constant D and, using Einstein's formula (2.6.3) (which Perrin like Einstein wrote as $D = RT/6\pi\eta a N_A$) he found[20] that $N_A = 7.15 \times 10^{23}$/mole, corresponding to $m_1 = 1.40 \times 10^{-24}$ g. The fair agreement of this result, which was obtained by direct observation of diffusing particles, with Perrin's earlier measurement based on equilibrium in a vertical column gave support to the view that diffusion is due to the motion of the balls in equilibrium with randomly moving molecules. Perrin was not hesitant in concluding that his work confirmed the reality of molecules – his results were summarized in a long article[21] titled "Brownian Movement and Molecular Reality." His measurements were not far off – with modern definitions of molecular weight, the value of Avogadro's number is 6.022142×10^{23}/mole.

Black Body Radiation

As we will see in the next chapter, in 1900 the early ideas of Max Planck[22] (1858–1947) about quantum theory led to a formula for the distribution with frequency of the radiation energy emitted by a totally absorbing body, which depended on the value of kT. Comparison of Planck's formula with observation gave $k \simeq 1.34 \times 10^{-16}$ erg/K.

---

19 J. Perrin, Comptes rendus **cxlvi**, 167 (1908) and **cxlvii**, 530 (1908).
20 J. Perrin, Comptes rendus **cxlvii**, 1044 (1908).
21 See Perrin, *Brownian Movement and Molecular Reality*, listed in the bibliography.
22 M. Planck, Verh. d. deutsche phys. Ges. **2**, 202, 237 (1900).

Consistency

The atomic theory underlying these measurements of microscopic parameters and the values found gained much credit from the consistency of the results obtained. For instance, in 1901 Planck used his measurement of k together with the known value $R = 8.27 \times 10^7$ erg/mole K of the gas constant to calculate a value for Avogadro's number $N_A = R/k = 6.17 \times 10^{23}$/mole, in fair agreement with Perrin's later result $N_A \simeq 7 \times 10^{23}$/mole. Planck also used this result together with the known value of the faraday, $F = eN_A = 9.63 \times 10^4$ coulombs/mole, to calculate the unit of charge, $e = 1.56 \times 10^{-19}$ coulombs, in very good agreement with Millikan's result.

This happy agreement of fundamental constants led to a widespread acceptance of the atomic theory of matter. For instance, the chemist F. W. Ostwald (1853–1932) had been a determined opponent of the atomic theory, but in 1908 he finally admitted that "I am now convinced that we have recently become possessed of experimental evidence of the discrete or grained nature of matter, which the atomic hypothesis sought in vain for hundreds and thousands of years."

An adverse voice remained. The physicist–philosopher Ernst Mach (1838–1916), who spoke of "the artificial hypothetical atoms of chemistry and physics," never accepted their existence. As late as 1916, shortly before his death, he declared that "I can accept the theory of relativity as little as I can accept the existence of atoms and other such dogmas." This goes to show that a scientist can maintain his own principles, bravely holding out against a wide consensus of the scientific establishment, and still be wrong.

Appendix: Einstein's Diffusion Constant Rederived

Einstein's derivation of Eq. (2.6.3) for the diffusion constant D relied on the introduction of an external force F acting on suspended particles, which prevents their diffusion from disturbing a time-independent equilibrium particle distribution. The presence of such an external force such as gravity is not uncommon, but it ought to be possible to obtain the same result where there is no external force, and where diffusion is actually taking place. Below is such a derivation, which indicates the presence of a correction for particles whose mass is not negligible.

The mean velocity $\mathbf{v}(\mathbf{x}, t)$ of diffusing suspended particles at position $\mathbf{x}$ and time t is given by setting the current (2.5.23) equal to $\nu\mathbf{v}$:

$$\mathbf{v}(\mathbf{x}, t) = -\frac{D\boldsymbol{\nabla}\,\nu(\mathbf{x}, t)}{\nu(\mathbf{x}, t)} \,. \tag{2.6.5}$$

According to Stokes' theorem, spherical balls of radius a with this mean velocity experience a mean viscous drag force:

$$\mathbf{F}_{\text{vis}} = -6\pi\eta a \mathbf{v} = \frac{6\pi\eta a D \boldsymbol{\nabla}\nu}{\nu} , \tag{2.6.6}$$

with the signs indicating that the viscous force is in a direction opposite to that of $\mathbf{v}$, and hence in the direction of the gradient of the particle number density. Diffusion occurs because this drag is overcome by osmotic pressure. Following the same reasoning that led to Eq. (2.6.1), if the gradient of ν is along the x-direction, then the force due to an environment at uniform temperature T on the particles in a small disk of area dA and thickness dx transverse to the x-direction is the osmotic pressure force $dAkT[\nu(x,t) - \nu(x+dx,t)] = -dA\,kT\,dx\,d\nu(x,t)/dx$ on the disk. Dividing this by the number $dA\,dx\,\nu(x,t)$ of suspended particles in the disk gives the osmotic pressure force on each particle:

$$F_{\text{osm}} = \frac{-dAkT\,dx\,d\nu(x,t)/dx}{dA\,dx\nu(x,t)} = -\frac{kTd\nu(x,t)/dx}{\nu(x,t)} .$$

Since in the absence of external forces there is nothing special about the x-direction, for a gradient in a general direction we have

$$\mathbf{F}_{\text{osm}} = -\frac{kT\boldsymbol{\nabla}\nu}{\nu} . \tag{2.6.7}$$

Assuming that the viscous drag is cancelled by the osmotic pressure, we have

$$0 = \mathbf{F}_{\text{vis}} + \mathbf{F}_{\text{osm}} , \tag{2.6.8}$$

which gives Einstein's formula (2.6.3):

$$D = \frac{kT}{6\pi\eta a} . \tag{2.6.9}$$

More generally, we should take into account the possibility that the viscous drag is not precisely cancelled by the osmotic pressure. In this case, Newton's law gives

$$m\frac{d\mathbf{v}}{dt} = \mathbf{F}_{\text{vis}} + \mathbf{F}_{\text{osm}} , \tag{2.6.10}$$

where m is the mass of the balls and the acceleration $d\mathbf{v}/dt$ is the total time derivative of the mean velocity, due both to the change in mean velocity at a fixed position and to the change in mean velocity of the particles carried from one point to another at the mean velocity:

$$\frac{d\mathbf{v}}{dt} = \frac{\partial\mathbf{v}}{\partial t} + \mathbf{v}\cdot\boldsymbol{\nabla}\mathbf{v} . \tag{2.6.11}$$

Inspection of Eqs. (2.6.5), (2.6.11), (2.5.22), and (2.5.23) shows that the magnitude of the acceleration can depend only on D and on L, the scale of distances over which ν varies appreciably. Dimensional analysis then tells us that it must be of order

$$\left|\frac{d\mathbf{v}}{dt}\right| \approx \frac{D^2}{L^3}\ . \tag{2.6.12}$$

This shifts the value of $|\mathbf{F}_{\rm vis}|$ for a given $\mathbf{F}_{\rm osm}$ by an amount of order mD^2/L^3, and hence shifts the value of the diffusion constant derived from Eq. (2.6.6) by a fractional amount of order

$$\frac{\Delta D}{D} \approx (mD/L^3) \times (L/6\pi\eta a) \approx \frac{mkT}{L^2(6\pi\eta a)^2}\ . \tag{2.6.13}$$

Einstein's formula for D is valid only if this is much less than one.

Einstein did not see this correction, because he was assuming that an external force was preventing any mean motion, so that there were no inertial forces. But the correction would affect the *horizontal* diffusion of suspended balls in Perrin's measurement of the diffusion constant. I do not know the parameters in Perrin's experiment, but the fact that he obtained close values for Avogadro's number from the measurement of horizontal diffusion and from the measurement of the vertical distribution of suspended balls indicates that in his experiment the correction (2.6.13) was not very large.

This is reassuring regarding the derivation of the Navier–Stokes equation (2.5.16), in which it was assumed that terms of second order in the inverse of the scale L over which properties of the fluid vary can be neglected. Using $mv^2 \approx kT$, where v is a typical particle velocity, we see that the fractional correction (2.6.13) is of order $(\mathcal{L}/L)^2$, where $\mathcal{L} \equiv kT/6\pi\eta a v$ is approximately the distance in which viscous forces will bring a particle with radius a and velocity v to rest. It is the ratios of just such microscopic lengths as $\mathcal{L}$ to the scale L of macroscopic variation whose second and higher powers are dropped in the derivation of the Navier–Stokes equation.

3
Early Quantum Theory

The early years of quantum theory were a time of guesswork, inspired by problems presented by the properties of atoms and radiation and their interaction. This is the subject of the present chapter. Later, in the 1920s, this struggle led to the systematic theory known as quantum mechanics, the subject of Chapter 5.

3.1 Black Body Radiation

Quantum mechanics started with the problem of understanding radiation in thermal equilibrium at a non-zero temperature. We define $\mathcal{E}(\nu,\ T)\,d\nu$ as the energy per volume of radiation with frequency between ν and $\nu + d\nu$ in an enclosure with walls at uniform temperature T. As noted in 1859–1862 by Gustav Robert Kirchhoff (1824–1887), this distribution is independent of any property of the enclosure except for its temperature, because to change $\mathcal{E}(\nu,\ T)$ by changing the material or the shape of the enclosure would require taking energy from one frequency to another, while keeping the same temperature, which is impossible.

Radiation Absorption, Emission, and Energy Density

Kirchhoff called this "black body radiation." This term refers to a relation between the energy density and the rates at which radiation is emitted and absorbed from any black heated surface. Consider radiation in an enclosure whose walls are at a uniform temperature T, and think how to calculate the energy received by a small patch of area dA on the inner walls of the enclosure. At a point within the enclosure at a distance r from this patch, the patch subtends a solid angle $dA\cos\theta/r^2$, where θ is the angle between the line of sight from the point to the patch and the normal to the patch. Hence a fraction $dA\cos\theta/4\pi r^2$ of the radiation at this point is aimed at the patch. In a time t all the radiation at a distance $r < ct$ that is aimed at the patch will hit it (where c is the speed of light), so the total rate $\Gamma(\nu, T)$ per unit time, per unit area, and per unit frequency interval at which radiation energy at a frequency near ν hits the patch will be

$$\Gamma(\nu,T) = \frac{1}{t\,dA\,d\nu}\int_0^{ct} 2\pi r^2\,dr \int_0^{\pi/2} \sin\theta\,d\theta\,\frac{dA\cos\theta}{4\pi r^2}\mathcal{E}(\nu,T)\,d\nu$$
$$= \left(\frac{c}{4}\right)\mathcal{E}(\nu,T)\,. \tag{3.1.1}$$

Equilibrium requires that the rate per area of emission of radiation energy in a frequency interval $d\nu$ must equal the rate per area of absorption of radiation energy in that frequency interval, which is $(c/4)f(\nu,T)\mathcal{E}(\nu T)\,d\nu$, where $f(\nu,T) \leq 1$ is the fraction of energy of radiation of frequency ν that is absorbed when it hits the wall of the enclosure. The emission is evidently greatest for "black" walls, which absorb all the radiation that falls on them, so that $f(\nu,T)$ tales its maximum value, $f(\nu,T) = 1$.

In the 1890s Eq. (3.1.1) was used at the Physikalisch-Technische Reichsanstalt in Berlin to accurately measure $\mathcal{E}(\nu,T)$. This presented a challenge to theorists, to understand the measured distribution $\mathcal{E}(\nu,T)$.

Electromagnetic Degrees of Freedom

To use the equipartition of energy to calculate the radiation energy density $\mathcal{E}(\nu,T)$ from first principles it is necessary to identify the degrees of freedom of radiation among which energy is shared. The deepest understanding of radiation at the beginning of the twentieth century was based on Maxwell's equations. In unrationalized electrostatic units these are

$$\begin{aligned} &\nabla\times\mathbf{B} - \frac{1}{c}\frac{\partial\mathbf{E}}{\partial t} = \frac{4\pi}{c}\mathbf{J}\,, &&\nabla\cdot\mathbf{E} = 4\pi\rho\,,\\ &\nabla\times\mathbf{E} + \frac{1}{c}\frac{\partial\mathbf{B}}{\partial t} = 0\,, &&\nabla\cdot\mathbf{B} = 0\,, \end{aligned} \tag{3.1.2}$$

where $\mathbf{E}(\mathbf{x},t)$ and $\mathbf{B}(\mathbf{x},t)$ are the electric and magnetic fields, while $\mathbf{J}(\mathbf{x},t)$ and $\rho(\mathbf{x},t)$ are the electric current density and charge density. For empty space, $\rho = \mathbf{J} = 0$, and Maxwell's equations have solutions of the form

$$\begin{aligned} \mathbf{E}(\mathbf{x},t) &= \mathbf{e}\exp\left(i\mathbf{k}\cdot\mathbf{x} - i\omega t\right) + c.c.\\ \mathbf{B}(\mathbf{x},t) &= \mathbf{b}\exp\left(i\mathbf{k}\cdot\mathbf{x} - i\omega t\right) + c.c. \end{aligned} \tag{3.1.3}$$

where $\mathbf{k}$ and ω are real constants; $\mathbf{e}$ and $\mathbf{b}$ are complex constant three-vectors; and $c.c.$ denotes the complex conjugate of the preceding term. Since in Eq. (3.1.3) we are including terms proportional to both $\exp(-i\omega t)$ and $\exp(i\omega t)$, without loss of generality we can take $\omega > 0$. Inserting (3.1.3) into (3.1.2), we see that this is a solution for $\rho = \mathbf{J} = 0$ if and only if

$$
\begin{aligned}
\mathbf{k}\times\mathbf{b}+\frac{\omega}{c}\mathbf{e}=0\,,\quad \mathbf{k}\cdot\mathbf{e}=0\\
\mathbf{k}\times\mathbf{e}-\frac{\omega}{c}\mathbf{b}=0\,,\quad \mathbf{k}\cdot\mathbf{b}=0\,.
\end{aligned}
\tag{3.1.4}
$$

Combining these, we have

$$
\frac{\omega^2}{c^2}\mathbf{e}=-\frac{\omega}{c}\mathbf{k}\times\mathbf{b}=-\mathbf{k}\times[\mathbf{k}\times\mathbf{e}]=\mathbf{k}^2\mathbf{e}\,;
\tag{3.1.5}
$$

so $\omega=|\mathbf{k}|c$, and electromagnetic radiation therefore propagates at the speed c.

Now, we want to calculate the electromagnetic energy in a finite volume V. Since $\mathcal{E}$ is universal, we can take our enclosure to be a cube, with edges $L=V^{1/3}$ that lie along the 1-, 2-, and 3- directions. Whatever boundary conditions the material of the enclosure imposes on the phases of the waves, it must be the same on opposite sides of the cube, so the phase $\mathbf{k}\cdot\mathbf{x}$ can only change by an integer multiple of 2π when x_1, x_2, or x_3 is shifted by L. That is, the wave number $\mathbf{k}$ and frequency ω must take the form

$$
\mathbf{k}_\mathbf{n}=(2\pi/L)\mathbf{n},\qquad \omega_\mathbf{n}=c\,|\mathbf{k}_\mathbf{n}|\,,
\tag{3.1.6}
$$

where $\mathbf{n}$ is a vector with integer components n_1, n_2, and n_3. Hence the general electric and magnetic fields in the enclosure are

$$
\mathbf{E}(\mathbf{x},t)=\sum_\mathbf{n}\mathbf{e}(\mathbf{n})\exp\left(i\mathbf{k}_\mathbf{n}\cdot\mathbf{x}-i\omega_\mathbf{n}t\right)+c.c.
\tag{3.1.7}
$$

$$
\mathbf{B}(\mathbf{x},t)=\sum_\mathbf{n}\left(\frac{c}{\omega_n}\right)[\mathbf{k}\times\mathbf{e}(\mathbf{n})]\exp\left(i\mathbf{k}_\mathbf{n}\cdot\mathbf{x}-i\omega_\mathbf{n}t\right)+c.c.\,,
\tag{3.1.8}
$$

where $\mathbf{e}(\mathbf{n})$ is, for each $\mathbf{n}$ a three-vector orthogonal to $\mathbf{n}$, and *c.c.* denotes the complex conjugate of the previous term.

It is a well-known result of classical electrodynamics that the energy density in radiation is $(\mathbf{E}^2+\mathbf{B}^2)/8\pi$. To integrate this over the volume of the enclosure, we use the orthogonality relations

$$
\begin{aligned}
\int_V d^3x\; e^{i(\mathbf{k}_\mathbf{n}-\mathbf{k}_\mathbf{m})\cdot\mathbf{x}}&=\begin{cases}V & \mathbf{n}=\mathbf{m}\\ 0 & \mathbf{n}\neq\mathbf{m}\,,\end{cases}\\
\int_V d^3x\; e^{i(\mathbf{k}_\mathbf{n}+\mathbf{k}_\mathbf{m})\cdot\mathbf{x}}&=\begin{cases}V & \mathbf{n}=-\mathbf{m}\\ 0 & \mathbf{n}\neq-\mathbf{m}\,.\end{cases}
\end{aligned}
\tag{3.1.9}
$$

(For instance, in one dimension for $n\neq m$,

$$
\int_0^L dx\; e^{(2\pi i/L)(n-m)x}=\big(L/2\pi i(n-m)\big)[e^{2\pi i(n-m)}-1]=0\,,
$$

while for $n=m$ it is just L. In three dimensions, the integral is a product of similar factors.) It follows then that

$$\frac{1}{V}\int_V d^3x\, \mathbf{E}^2(\mathbf{x},t) = \sum_{\mathbf{n}} \mathbf{e}(\mathbf{n})\cdot\mathbf{e}(-\mathbf{n})e^{-2i\omega_{\mathbf{n}}t}$$
$$+\sum_{\mathbf{n}} \mathbf{e}^*(\mathbf{n})\cdot\mathbf{e}^*(-\mathbf{n})e^{+2i\omega_{\mathbf{n}}t}$$
$$+2\sum_{\mathbf{n}} \mathbf{e}(\mathbf{n})\cdot\mathbf{e}^*(\mathbf{n})\,,$$

$$\frac{1}{V}\int_V d^3x\, \mathbf{B}^2(\mathbf{x},t) = \sum_{\mathbf{n}} \left(\frac{c}{\omega_{\mathbf{n}}}\right)^2 (\mathbf{k}_{\mathbf{n}}\times\mathbf{e}(\mathbf{n}))\cdot(-\mathbf{k}_{\mathbf{n}}\times\mathbf{e}(-\mathbf{n}))e^{-2i\omega_{\mathbf{n}}t}$$
$$+\sum_{\mathbf{n}} \left(\frac{c}{\omega_{\mathbf{n}}}\right)^2 (\mathbf{k}_{\mathbf{n}}\times\mathbf{e}^*(\mathbf{n}))\cdot(-\mathbf{k}_{\mathbf{n}}\times\mathbf{e}^*(-\mathbf{n}))e^{+2i\omega_{\mathbf{n}}t}$$
$$+2\sum_{\mathbf{n}} \left(\frac{c}{\omega_{\mathbf{n}}}\right)^2 (\mathbf{k}_{\mathbf{n}}\times\mathbf{e}(\mathbf{n}))\cdot(\mathbf{k}_{\mathbf{n}}\times\mathbf{e}^*(\mathbf{n}))\,.$$

Noting that for $\mathbf{k}\cdot\mathbf{e} = \mathbf{k}\cdot\mathbf{e}' = 0$ we have $(\mathbf{k}\times\mathbf{e})\cdot(\mathbf{k}\times\mathbf{e}') = \mathbf{k}^2\mathbf{e}\cdot\mathbf{e}'$, and noting also that $\omega_{\mathbf{n}}^2 = c^2\mathbf{k}_n^2$, we see that the terms proportional to $\exp(-2i\omega_{\mathbf{n}}t)$ in the electric and magnetic energy cancel, as do the terms proportional to $\exp(+2i\omega_{\mathbf{n}}t)$, leaving us with the total energy:

$$E = \frac{1}{8\pi}\int_V d^3x\, [\mathbf{E}^2(\mathbf{x}) + \mathbf{B}^2(\mathbf{x})] = \frac{V}{2\pi}\sum_{\mathbf{n}} \mathbf{e}(\mathbf{n})\cdot\mathbf{e}^*(\mathbf{n})\,. \qquad (3.1.10)$$

There are two independent components of each $\mathbf{e}(\mathbf{n})$ orthogonal to $\mathbf{n}$, each with independent real and imaginary terms, all four quantities for each $\mathbf{n}$ contributing independently to E, so there are four degrees of freedom for each $\mathbf{n}$.

We will assume that $L = V^{1/3}$ is much larger than the wavelengths c/ν under consideration, so that the frequencies $\nu_{\mathbf{n}} \equiv \omega_n/2\pi$ are very close together and we can replace sums over $\mathbf{n}$ with integrals over ν. To count the number of integer-component vectors $\mathbf{n}$ in a given range of frequencies, note that, according to Eq. (3.1.6),

$$|\mathbf{n}| = |\mathbf{k}_{\mathbf{n}}|L/2\pi = \omega_{\mathbf{n}}L/2\pi c = \nu_{\mathbf{n}}L/c\,.$$

The number of allowed frequencies between ν and $\nu + d\nu$ therefore equals the number of integer-component vectors $\mathbf{n}$ in a shell with $|\mathbf{n}|$ between $\nu L/c$ and $(\nu + d\nu)L/c$. These vectors form a cubic lattice with lattice site width unity, so the number dN of these vectors in this shell just equals the volume of the shell:

$$dN = 4\pi|\mathbf{n}|^2 d|\mathbf{n}| = 4\pi(L/c)^3\nu^2\,d\nu = 4\pi V\nu^2\,d\nu/c^3\,. \qquad (3.1.11)$$

With two polarizations for each $\mathbf{n}$, the total energy density per frequency interval is then

$$\mathcal{E}(\nu, T) = \frac{8\pi}{c^3}\nu^2 \overline{E}(\nu, T) , \tag{3.1.12}$$

where $\overline{E}(\nu, T)$ is the mean energy for each of the two complex polarization vectors orthogonal to a wave vector $\mathbf{k}$ with a given value of $\nu = |\mathbf{k}|c$.

The Rayleigh–Jeans Distribution

In 1900 a calculation along these lines was presented by John William Strutt (1842–1919), better known as Lord Rayleigh.[1] He used the result of classical thermodynamics, described here in Section 2.4, that for systems whose total energy can be expressed as a sum over degrees of freedom of squared amplitudes, as in Eq. (3.1.10), each degree of freedom such as Re($\mathbf{e}$) and Im($\mathbf{e}$) for a given polarization and wave vector contributes an energy $kT/2$, so the mean total energy for a given polarization and wave vector is $\overline{E} = kT$. Using this in Eq. (3.1.12) gives an energy density

$$\mathcal{E}(\nu, T) = \frac{8\pi kT}{c^3}\nu^2 . \tag{3.1.13}$$

A more detailed derivation is given in the next section, to serve as a basis for the modification introduced by Einstein.

Rayleigh had made a mistake of a factor 8, which was corrected in 1905 by James Jeans[2] (1877–1946); the result (3.1.13) is therefore known as the *Rayleigh–Jeans* formula. Unfortunately a mere factor 8 was the least of Rayleigh's problems. If Eq. (3.1.13) held at all frequencies, however high, then the total energy E in a volume V at any temperature $T \neq 0$ would be given by a divergent integral:

$$E = \frac{8\pi kTV}{c^3}\int_0^\infty \nu^2\, d\nu$$

a result that became known as the ultraviolet catastrophe.

The Planck Distribution

Meanwhile, back in Berlin, a different approach was being followed by Max Planck (1868–1947). Measurements indicated that $\mathcal{E}(\nu, T)$ increases as ν^2 for small ν, reaches a maximum at a frequency proportional to temperature, and decreases more or less exponentially for large ν. To fit this behavior, it would be natural to guess that $\mathcal{E}(\nu, T) = CT\nu^2\exp(-C'\nu/T)$, with C and C' some temperature-independent constants, which would give a total energy $\int d\nu \mathcal{E}(\nu, T)$ proportional to T^4, which as we saw in Section 2.3 is required by

---

[1] Lord Rayleigh, Phil. Mag. **49**. 539 (1900); Nature **72**, 54 (1905).

[2] J. Jeans, Phil. Mag. **10**, 91 (1905).

thermodynamics. But this formula would not agree with a more detailed result of classical thermodynamics, known as the *Wien displacement law*.[3] Planck in 1900 guessed the formula[4]

$$\mathcal{E}(\nu, T)\, d\nu = \frac{8\pi h}{c^3} \frac{\nu^3\, d\nu}{\exp(h\nu/kT) - 1}\,, \tag{3.1.14}$$

where h and k again are constants.

A little later in the same year, Planck published an attempted derivation[5] of Eq. (3.1.14), which indicated that k is Boltzmann's constant, while h is a new constant, known ever since as *Planck's constant*. To derive this formula, he adopted a model of the wall of the enclosure whereby it consists of electrically charged harmonic oscillators with a wide range of frequencies, with the oscillators of frequency ν coming into equilibrium with the electromagnetic radiation of frequency ν. Planck assumed that the energies of oscillators of frequency ν can only take the form $E = nh\nu$, with n a positive integer. Planck calculated the radiation emitted by these oscillators when they are in thermal equilibrium at temperature T, and found that in order for them to absorb just as much radiation as they emit, the radiation in the enclosure must have the energy density distribution given by Eq. (3.1.14). We will not go into Planck's derivation because it was superseded a few years later with the modern derivation, due to Albert Einstein, described in the next section.

Finding the Boltzmann Constant

By comparing Eq. (3.1.14) with the Reichanstalt data, Planck was able to infer values for the Boltzmann and Planck constants. One set of early results was

$$k = 1.4 \times 10^{-16}\ \text{erg/K}, \quad h = 6.6 \times 10^{-27}\ \text{erg sec}\,,$$

which compare well with the modern values,

$$k = 1.38062 \times 10^{-16}\ \text{erg/K}, \quad h = 6.62620 \times 10^{-27}\ \text{erg sec}\,.$$

As described in Section 2.6, from his value of k and the known gas constant $R = kN_A$, Planck calculated a value for the Avogadro number N_A (or equivalently, for the mass $m_1 = 1/N_A$ of unit atomic weight) and from N_A and the known value of the faraday, $F = eN_A$, he calculated the electric charge e carried by singly charged ions in electrolysis.

[3] This result was derived by Wilhelm Wien (1864–1926) in 1893. It requires that the energy density distribution must take the form $\mathcal{E}(\nu, T) = \nu^3\, \mathcal{F}(\nu/T)$ where $\mathcal{F}$ is some function, of only the ratio ν/T, that is not dictated by thermodynamics alone. For a proof, see Appendix XXXIII of Born, *Atomic Physics*, listed in the bibliography. We will not be relying here on this result.

[4] M. Planck, Verhand. deutsch. phys. Ges. **2**, 202 (1900).

[5] M. Planck, Verhand. deutsch. phys. Ges. **2**, 237 (1900).

The Rayleigh–Jeans formula (3.1.13) agrees with Planck's for $h\nu \ll kT$, and in fact gives the correct low-frequency limit of the energy density distribution. It is an irony of history that in principle Rayleigh could have used the comparison of his formula with the data for low frequency to find the value of k, and then like Planck calculated the values of m_1 and e. For this, the quantum hypothesis is unnecessary. This would have been difficult, for it is not easy to fit experimental data for $\mathcal{E}(\nu, T)$ at low frequencies with a formula that is only supposed to be valid at these frequencies, when the form of the distribution at higher frequencies is not known. Anyway, it is just as well that Rayleigh did not do this, as his factor of 8 mistake in $\mathcal{E}(\nu, T)$ would have led to the wrong results for Avogadro's number and the fundamental electric charge.

Radiation Energy Constant

Unlike the Rayleigh–Jeans distribution, the Planck distribution gives a finite total energy density:

$$\mathcal{E}_\gamma(T) = \int_0^\infty \mathcal{E}(\nu, T)\, d\nu = aT^4 \,. \tag{3.1.15}$$

This was in agreement with the known temperature dependence Eq. (2.3.13), which as we saw had been derived thermodynamically using the result of classical electrodynamics that radiation pressure is one-third of the energy density. But now there was a value for the radiation energy constant:

$$a = 16\pi^8 k^4/15h^3c^3 \,.$$

(Using modern values for h, c, and k, the constant a has the value $7.56577(5) \times 10^{-15}$ erg/cm^3 K^4.) According to Eq. (3.1.1), this also tells us that the total rate, $\Gamma \equiv \int \Gamma(\nu, T)\, d\nu$, per unit area and per unit time at which a black surface at temperature T emits radiation energy is $\Gamma = \sigma T^4$, where $\sigma = ca/4$ is another constant, known as the Stefan–Boltzmann constant.

3.2 Photons

Quantization of Radiation Energy

The modern interpretation of the Planck distribution (3.1.14) emerged from a heuristic conjecture of Albert Einstein[6] in 1905. Planck had assumed a quantization of the energies of the charged harmonic oscillators that he supposed made up the walls of an enclosure. Einstein instead imposed the quantization on the radiation itself.

[6] A. Einstein, Ann. Phys. **17**, 132 (1905).

Confusingly, Einstein was not actually dealing with the Planck distribution, but with an attempted fit to the data given earlier by Wilhelm Wien:

$$\mathcal{E}(\nu, T) \propto \nu^3 \exp(-\beta\nu/T)$$

where β is a constant. Einstein used thermodynamic arguments to show that this distribution would require that the energy of radiation at frequency ν must be a whole number multiple of $\beta R\nu/N_A$. Physicists soon learned that $\mathcal{E}(\nu, T)$ is really given by the Planck distribution (3.1.14) and could interpret the Wien distribution as the high-frequency limit of the Planck distribution, which for large ν is proportional to $\nu^3 \exp(-h\nu/kT)$. Thus β in Einstein's quantization condition could be identified as $\beta = h/k$. With the gas constant R equal to kN_A, this means that the energy of the radiation at frequency ν must be a whole number multiple of $(h/k)(R\nu/N_A) = h\nu$, the same rule as for Planck's mythical oscillating charges.

Derivation of Planck Distribution

To see how Einstein's assumption leads to the Planck distribution (*not* the Wien distribution), it is helpful to follow the reasoning a few years later of Hendrik Lorentz[7](1853–1928). For this, we will go back in more detail to the use by Rayleigh and Jeans of the principle of equipartition of energy, which will make it easy to see the difference made by Einstein's assumption of the quantization of radiation energy.

Recall that by counting degrees of freedom, we found that the energy density per frequency interval is given by Eq. (3.1.12):

$$\mathcal{E}(\nu, T) = \frac{8\pi\nu^2}{c^3}\overline{E}(\nu, T) \,, \tag{3.2.1}$$

where $\overline{E}(\nu, T)$ is the mean energy of each polarization state of electromagnetic waves of frequency ν. According to Eq. (3.1.10) the energy $E_{\mathbf{k},\mathbf{e}}$ for a given wave vector $\mathbf{k} = 2\pi\mathbf{n}/L$ and polarization vector $\mathbf{e}$ orthogonal to $\mathbf{n}$ in a cubical box of volume L^3 is a sum of squares:

$$E_{\mathbf{n},\mathbf{e}} = \frac{L^3}{2\pi}\left[(\mathrm{Re}\,\mathbf{e}(\mathbf{n}))^2 + (\mathrm{Im}\,\mathbf{e}(\mathbf{n}))^2\right] .$$

Therefore, in the same way as in Eq. (2.4.17), in classical statistical mechanics we would have a mean energy for each frequency and polarization, given by

[7] H. A. Lorentz, Phys. Z. **11m**, 1234 (1910).

$$\overline{E}(\nu, T)_{\text{class}} = \frac{\int_{-\infty}^{\infty} dX \int_{-\infty}^{\infty} dY \; (X^2 + Y^2) \; \exp\big(- (X^2 + Y^2)/kT\big)}{\int_{-\infty}^{\infty} dX \int_{-\infty}^{\infty} dY \; \exp\big(- (X^2 + Y^2)/kT\big)} \,,$$

where

$$X \equiv \sqrt{\frac{L^3}{2\pi}} \mathrm{Re}\, \mathbf{e}(\mathbf{n}) \,, \quad Y \equiv \sqrt{\frac{L^3}{2\pi}} \, \mathrm{Im}\, \mathbf{e}(\mathbf{n}) \,.$$

(The factor $L^3/2\pi$ in $dX\, dY$ is irrelevant, as it cancels between numerator and denominator.) Defining θ and E by $X = \sqrt{E} \cos\theta$ and $Y = \sqrt{E} \; \sin\theta$ and integrating over θ gives

$$\begin{aligned}\overline{E}(\nu, T)_{\text{class}} &= \frac{\int_0^\infty 2\pi\sqrt{E} d\sqrt{E}\, E \exp(-E/kT)}{\int_0^\infty 2\pi\sqrt{E} d\sqrt{E} \exp(-E/kT)} \\ &= \frac{\int_0^\infty dE\, E \exp(-E/kT)}{\int_0^\infty dE \exp(-E/kT)} = kT \,. \end{aligned} \tag{3.2.2}$$

This is the classical equipartition result used by Rayleigh and Jeans, leading to the Rayleigh–Jeans energy density distribution (3.1.13).

According to Einstein's conjecture, the energy E (*not* X^2 or Y^2) of each polarization state can only take the values $nh\nu$, with $n = 0, 1, 2, \ldots$, so the integrals in Eq. (3.2.2) must be replaced with sums. That is, according to Einstein,

$$\begin{aligned}\overline{E}(\nu, T) &= \frac{\sum_{n=0}^\infty nh\nu \exp(-nh\nu/kT)}{\sum_{n=0}^\infty \exp(-nh\nu/kT)} = -\frac{d}{d(1/kT)} \ln\left[\sum_{n=0}^\infty \exp(-nh\nu/kT)\right] \\ &= -\frac{d}{d(1/kT)} \ln\big[1 - \exp(-h\nu/kT)\big]^{-1} = \frac{h\nu \exp(-h\nu/kT)}{1 - \exp(-h\nu/kT)} \\ &= \frac{h\nu}{\exp(h\nu/kT) - 1} \,. \end{aligned} \tag{3.2.3}$$

Using this in Eq. (3.2.1) gives an energy density distribution

$$\mathcal{E}(\nu, T) = \frac{8\pi h}{c^3} \frac{\nu^3}{\exp(h\nu/kT) - 1} \,. \tag{3.2.4}$$

This is the same as the Planck distribution (3.1.14), but derived from quite different assumptions.

Einstein's interpretation of his quantization assumption $E = nh\nu$ was that the energy in radiation of frequency ν comes in individual bundles, or "quanta," each with energy $h\nu$. A state of this radiation with energy $nh\nu$ is simply one containing n such quanta. This interpretation was soon confirmed by data on the photoelectric effect.

Photoelectric Effect

Several physicists had observed in the late nineteenth century that electric charge is expelled from metal surfaces when the surfaces are exposed to ultraviolet light. After Thomson's discovery of the electron in 1897, it was generally assumed that this charge was carried by electrons. A metal is a lattice of positively charged ions that have each lost one or more electrons, which circulate freely through the metal, accounting for the good electrical and thermal conductivity of metals. The positively charged metal ions produce an electrostatic potential, so that in normal circumstances it takes a definite energy (called the "work function") ϕ to pull the negatively charged electrons out of the metal. One might think that the more intense the radiation, the more energy is given to these electrons. In 1902 experiments by Philipp Lenard (1862–1947) showed that this is not the case. Instead, no matter how intense the radiation, no electrons are ejected from the metal unless the frequency exceeds a certain minimum (which is why photoelectricity was discovered using ultraviolet rather than visible light), and when that condition is met, the energy of each expelled electron increases with the frequency. Only the number of photoelectrons depends on the intensity of the radiation, not their individual energies.

Einstein in his 1905 paper seized on these phenomena as evidence for his quantization assumption. Any electron expelled from the metal was assumed to have been struck by one of Einstein's quanta. In order to get out of the metal, the energy $h\nu$ of the radiation quantum must at least equal the work function ϕ, so no electrons can be emitted unless $\nu \geq \phi/h$. If this condition is satisfied, then the kinetic energy E_e of the emitted electron will be given by the excess

$$E_e = h\nu - \phi \,. \tag{3.2.5}$$

These energies could be measured by observing how strong an electric field is needed to stop the electron emission by exerting a force toward the surface. In this way, Millikan at Chicago in 1914–1916 (while Europeans had other things on their minds) confirmed the form of the Einstein relation (3.2.5), and found a value for h, which turned out to agree with the value measured in studies of black body radiation.

Particles of Light

As we shall see in Chapter 4, with the advent of special relativity it became clear that since Einstein's quanta would have to travel at the speed of light, as particles they would have to have momenta equal to the energy divided by c, or $h\nu/c$. As discussed in Section 4.5, this was confirmed in experiments of

Arthur Holly Compton (1892–1962) in 1922–1923 on the scattering of X-rays by electrons in atoms; Compton's measurements removed the last doubt about the existence of Einstein's radiation quanta. A few years later, they were given their present name, *photons*.

3.3 The Nuclear Atom

It was not possible to make progress in applying quantum ideas to atoms without some understanding of what atoms are. The growth of this understanding began with the discovery of radioactivity.

Radioactivity

In 1896 Antoine Henri Becquerel (1852–1908) was trying to find whether various crystals that had been exposed to sunlight would emit energetic radiation, like the X-rays that had been discovered a few months earlier. He put these crystals next to photographic plates wrapped in dark paper that would block sunlight but might not block rays emitted by the crystal that had earlier been exposed to the Sun. A wire mesh was inserted between the crystal and the paper, so that any exposure of the plate by these rays would show an image of the mesh. One of the crystals Becquerel intended to study was uranium potassium bisulphate, because it exhibits the phenomenon of phosphorescence, the delayed emission of light by substances such as the luminous paint on clock dials that have been exposed to bright light. At first, in February 1896, the skies in Paris were too cloudy to provide the needed sunlight, so Becquerel left his crystals and photographic plates in a drawer. When he took them out in early March, he found that the plates that had been left near the crystals containing uranium were exposed, showing clear images of the wire mesh, even though they had never been put in sunlight. In the following months he found that some sort of ray from various compounds of uranium would expose photographic plates, even when the crystals and plates were put together in lead-lined boxes.

It was soon realized that this phenomenon was not limited to uranium compounds. In 1898 Marie Curie (1867–1934) showed that similar phenomena are produced by compounds of thorium, and she and Pierre Curie (1859–1906) were able to isolate a previously unknown element, radium, that was millions of times more active than uranium or thorium. The Curies gave this phenomenon the name *radioactivity*.

Two different kinds of radioactivity were distinguished in the next few years by Ernest Rutherford (1871–1937). There are beta rays, which are about as penetrating as X-rays, and alpha rays, which cannot penetrate even very thin sheets of foil. (Gamma rays, which are energetic photons, were discovered later.) In 1898 Becquerel discovered that beta rays could be deflected by magnetic and electric fields. From the amount of the deflection he concluded that these rays are composed of particles with the same ratio of charge to mass as the cathode ray particles whose deflection had been measured by J. J. Thomson shortly before. Beta rays are in fact what later became known as electrons, but moving much faster than the electrons in cathode rays. It was harder to deflect alpha rays, but this was eventually accomplished by Rutherford. From the amount and direction of deflection, Rutherford concluded that these rays consist of positively charged particles, with a ratio of charge e_α to mass m_α equal to half the ratio of charge to mass for hydrogen ions, as had been measured in electrolysis. The lightest element heavier than hydrogen is helium, with atomic weight about four times greater than hydrogen, so Rutherford guessed that alpha rays are helium ions, with charge twice that of hydrogen ions. That is,

$$\frac{e_\alpha}{m_\alpha} = \frac{2e}{4m_1} = \frac{1}{2}\frac{e}{m_1} .$$

This was finally confirmed in 1907 when Rutherford together with T. D. Royds was able to collect enough alpha particles from radioactive decay to show that the atoms that they form absorb light at the same spectral frequencies as helium.

Once the particles in alpha and beta rays were identified respectively as helium ions and electrons, it became possible to use measurements of their deflection to find the particles' energies. These energies were enormous, typically about a million times larger than the energies of photons emitted in ordinary chemical reactions such as burning. Studies of radioactivity by Rutherford with the chemist Frederick Soddy (1877–1956) at McGill University showed that this energy is released when elements like radium and thorium spontaneously change to other elements, such as radon. But, for the understanding of atoms, the most important consequence of the discovery of radioactivity was that it provided highly energetic charged particles that could be used as probes of atomic structure.

Discovery of the Atomic Nucleus

After Thomson's discovery of the electron, it was widely supposed that atoms are like puddings, in which negatively charged electrons swim like raisins in a smooth background of positive charge. This seemed at first to be verified by experiments at the laboratory of Ernest Rutherford at the University of Manchester, to which Rutherford had moved in 1907. Rutherford's assistant Hans Geiger (1882–1945) used a beam of alpha particles from what he called "radium emanation" (radon 222, a product of the alpha decay of radium 226), collimated by letting the alpha particles pass through a small slit in a metal sheet through which the alpha particles emerged in a narrow beam. The beam was directed at a gold foil, thin enough that the alpha particles could penetrate the foil.

The beam then struck a screen covered with zinc sulfide, which emits a flash of light when hit by an energetic charged particle such as an alpha particle. If the gold atom really consisted only of the very light electrons in a continuum of positive charge, it would scatter the alpha particles only weakly. Geiger at first found flashes of light from an area only slightly larger than the geometric image of the slit, where the unscattered beam would have struck the screen, indicating the expected slight scattering.[8]

A better model was suggested in 1911 by Rutherford,[9] on the basis of further experiments in 1910 in his laboratory. Geiger and Ernst Marsden (1889–1970) again used alpha rays emitted from a glass tube containing radon 222 gas.[10] Rutherford for some reason asked Geiger and Marsden to see whether any alpha particles could be deflected at large angles, more than 90°, so that the particles would be reflected backwards from surfaces of gold or other metals, producing flashes of light in a zinc sulfide screen on the same side of the metal surface as the alpha particle source. To his surprise Rutherford learned that some alpha particles were scattered almost straight back from various metal surfaces: gold, lead, platinum, etc.[11]

Nuclear Mass

The observation of backward scattering immediately indicated that the alpha particles were repelled by something much heavier than an electron, heavier indeed than an alpha particle. Suppose two particles with masses m_A and m_B and initial velocities v_A and v_B along some line collide head on, emerging with velocities v'_A and v'_B along the same line. (These vs can be positive or negative; when two vs have the same sign the particles are going in the same direction; if opposite signs, they are going in opposite directions.) The conservation of momentum requires that

$$m_A v'_A + m_B v'_B = m_A v_A + m_B v_B$$

while (as long as velocities are measured when the particles are sufficiently far apart that they exert no force on each other) the conservation of energy requires that

---

[8] H. Geiger, Proc. Roy. Soc. A **81**, 141 (1908). This reference gives citations to earlier work of Rutherford and others along the same lines.

[9] E. Rutherford, Phil. Mag. **21**, 669 (1911); **27**, 488 (1914). The first article is reprinted in Beyer, *Foundations of Nuclear Physics*, listed in the bibliography.

[10] It is not clear whether these alpha particles were produced by the direct alpha decay of radon or in the alpha decay of radon's decay products. Without explanation Rutherford's 1911 paper cited an alpha particle velocity of 2.09×10^9 cm/sec. If this is accurate, then these alpha particles could not have been those that are emitted in the decay of radon 222 to polonium 218, which have a velocity of 1.6×10^9 cm/sec. Polonium 218 decays into lead 214 with a half life of 3.1 minutes, producing an alpha particle with velocity 1.7×10^9 cm/sec, and lead 214 then undergoes further decays. Rutherford's estimate of an alpha particle velocity of 2.09×10^9 cm/sec may have been just a guess.

[11] H. Geiger and E. Marsden, Proc. Roy. Soc. A **82**, 495 (1910).

$$m_A v_A'^2 + m_B v_B'^2 = m_A v_A^2 + m_B v_B^2 \,.$$

We can use the first equation to express v'_B in terms of the other velocities. Using this in the second equation, we then have a quadratic equation for v'_A, with coefficients depending on v_A and v_B. Like any quadratic equation, this has two solutions. If nothing changes in the collision then the conservation laws are automatically satisfied, so one solution is obvious; without even writing down the equation, we know that $v'_B = v_B$, $v'_A = v_A$ is a solution. Since there are only two solutions, the other solution, for which something does happen in the collision, is unique. Here it is:

$$v'_A = \frac{(m_A - m_B)v_A + 2m_B v_B}{m_A + m_B} \tag{3.3.1}$$

and, just interchanging the As and Bs,

$$v'_B = \frac{(m_B - m_A)v_B + 2m_A v_A}{m_A + m_B.} \,.$$

It is easy to check that these do satisfy the conservation laws.

In particular, suppose that particle B is initially at rest, so $v_B = 0$. Then

$$v'_A = \left(\frac{m_A - m_B}{m_A + m_B}\right) v_A \,.$$

So it is only possible for particle A to be reflected backward (that is, with the sign of v'_A opposite to that of v_A) in the collision with a particle B at rest if $m_A - m_B$ is negative. Taking A to be the alpha particle, B to be whatever it is in the atom encountered by the alpha particle, we see that whatever the forces between them may be, the alpha particle must be repelled by something in the atom heavier than itself.

Nuclear Size

The observations of alpha particles reflected backward also shows that they are repelled by something small. Here it is necessary to assume that at the separations reached in these collisions, the force between the alpha particle and whatever it is encountering is purely electrostatic. If we assume that the alpha particle has charge e_α and mass m_α, and is repelled by something heavy with charge Ze, then the potential energy of the alpha particle at separation r is Zee_α/r. In order for the alpha particle to be brought momentarily to rest before reversing direction, its initial kinetic energy $m_\alpha v_\alpha^2/2$ must be entirely converted to potential energy, so it must at that moment reach a separation r satisfying

$$Zee_\alpha/r = m_\alpha v_\alpha^2/2$$

or, in other words,

$$r = 2Ze(e_\alpha/m_\alpha)v_\alpha^2 \, . \tag{3.3.2}$$

Both v_α and e_α/m_α could be measured, as Thomson had done for electrons, by measuring the electric and magnetic deflection of the beam of alpha particles. Rutherford cited a velocity $v_\alpha \simeq 2.09 \times 10^9$ cm/sec, and, as already mentioned, $e_\alpha/m_\alpha \simeq e/2m_1 = 1/2$ faraday. Using these values in Eq. (3.3.2) gives the separation when the alpha particle comes to rest as $r = 3Z \times 10^{-14}$ cm. Even for Z as large as 100, this is much smaller than the diameter $>10^{-8}$ cm of heavy atoms, estimated from the density of the metals and the mass Am_1 of their atoms.

Rutherford jumped to the conclusion that the positive charge of any atom and most of its mass is concentrated in a small heavy nucleus, which was repelling the alpha particles in his experiment. Whether or not by chance, Rutherford announced this discovery at a session of the Manchester Literary and Philosophical Society, the same organization at whose meeting Dalton had announced the law of combining weights a little more than a century earlier.

Scattering Pattern

Further experiments in Rutherford's laboratory measured the rate $d\Gamma$ at which alpha particles in a beam with flux Φ (in particles per unit time and per unit area transverse to the beam) are scattered into any solid angle $d\Omega$ (that is, into ranges $d\theta$ of angles to the initial direction and $d\phi$ of angles around the initial direction, with $d\Omega = \sin\theta \, d\theta \, d\phi$). Rutherford compared the result with a calculation using Newtonian dynamics to follow the hyperbolic orbits of alpha particles in the electric field of a single charged nucleus and find into what area $d\sigma$ transverse to the beam the alpha particles must be directed in order to be scattered into the solid angle $d\Omega$. This gave the ratio of $d\sigma$ to $d\Omega$, known as the *differential cross section*:

$$\frac{d\sigma}{d\Omega} = \frac{Z_\alpha^2 Z^2 e^4}{16 E_\alpha^2 \sin^4(\theta/2)} \, , \tag{3.3.3}$$

where $Z_\alpha e$ and Ze are the electric charges respectively of the alpha particle and the nucleus, and E_α is the initial alpha particle kinetic energy. Since any given alpha particle can be anywhere in the beam, for a beam of transverse area $\mathcal{A}$ the probability that a particular alpha particle will be aimed at the area $d\sigma$ for scattering into $d\Omega$ by a single nucleus is $d\sigma/\mathcal{A}$. If there are N atomic nuclei in the part of a metal surface within the area of the beam of alpha particles, then the probability that a given alpha particle will be scattered into the solid angle $d\Omega$ will be $N d\sigma/\mathcal{A}$. With a flux Φ, the number of alpha particles per second hitting the metal surface is $\Phi\mathcal{A}$, so the rate at which alpha particles are scattered into the solid angle $d\Omega$ is

$$\Phi\mathcal{A} \times N d\sigma/\mathcal{A} = \Phi N \frac{d\sigma}{d\Omega} d\Omega \,.$$

The observed pattern of scattering at angles greater than 90° agreed with the proportionality to $1/\sin^4(\theta/2)$ indicated by Eq. (3.3.3), confirming to Rutherford that this was indeed Coulomb scattering by a heavy point charge.

Rutherford was lucky. He was calculating these probabilities using classical mechanics and got the right answer, even though at these velocities and separations quantum mechanics would normally be needed. Scattering by inverse square law forces is special; it allows the use of classical mechanics in some circumstances where for any other force it would be necessary to use quantum mechanics. Equation (3.3.3) will be derived using quantum mechanics in Section 5.6, so we will not trouble to repeat Rutherford's classical calculation here.

Of course, Rutherford's discovery was made before the development of quantum mechanics. The agreement of his experimental results with theory generally convinced physicists of a new picture of the atom, that it consists of a small heavy positively charged nucleus, around which electrons revolve like planets around the Sun, held in orbit by electrostatic attraction, which in part had already been guessed at in 1904 by Hantaro Nagaoka (1854–1950).

Nuclear Charge

In order for atomic theory to make contact with chemistry, it was essential to know the precise number of electrons in the atoms of various elements. For instance, as we shall see in Chapter 5, the dramatic difference in the chemical properties of chlorine and argon is almost entirely due to the fact that chlorine atoms contain 17 electrons while argon atoms contain 18. Because atoms are electrically neutral, knowing the electric charge of the nucleus tells us the number of electrons: if the nuclear charge is Ze, the atom must contain Z electrons.

Almost immediately after Rutherford in 1911 announced his conclusion about the existence of the nucleus, Antonius van den Broek (1870–1926) argued in a brief note[12] (apparently on the basis of the steady progression of chemical properties with increasing atomic weight) that the nuclear charge in units of e equals the *atomic number*, defined as the position in the catalog of elements when they are listed in order of increasing atomic weight – that is, hydrogen, helium, lithium, and so on – but he had no experimental evidence for this hypothesis.

Rutherford offered no opinion about this. In his 1911 article cited in footnote 9 he had used Eq. (3.3.3) (with $Z_\alpha = 2$, known from previous measurements of the deflection of alpha particles by electric and magnetic fields) together with several measurements by Geiger of the scattering by small angles

[12] A. van den Broek, Nature **87**, 78 (1911). He later published a longer paper, Phys. Zeit. **14**, 32 (1913).

of alpha particles in thin gold foil to derive a value of $97e$ or $114e$ for the charge of the gold nucleus. The atomic number of gold is 79, so if Rutherford's value for the charge of the gold nucleus had been correct it would have ruled out the equality of atomic number and the nuclear charge in units of e.[13] As we shall see in the next section, this equality was established in 1913 by measurements of the wavelengths of X-rays from various elements.

3.4 Atomic Energy Levels

Spectral Lines

In Munich in 1814–1815 the optician Joseph Fraunhofer (1787–1826) observed that when light from the Sun is passed through a slit, focussed by a telescope, and then dispersed by a prism into a spectrum of colors, the spectrum is crossed with hundreds of dark lines, each an image of the slit. These lines were always found in the same places in the spectrum, each corresponding to a definite wavelength of light. It was realized that these dark lines must be caused by selective absorption of light as it passes from the hot solar surface through the cooler part of the Sun's atmosphere. The same dark lines were seen in the spectrum of the Moon and bright stars. Similar observations of the light from flames and other terrestrial sources showed lines in the same places, sometimes dark and sometimes bright, so it became possible to identify the elements producing these lines: sodium, iron, magnesium, calcium, etc. Some elements, such as helium, were discovered in this way on the Sun before they were found on Earth.

By the end of the nineteenth century large books had been published for physicists and chemists, giving vast numbers of wavelengths for the spectral lines of various elements. The observation of spectra became a standard tool of astronomy and chemical analysis. But what could cause the atoms of a given element preferentially to emit and absorb light at only certain definite wavelengths? Answering this question had to wait for a realistic model of atoms.

Electron Orbits

In classical electrodynamics the simple harmonic oscillation of a charged body produces electromagnetic radiation with the same frequency as the oscillating

[13] Rutherford's over-estimate of the charge of the gold nucleus may have arisen because he was using a wrong value for the velocity of the alpha particle in these experiments. As mentioned in footnote 10, in the same paper Rutherford had given a value 2.09×10^9 cm/sec for the alpha particle velocity, while the alpha particles from the decay of radon 222 actually have a velocity of 1.6×10^9 cm/sec. According to Eq. (3.3.3) the scattering cross section depends on Z/E_α, so by over-estimating the velocity of the alpha particles he would be over-estimating the electric charge of the gold nucleus.

charge, and the charged body is also effective at absorbing radiation at that frequency. After the discovery of the electron in 1897, as mentioned earlier it was widely supposed that atoms consist of electrons trapped in a smooth background of positive charge, and it was natural to assume that the characteristic frequencies observed in atomic spectra are the frequencies with which these electrons can oscillate back and forth around their normal positions.

Then, with the discovery of the nucleus discussed in the previous section, this picture was replaced with a planetary model of the atom, in which electrons circulate in orbits around the nucleus, like planets around the Sun only held in orbit by electrostatic rather than gravitational attraction. In classical electrodynamics the periodic motion of the electrically charged electrons would produce electromagnetic radiation, with a frequency for circular orbits equal to the frequency with which the electron goes around its orbit.

For elliptical orbits matters are more complicated. While the Cartesian coordinates of an electron traveling at constant speed in a circular orbit are simple harmonic functions of time, and in classical electrodynamics the electron radiates at the corresponding frequency, for elliptical orbits the motion is periodic though not simple harmonic. The Cartesian coordinates for an orbit of period $1/\nu$ can still be expressed as Fourier series of simple harmonic terms proportional to $\sin 2\pi n\nu t$ and $\cos 2\pi n\nu t$ with n an integer, so the electron classically radiates at all frequencies equal to whole number multiples of the frequency ν of revolution. No such pattern is seen in actual spectra.

Even if the orbits were all circular, this view of atomic spectra would have problems. One trouble with this picture is that classically the electrons would continually lose energy to radiation, bringing them closer to the nucleus and thereby speeding up its revolution, hence replacing the discrete spectral line with a continuum of frequencies. Even worse, classically there would be nothing to prevent electrons from spiraling onto the nucleus, so that there would be no stable atoms. Of course, one could simply assume that only certain orbits are possible, and that these are all stable. The frequencies of these allowed orbits would then correspond to the observed spectral lines. But there was another trouble even with this picture: it offered no explanation of a systematic property of observed spectral frequencies, known as the *Ritz combination principle*.

The Combination Principle

In 1908 the spectroscopist Walther Ritz (1878–1909) noticed a peculiar property of the observed wavelengths of spectral lines:[14] in any one atom, the frequencies corresponding to the observed wavelengths of spectral lines are differences of a smaller number of quantities, which he called *terms*. That is, if we label the nth term as ν_n, then the observed spectral frequencies are all of the form

[14] W. Ritz, Phys. Z. **9**, 521 (1908).

$$\nu_{nm} = \nu_n - \nu_m , \tag{3.4.1}$$

with n and m equal to $1, 2, 3, \ldots$ (This was traditionally expressed in terms of inverse wavelengths instead of frequencies, but the frequency of any wave is just the speed of light times the inverse wavelength, so this makes no difference.) Ritz could offer no explanation of this principle.

The explanation of the Ritz principle and much else was provided in the visit in 1913 to Rutherford's Manchester laboratory of a young Danish theorist, Niels Bohr[15] (1885–1962). He assumed that the states of an atom have energies in a discrete set, labeled E_n with n running from one to infinity. These states are stable, except for radiative transitions among them, whose rates are typically much slower than the frequencies of spectral lines. When an atom makes a transition from an energy E_n to a smaller energy E_m, it emits a photon with energy $E_n - E_m$ and hence with frequency

$$\nu_{nm} = (E_n - E_m)/h .$$

Similarly, for an atom to make a transition from an energy E_m to a higher energy E_n, it must absorb a photon with the same energy and frequency. These are the transitions that produce the bright and dark lines observed in spectrographs. Their frequencies match the results (3.4.1) given by the Ritz principle, if we identify the "terms" ν_n as simply the energies E_n of the various states, divided by h.

Bohr's Quantization Condition

But what determines the energies E_n? Casting about for something to quantize, Bohr noted that h has the units of energy per frequency, which is the same as the units of angular momentum, so Bohr guessed that the angular momenta of atomic states are integer multiples of some quantity $\hbar$, similar in magnitude to h. (Some readers may already know what $\hbar$ turned out to be. At first, Bohr had no idea what it was, so until we see how Bohr figured this out, please forget whatever you know about $\hbar$.)

The applications of Bohr's quantization principle are simplest for one-electron atoms, such as neutral hydrogen, singly ionized helium, etc. An electron with velocity v_n in a circular orbit of radius r_n about a nucleus of charge Ze has angular momentum $m_e v_n r_n$, so Bohr's quantization condition was

$$m_e v_n r_n = n\hbar , \tag{3.4.2}$$

with n an integer running from one to infinity. A second relation between v_n and r_n is given by equating the electrostatic attraction Ze^2/r_n^2 to m_e times the centripetal acceleration v_n^2/r_n:

$$Ze^2/r_n^2 = m_e v_n^2/r_n , \tag{3.4.3}$$

[15] N. Bohr, Phil. Mag. **26**, 1, 476, 857 (1913); Nature **92**, 231 (1913).

just as for planets in the solar system, but of course with different constant factors on each side of the equation. We can solve these two equations for radius and velocity. Multiplying Eq. (3.4.3) with r_n^3/Ze^2 gives $r_n = m_e v_n^2 r_n^2/Ze^2$, so

$$r_n = \frac{n^2\hbar^2}{Ze^2 m_e} . \tag{3.4.4}$$

Using this back in Eq. (3.4.2) then gives

$$v_n = \frac{Ze^2}{n\hbar} . \tag{3.4.5}$$

The electron has total energy

$$E_n = \frac{m_e v_n^2}{2} - \frac{Ze^2}{r_n} = -\frac{Z^2 e^4 m_e}{2n^2\hbar^2} . \tag{3.4.6}$$

(By the way, it immediately follows from Eq. (3.4.3) that the kinetic energy is $-1/2$ times the potential energy. One consequence, already mentioned in the previous section, is that classically when an electron in orbit loses energy the potential energy decreases, becoming more negative, so that the kinetic energy *increases*.)

The Correspondence Principle

Now, what is $\hbar$? To answer this, Bohr invoked what he called the *correspondence principle*, that the larger a system is, the more closely it obeys classical mechanics. From Eq. (3.4.4) we see that the large orbits are those with large n. (Atoms with n of order 100 have actually been studied experimentally.) For $n \gg 1$, the energy emitted when a single-electron atom goes from state n to state $n - 1$ is

$$E_n - E_{n-1} = \frac{Z^2 e^4 m_e}{2\hbar^2}\left(\frac{1}{(n-1)^2} - \frac{1}{n^2}\right) \simeq \frac{Z^2 e^4 m_e}{2\hbar^2} \times \frac{2}{n^3} ,$$

so the frequency of the photon emitted in this transition must be

$$\nu_{n\to n-1} \simeq \frac{Z^2 e^4 m_e}{n^3\hbar^2 h} .$$

On the other hand, classically the frequency with which the electron goes around its orbit is

$$\nu_n = \frac{v_n}{2\pi r_n} = \frac{Ze^2/n\hbar}{2\pi n^2\hbar^2/Ze^2 m_e} = \frac{Z^2 e^4 m_e}{2\pi n^3\hbar^3} .$$

In order for these two frequencies to be equal, as required by the correspondence principle, we must have $\hbar^2 h = 2\pi\hbar^3$, and therefore

$$\hbar = h/2\pi \simeq 1.054 \times 10^{-27}\ \text{erg sec} \simeq 6.582 \times 10^{-16}\ \text{eV sec} . \tag{3.4.7}$$

Bohr could then give numerical values of parameters for one-electron atoms:

$$v_n \simeq \frac{Z}{137n} \times c\,, \quad r_n \simeq \frac{n^2}{Z} \times 0.5292 \times 10^{-8}\,\text{cm}\,, \qquad E_n = -\frac{Z^2}{n^2} \times 13.6\,\text{eV}\,. \tag{3.4.8}$$

Comparison with Observed Spectra

Bohr's result for E_n was in good agreement with measurements of spectral wavelengths. In 1885 the Swiss mathematician Johann Balmer (1825–1898) had noticed that the wavelengths of many of the lines in the visible spectrum of hydrogen are well fit by the formula

$$\lambda^{-1}_{\text{Balmer},n} \propto \left(\frac{1}{4} - \frac{1}{n^2}\right)\,, \quad n = 3, 4, \ldots$$

This was generalized in 1888 by the Swedish physicist Johannes Rydberg (1854–1919), to a general formula for the wavelengths of lines in the spectrum of neutral hydrogen:

$$\lambda^{-1}_{m,n} = R_H \left(\frac{1}{m^2} - \frac{1}{n^2}\right)\,, \quad m = 1, 2, 3, \ldots, \quad n = m+1, m+2, \ldots\,,$$

where $R_H \simeq 1.1 \times 10^5\ \text{cm}^{-1}$ is a constant, later named the Rydberg constant for hydrogen. The visible Balmer series is the case $m = 2$, while the infrared series $m = 3$, $m = 4$, $m = 5$, etc. became named for Paschen, Brackett, Pfund, etc. The lines of the $m = 1$ series were predicted by Rydberg's formula to be in the ultraviolet, with wavelengths from 121.7 to 91.1 nm. It was not until 1903 that they were measured, studying hydrogen excited by electric currents, by Theodore Lyman (1833–1897) at Harvard. These results for hydrogen may have provided Ritz with inspiration to formulate his combination principle for all elements.

For comparison with Rydberg's formula, Bohr's formula (3.4.6) for E_n gave the inverse wavelength of the photon emitted in a transition from energy level n to energy level m: in an atom whose nucleus has charge Ze,

$$\lambda^{-1}_{n\to m} = \frac{\nu_{n\to m}}{c} = \frac{1}{2\pi\hbar c}(E_n - E_m) = \frac{Z^2 e^4 m_e}{4\pi\hbar^3 c}\left(\frac{1}{m^2} - \frac{1}{n^2}\right)\,, \tag{3.4.9}$$

which is the same for $Z = 1$ as Rydberg's formula if we identify the Rydberg constant for hydrogen as

$$R_H = \frac{e^4 m_e}{4\pi\hbar^3 c}\,. \tag{3.4.10}$$

Using the best values then available for the fundamental constants, Bohr obtained a value for R_H in agreement with the results from contemporary spectroscopic measurements. (Using modern values for fundamental constants gives $R_H = 13.605693009(84)\,\text{eV}/hc = 1.0968 \times 10^5\ \text{cm}^{-1}$.)

Reduced Mass

But the agreement was not perfect. According to Eq. (3.4.6), all energies and hence all frequencies in the spectrum of once-ionized helium should be $Z_{\text{He}}^2 = 4$ times larger than for neutral hydrogen, but experiment showed that the ratio was actually larger than 4 by about 0.04%. Bohr realized that the source of this discrepancy was that in order to take account of the motion of the nucleus the formulas for energy and angular momentum of an electron in orbit around a nucleus of mass M should contain the *reduced mass* $\mu = m_e/(1 + m_e/M)$ in place of the electron mass itself. It is therefore the reduced mass that should appear in Bohr's formulas for energies and frequencies in place of m_e. All energies and frequencies are thus larger for singly ionized helium than for hydrogen by a factor

$$Z_{\text{He}}^2/Z_{\text{H}}^2 \times \mu_{\text{He}}/\mu_{\text{H}} = 4 \times \frac{1 + m_e/m_{\text{H}}}{1 + m_e/m_{\text{He}}} = 4 \times 1.00041\ ,$$

in agreement with observation. Bohr's success in getting this factor right was a key factor in convincing physicists of the correctness of his assumptions.

Incidentally, although Bohr's formula (3.4.10) for hydrogen energy levels (with the reduced mass in place of m_e) worked very well, the n in this formula is not quite equal to the angular momentum in units of $\hbar$, as Bohr had assumed. We will see in Section 5.2 that in general there are several hydrogen states with energies given by this formula with the same n, in which the electron has orbital angular momenta $(n-1)\hbar, (n-2)\hbar, \ldots, 0$, but not $n\hbar$. The electrostatic attraction exerted on electrons by the nucleus is not balanced solely by the centrifugal force of motion in closed orbits, but by motions implicit in the wave nature of the electron. Although Bohr's calculation of the energy levels in hydrogen has not survived as a correct derivation of the formula for these energies, Bohr made a contribution of permanent importance in using a hypothesis of discrete energy levels for electrons in all atoms to explain the existence of bright and dark lines in atomic spectra.

Atomic Number

The alpha particle scattering experiments in Rutherford's laboratory had not settled the crucial question of the electric charge Ze of the atomic nucleus and its possible relation to the atomic number, which gives the order of an element in the list of elements in order of increasing atomic weight. One of the great achievements of the Bohr theory is that it made possible precise measurements of nuclear charge.

Of course, Bohr's formula (3.4.8) was strictly applicable only to one-electron atoms, but under the approximation of spherical symmetry the electric field felt by the innermost electrons in any atom arises entirely from the nucleus, not from electrons farther out. Hence the energy of the photon emitted when an electron falls from a state in which it is more or less at rest far from the atom and has essentially zero energy to the innermost $n = 1$ orbit of any atom is given by Bohr's formula (3.4.8) as $-E_1 = 13.6Z^2$ eV. For $Z > 10$ this is an X-ray energy.

After publication of Bohr's work in 1913, a young physicist at Manchester, Henry G. J. Moseley (1887–1915), set out to measure these energies. Instead of a prism he used a crystal that (as described in Section 5.1) preferentially reflects X-rays at certain angles that depend on the wavelength. His results[16] for the nuclear charge Ze are shown in the following table, along with the values then known for atomic weight A, which are close to the values accepted now:

Element	Z	A
calcium	20.00	40.09
scandium	—	44.1
titanium	21.99	48.1
vanadium	22.96	51.06
chromium	23.98	52.0
manganese	24.99	54.93
iron	25.99	55.85
cobalt	27.00	58.97
nickel	28.04	58.68
copper	29.01	63.57
zinc	30.01	65.37

Two aspects of this table stand out dramatically. The first is that Z always turns out to be very close to an integer; the small discrepancies can be easily blamed on experimental uncertainties. That of course is what one expects, if Z is the number of electrons in the atom, but it reassured everyone that Moseley's measurements were reliable. The second remarkable feature is that Z goes up by one unit as you go up one step in the list of elements according to atomic weight; there are no elements with atomic weights between 40 and 65 other than those listed here. (Nickel is an exception to the steady increase of A with Z, understood today as due to forces in the nucleus of nickel that make it unusually strongly bound, for a reason discussed in Section 6.3.) This tight

[16] H. G. J. Moseley, Phil. Mag. **26**, 1024 (1913).

correspondence between atomic number and atomic weight goes beyond the elements in the table. For instance, there are just 19 elements with atomic weights less than calcium, which has $Z = 20$. Thus with a few exceptions, one can find Z for any element just by making a list of all elements in order of increasing atomic weight; the atomic number, defined as the place of the element in that list, gives the number Z of electrons in the atom and the positive charge Ze of the nucleus.

Incidentally, the Bohr theory also provides a rough idea of the sizes of all atoms. The electric field felt by the outermost electron in any atom is largely shielded by the $Z - 1$ electrons closer to the nucleus, so the radius of its orbit is very crudely given by the Bohr result (3.4.8), only with $Z \simeq 1$. This is why the sizes of the atoms of heavy elements are not very much larger than that of the hydrogen atom, of order 10^{-8} cm. They are in fact somewhat larger, because the radius r_n increases with n, and for reasons we will learn in Chapter 5 the outermost electrons in heavy atoms have n greater than 1.

Outstanding Questions

Successful as it was, the Bohr theory raised a number of new questions.

1. Why should angular momentum (or anything else) be quantized?
2. How many atomic states are there for each energy? (It was already known that spectral lines could be split by exposing atoms to external electric and magnetic fields.)
3. Above all, how should quantum theory be applied to states that cannot be approximated as consisting of electrons moving in a fixed Coulomb potential. This includes all molecules.

The solution of these problems had to wait until the advent of modern quantum mechanics in the 1920s. This is the subject of Chapter 5.

3.5 Emission and Absorption of Radiation

A and B Coefficients

In 1917 Einstein returned to the theory of black body radiation,[17] this time combining it with the Bohr idea of quantized atomic energy states. Einstein defined a quantity A^n_m as the rate at which an atom will spontaneously make a transition from a state m of energy E_m to a state n of lower energy E_n, emitting a photon of energy $E_m - E_n$. He also considered the absorption of photons

[17] A. Einstein, Phys. Z. **18**, 121 (1917), reprinted in English translation in Van der Waerden, *Sources of Quantum Mechanics*, listed in the bibliography.

from radiation (not necessarily black body radiation) with an energy density $\mathcal{E}(\nu)\,d\nu$ at frequencies between ν and $\nu + d\nu$. The rate at which an individual atom in such a field makes a transition from a state n to a state m of higher energy is written as $B_n^m \mathcal{E}(\nu_{nm})$, where $\nu_{nm} \equiv (E_m - E_n)/h$ is the frequency of the absorbed photon. As we will see, Einstein also found it necessary to take into account the possibility that the radiation would stimulate the emission of photons of frequency ν_{nm} by the atom in transitions from a state m to a state n of lower energy, at a rate written as $B_m^n \mathcal{E}(\nu_{nm})$. The coefficients B_n^m, and B_m^n like A_m^n, were assumed to depend only on the properties of individual atoms, not on their temperature or any properties of the radiation.

Now, suppose the radiation is black body radiation, at a temperature T, with which the atoms are in equilibrium. The energy density per frequency interval of the radiation will be the function $\mathcal{E}(\nu, T)$ given by Eq. (3.2.4):

$$\mathcal{E}(\nu, T) = \frac{8\pi h}{c^3} \frac{\nu^3}{\exp(h\nu/kT) - 1} .$$

In equilibrium the rate at which atoms make a transition $m \to n$ from higher to lower energy must equal the rate at which atoms make the reverse transition $n \to m$:

$$N_m \left[A_m^n + B_m^n \mathcal{E}(\nu_{nm}, T)\right] = N_n\, B_n^m \mathcal{E}(\nu_{nm}, T) , \tag{3.5.1}$$

where N_n and N_m are the numbers of atoms in states n and m. According to the Boltzmann rule of classical statistical mechanics, at temperature T the number of atoms in a given state of energy E is proportional to $\exp(-E/kT)$, so

$$N_m/N_n = \exp\left(-(E_m - E_n)/kT\right) = \exp\left(-h\nu_{nm}/kT\right) . \tag{3.5.2}$$

(It is important here to take the various N_n as the numbers of atoms in the individual states n, some of which may have precisely the same energy, rather than the numbers of atoms in all states with energies E_n.) Putting this together, we have

$$A_m^n = \frac{8\pi h}{c^3} \frac{\nu_{nm}^3}{\exp(h\nu_{nm}/kT) - 1} \left(\exp(h\nu_{nm}/kT)\, B_n^m - B_m^n\right) . \tag{3.5.3}$$

For this to be possible at all temperatures for temperature-independent A and B coefficients, these coefficients must evidently be related by

$$B_m^n = B_n^m , \quad A_m^n = \left(\frac{8\pi h \nu_{nm}^3}{c^3}\right) B_m^n . \tag{3.5.4}$$

Hence, knowing the rate at which a classical light wave of a given energy density is absorbed or stimulates emission by an atom, we can calculate the rate at which it spontaneously emits photons, an explicitly quantum process.

Lasers

The phenomenon of stimulated emission makes possible the amplification of beams of light in a laser. (This is an acronym for "light amplification by stimulated emission of radiation." Before lasers there were masers, in which it was microwave radiation rather than visible light that was amplified by stimulated emission.) Suppose a beam of light with energy density distribution $\mathcal{E}(\nu)$ passes through a medium consisting of N_n atoms at energy level E_n. Stimulated emission from the first excited state $n = 2$ to the ground state $n = 1$ adds photons of frequency $\nu_{12} \equiv (E_2 - E_1)/h$ to the beam at a rate $N_2\mathcal{E}(\nu_{12})B_2^1$, but absorption from the ground state removes photons at a rate $N_1\mathcal{E}(\nu_{12})B_1^2$, and since $B_2^1 = B_1^2$ there will be a net addition of photons only in the case $N_2 > N_1$. Unfortunately, such a population inversion never occurs in thermal equilibrium, and cannot even be produced by exposing the atoms in their ground state to light at the resonant frequency ν_{12}. The *net* rate of change in the population of the first excited state, labeled $n = 2$, due to spontaneous and stimulated emission from the excited state and absorption from the ground state will be

$$\frac{dN_2}{dt} = -N_2\mathcal{E}(\nu_{12})B_2^1 - N_2A_2^1 + N_1\mathcal{E}(\nu_{12})B_1^2 ,$$

or, using the Einstein relation $B_1^2 = B_2^1$,

$$\frac{dN_2}{dt} = B_2^1\Big[- N_2\big[\mathcal{E}(\nu_{12}) + 8\pi\nu_{12}^3 h/c^3\big] + N_1\mathcal{E}(\nu_{12})\Big] . \tag{3.5.5}$$

If we start with $N_2 = 0$, then N_2 increases until it approaches a value $N_1/(1+\xi)$, where $\xi \equiv 8\pi\nu_{12}^3 h/\mathcal{E}(\nu_{12})c^3$, when N_2 becomes constant. Not only can this process not produce a population inversion; because of spontaneous emission it cannot even make N_2 as large as N_1.

A population inversion can be produced in other ways, for instance by optical pumping, in which atoms are excited to some state, say $n = 3$, by absorption of light with frequency $\nu_{31} = (E_3 - E_1)/h$, and then spontaneously decay to the state $n = 2$. This can also happen naturally. Masers have been observed in the accretion disks surrounding the centers of several galaxies, including NGC 4258 and M33.

Suppressed Absorption

Stimulated emission can not only intensify emission lines, such as those from masers – it can also suppress absorption lines. Consider a steady beam with area A of radiation moving in the $+x$-direction, with local energy density per unit frequency interval $\mathcal{E}(\nu, x)$ at x. In the steady state, the rate of change of energy per unit frequency interval $\mathcal{E}A\,dx$ in the slab between x and $x + dx$ due to atomic transitions $n \to m$ and $m \to n$ with $E_m - E_n = h\nu > 0$ must be balanced by the difference in the rates at which radiation energy enters and leaves the slab:

$$c[\mathcal{E}(\nu,x) - \mathcal{E}(\nu,x+dx)]A = h\nu\mathcal{E}(\nu,x)[-n_n B_n^m + n_m B_m^n]A\,dx$$

where n_m and n_n are the number densities of atoms in states m and n, respectively. The two terms in square brackets on the right arise respectively from absorption and stimulated emission; we do not include a term for spontaneous emission because the photons it produces leave the beam. If the medium is in thermal equilibrium at temperature T then $n_m/n_n = \exp(-h\nu/kT)$; so, since $B_m^n = B_n^m$, the energy density per unit frequency interval along the beam must satisfy

$$\frac{d}{dx}\mathcal{E}(\nu,x) = -\frac{h\nu}{c}\mathcal{E}(\nu,x)n_n B_n^m\big[1 - \exp(-h\nu/kT)\big]\,. \tag{3.5.6}$$

Thus, if $h\nu \ll kT$, stimulated emission suppresses the intensity of the absorption line by a factor $h\nu/kT$. This is important for radio and microwave frequency lines, like the famous "21-cm" line in hydrogen discussed in Section 5.4. It has $h\nu/k = 0.068$ K, which is less even than the temperature of the cosmic microwave background, so this absorption line is strongly suppressed by stimulated emission everywhere. Nevertheless, the absorption line is observed. Its intensity and Doppler shifts provide valuable information about the temperature and motion of hydrogen gas in galactic disks.

4
Relativity

We now turn to the special theory of relativity, introduced by Einstein in a pair of papers in 1905, the same year in which he postulated the quantization of radiation energy and showed how to use observations of diffusion to measure constants of microscopic physics. Special relativity revolutionized our ideas of space, time, and mass, and it gave the physicists of the twentieth century a paradigm for the incorporation of conditions of invariance into the fundamental principles of physics.

4.1 Early Relativity

Motion of the Earth

The idea of the relativity of motion first appeared in medieval arguments over whether or not the Earth can be in motion. For no good reason, it had been proposed by the followers of the cult of Pythagoras in the fifth century BC that the Earth along with the Sun and planets was in orbit about some sort of central fire. A more sober proposal was made in the third century BC by the Hellenistic astronomer Aristarchus of Samos (ca. 310–230 BC).[1] From observations of the Sun and Moon, he calculated that the Sun is much larger than the Earth. According to a later book of Archimedes, Aristarchus concluded from the difference in their sizes that instead of the Sun going around the Earth it was more plausible to suppose that the Earth goes around the Sun.

Better motivated was the idea that the Earth is rotating. It was not hard to see that the apparent rotation once a day from east to west of the Sun, Moon, planets, and stars could be neatly explained if instead the Earth were rotating on an axis once a day from west to east. At least one astronomer suggested this as

[1] Aristarchus, "On the Sizes and Distances of the Sun and Moon," translated by T. L. Heath, in *Aristarchus of Samos* (Clarendon Press, Oxford, 1923). The calculations of Aristarchus are described in S. Weinberg, *To Explain the World* (HarperCollins Publishers, New York, 2015).

early as the fourth century BC; it was Heraclides of Pontus (ca. 388–310 BC), a student at Plato's Academy at Athens.

There is a classic argument against both the rotation and motion of the Earth, given originally by Aristotle, and picked up around 150 AD by the astronomer Claudius Ptolemy of Alexandria (ca. 100–170 AD). Ptolemy argued that if the surface of the Earth were in motion then an arrow shot straight up would not fall back to the same spot from which it was shot, as is observed, because while the arrow was in flight that spot would have moved some distance under the arrow. This argument was first countered in the mid-1300s AD by Nicole Oresme (1321–1382), bishop of Lisieux. Relying on the concept of *impetus* introduced by his teacher at the University of Paris Jean Buridan (1300–1358), Oresme argued that an arrow on the surface of the Earth would pick up an impetus from the Earth's motion, which would keep it moving with the same horizontal component of velocity while going up and down in the air, so it would fall back to the same spot on Earth, despite the Earth's motion. Sadly, whether from respect for the teachings of the Church or fear of its discipline, Oresme never publicly adopted the notion that the Earth really is in motion. But he had established that purely terrestrial observations cannot detect a possible motion of the Earth.

It was not so obvious that the peculiar motion of the planets around the constellations of the zodiac, sometimes even seeming to reverse their motion, could be explained if the Earth were in orbit about the Sun, sometimes passing Mars or some other outer planet, and sometimes being passed by Venus or Mercury. As everyone knows, this was finally made clear in the 1540s by Nicolaus Copernicus (1473–1543).

Relativity of Motion

I don't know if it was the writings of Oresme or similar ideas of their own, but Johannes Kepler (1571–1630) and Galileo Galilei (1564–1642) in their defense of Copernicanism were comfortable with the conclusion that there is no way that a uniformly moving observer without observing the surroundings can tell that he or she is in motion. It was generally understood that (in modern notation) if a first observer describes any event as having Cartesian space coordinates x^i (with $i = 1, 2, 3$ or x, y, z) and time coordinate t, then a second observer who moves with velocity $-\mathbf{u}$ with respect to the first will see the same event with coordinates

$$x'^i = x^i + u^i t, \quad t' = t \,, \tag{4.1.1}$$

because an object seen by the first observer with any time-independent coordinates $x^i = a^i$ will seem to the second observer to be moving with velocity $+\mathbf{u}$, with coordinates $x'^i = a^i + u^i t$.

Invariance under these transformations was built into Newton's theory of motion and gravitation. In a system of bodies acted on by their mutual gravitational attraction, the equations of motion obeyed by the Cartesian coordinates x_N^i of the Nth body are

$$\frac{d^2}{dt^2}x_N^i = \sum_{M \neq N} G m_M \frac{x_M^i - x_N^i}{|\mathbf{x}_M - \mathbf{x}_N|^3} \tag{4.1.2}$$

where G is Newton's gravitational constant, and $|\mathbf{x}_M - \mathbf{x}_N|^2 \equiv \sum_i (x_M^i - x_N^i)^2$. These equations are invariant under the transformation (4.1.1), which here is

$$x_N^i \to x_N'^{\,i} = x_N^i + u^i t \,, \quad t \to t' = t \,, \tag{4.1.3}$$

because the term $u^i t$ drops out in the second time derivative on the left-hand side of Eq. (4.1.2) and does not appear in the differences of spatial coordinates on the right-hand side. The principle that the laws of nature are invariant under the transformations (4.1.1) is known as the *principle of Galilean relativity*. It is a good approximation for bodies moving at speeds much less than that of light. For instance, we saw in Section 2.5 how invariance under Galilean transformations is used to infer the equations of motion for imperfect fluids.

The equations of motion (4.1.2) are of course also invariant under constant rotations of space coordinates and constant translations of space and time coordinates. The set of all these transformations and all their combinations is known as the *Galileo group*.

Speed of Light

It is obvious that Maxwell's equations are not invariant under the Galilean transformations (4.1.1). Maxwell's equations tell us that light always travels at the same speed, which we call c. If a light wave moves along the 1-direction, the 1-coordinate of the wave front must have the time dependence

$$x^1(t) = x^1(0) + ct \,. \tag{4.1.4}$$

But then if a second observer who moves in the -1-direction with speed u uses the coordinates (4.1.1), she will see the 1-coordinate of the wave front as

$$x'^1(t) = x^1(0) + (c + u)t \,, \tag{4.1.5}$$

so the wave would seem to travel faster or slower than the speed of light according to whether u is positive or negative. Observers can use any coordinate systems they like, but Eq. (4.1.5) shows that if Maxwell's equations in the form (3.1.2) are found to hold when an observer uses coordinates x^i, t then they cannot hold in that form when she uses coordinates x'^i, t'.

Einstein worried about this as a young man. He was particularly concerned with what a light wave would look like to an observer with $u = -c$ in our example – that is, an observer moving with the light wave. He concluded that the electric and magnetic fields would appear frozen in time, though varying with position along the ray. Needless to say, this is not a solution of the Maxwell equations.

This problem did not worry Maxwell. In formulating his equations, he regarded the electric and magnetic fields as vibrations in an elastic medium, the *aether*. In this case one would not expect the equations to hold for observers moving with respect to the aether, any more than the equations for a sound wave traveling up and down in an organ pipe would seem the same to an observer flying up the pipe as to an observer at rest with respect to the pipe. Maxwell thought that his equations would apply only for observers at rest in the aether.

Michelson–Morley Experiment

So, if electromagnetic waves are vibrations in the aether, can we measure the velocity of the Earth through the aether? The Earth's orbital motion gives it a speed of 30 km/sec relative to the Sun, and the rotation of our galaxy gives the solar system a speed of about 200 km/sec relative to the galaxy's center. These speeds are much less than the speed of light, 300 000 km/sec, but not too small to be measured with a device known as a Michelson interferometer, invented by the American physicist Albert Michelson (1852–1931). (Michelson interferometers have been used for many purposes since then, most recently in the detection of gravitational waves from distant coalescing black holes and/or neutron stars.)

In 1886 Michelson and Edward Morley (1838–1923) set out to measure the speed of the Earth through the aether in observations at the US Naval Academy, where Michelson had been a midshipman. As a base for their interferometer, they used a large stone disk floating on mercury, to allow an easy change in its orientation and also to give it some insulation from vibrations in the Earth. On this disk they placed a strong source of light, which sent a beam of light toward a half-silvered mirror set at 45° to the beam. (See Figure 4.1.) Half the beam went straight ahead to an ordinary mirror A at distance L_A from the half-silvered mirror, and half went at a right angle to another ordinary mirror B at a distance L_B. From both these two mirrors the beam was reflected back to the half-silvered mirror. Some of the two reflected beams went together in the direction opposite to the direction to mirror B, to a detector which measured the intensity of the recombined beam. If it takes times t_A and t_B for the light to travel from the half-silvered mirror M along the paths to mirrors A and B and back again, then the intensity observed at the detector is proportional to

$$\begin{aligned}&\left|\mathcal{A}_A e^{-2\pi i \nu t_A} + \mathcal{A}_B e^{-2\pi i \nu t_B}\right|^2 \\ &\qquad = |\mathcal{A}_A|^2 + |\mathcal{A}_B|^2 + 2|\mathcal{A}_A||\mathcal{A}_B| \cos(2\pi \nu(t_A - t_B) + \alpha)\ , \qquad (4.1.6)\end{aligned}$$

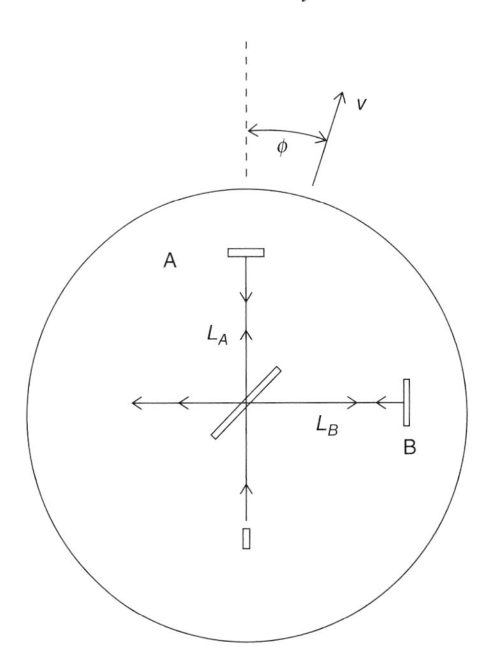

Figure 4.1 The interferometer used in the Michelson–Morley experiment, seen from above.

where $\mathcal{A}_A$ and $\mathcal{A}_B$ are the amplitudes that would be received from mirrors A and B if t_A and t_B were negligible, α is the relative phase of these amplitudes, and ν is the light frequency. It is easy to arrange that $|\mathcal{A}_A|$ and $|\mathcal{A}_B|$ are approximately equal, in which case the intensity (4.1.6) is quite sensitive to the argument of the cosine. So we need to calculate the times t_A and t_B for various orientations of the interferometer.

Adopting the idea of an aether for the sake of argument, let us assume that the Earth is traveling through the aether with a speed v, at an angle ϕ to the direction of the interferometer's incident light beam. To calculate t_A and t_B it is easiest to work in the frame of reference at rest in the aether, in which the speed of light according to Maxwell is c in all directions. If the light takes a time t_A^+ to travel from the half-silvered mirror M to mirror A and a time t_A^- to travel back from A to M, then in the time intervals $t_A^\pm$ it travels a distance $L_A \pm t_A^\pm v\cos\phi$ along its original direction (because during time t_A^+ the mirror A moves in a direction away from M by a distance $t_A^+ v\cos\phi$ while in the time t_A^- the half-silvered mirror M moves in a direction toward A by a distance $t_A^- v\cos\phi$). In both time intervals the light beam also moves at right angles to its original direction by a distance $t_A^\pm v\sin\phi$. The total distance traveled in these time intervals is then the hypotenuse of a right triangle with sides $L_A \pm t_A^\pm v\cos\phi$ and $t_A^\pm v\sin\phi$, so

$$ct_A^{\pm} = \sqrt{(L_A \pm t_A^{\pm} v \cos\phi)^2 + (t_A^{\pm} v \sin\phi)^2} = \sqrt{L_A^2 \pm 2L_A t_A^{\pm} v \cos\phi + (t_A^{\pm} v)^2}\,.$$

Because v is presumably much less than c, it will be enough to keep only terms up to second order in v. We can then use the familiar expansion

$$\sqrt{1+x} = 1 + \frac{x}{2} - \frac{x^2}{8} + \cdots\,,$$

so that

$$\begin{aligned} ct_A^{\pm} &\simeq L_A \pm t_A^{\pm} v \cos\phi + \frac{1}{2L_A}(t_A^{\pm} v)^2(1 - \cos^2\phi) \\ &\simeq L_A \pm t_A^{\pm} v \cos\phi + \frac{1}{2} L_A (v^2/c^2)(1 - \cos^2\phi) \end{aligned}$$

and therefore

$$t_A^{\pm} \simeq \frac{L_A}{c \mp v\cos\phi}\left(1 + \frac{v^2}{2c^2}(1 - \cos^2\phi)\right).$$

Adding these results for t_A^+ and t_A^-, we see that the terms of first order in v/c cancel, leaving us with the second-order correction

$$\begin{aligned} t_A = t_A^+ + t_A^- &\simeq \frac{2L_A c}{c^2 - v^2\cos^2\phi}\left(1 + \frac{v^2}{2c^2}(1 - \cos^2\phi)\right) \\ &\simeq \frac{2L_A}{c}\left(1 + \frac{v^2}{2c^2}(1 + \cos^2\phi)\right)\,. \end{aligned}$$

Since we assumed that the line from the half-silvered mirror M to mirror A is at an angle ϕ to the Earth's velocity through the aether, the line from M to mirror B is at an angle $90^\circ - \phi$ to the Earth's velocity. We can therefore find t_B by simply replacing ϕ with $90^\circ - \phi$ and of course replacing L_A with L_B:

$$t_B \simeq \frac{2L_B}{c}\left(1 + \frac{v^2}{2c^2}(1 + \sin^2\phi)\right)\,.$$

The difference, which appears in Eq. (4.1.6), is then

$$t_A - t_B \simeq \frac{2(L_A - L_B)}{c}\left(1 + \frac{3v^2}{4c^2}\right) + \frac{(L_A + L_B)}{2c}\left(\frac{v^2}{c^2}\right)\cos 2\phi\,. \quad (4.1.7)$$

There is no way that Michelson and Morley could know $L_A - L_B$ and α accurately enough to allow them to detect the presence of corrections proportional to v^2/c^2 by measuring the intensity with a fixed orientation of their interferometer, even if they knew the value of ϕ for that orientation, which of course they did not since no one knew the direction of the Earth's motion through the aether. But if they rotated the interferometer through 180°, then $\cos 2\phi$ would vary through the whole range from -1 to $+1$, so $t_A - t_B$ in Eq. (4.1.7) would vary by an amount $(L_A + L_B)v^2/c^3$ and the argument of the cosine in Eq. (4.1.6) would change by an amount $2\pi\nu(L_A + L_B)v^2/c^3$. This predicts an observable change in the intensity (4.1.6) as the interferometer is rotated through 180°, provided that $2\pi\nu(L_A + L_B)v^2/c^3$ is not much less than 2π, or in other words provided that v^2/c^2 is not too small compared with $c/\nu(L_A + L_B) = \lambda/(L_A + L_B)$, where $\lambda = c/\nu$ is the light wavelength. In the Michelson–Morley experiment (taking account of repeated reflections between the half-silvered mirror and the other mirrors) $L_A + L_B$ was of order 10^3 cm, while the wavelength λ was a few times 10^{-5}cm, so $\lambda/(L_A + L_B)$ was of order 10^{-8}, and velocities roughly of order $10^{-4}c = 30$ km/sec could be easily detected.

Finding no change in the intensity (4.1.6) as the interferometer was rotated, Michelson and Morley concluded in 1887 that the velocity of light as observed from the moving Earth is the same in all directions to within 5 km/sec.[2] That is, within the aether theory of that time, the speed v of the Earth relative to the aether would have to be less than 5 km/sec, as compared with the undoubted orbital velocity of the Earth relative to the Sun of 30 km/sec. By 1964, with the use of a laser instead of an incoherent light source, the upper limit on this velocity had been reduced to about 1 km/sec.[3] Even if one imagined that on a particular day the Earth happened to be more or less at rest in the aether, six months later the Earth would be moving in the opposite direction, with the same speed relative to the Sun, and hence with a speed of 60 km/sec relative to the aether.

This surprising result evoked various explanations. H. A. Lorentz[4] in 1892 and George Francis Fitzgerald (1851–1901) at about the same time proposed that motion through the aether causes a contraction of the dimension of the interferometer along the direction of motion, just such as to hide the effect of motion on the speed of light. Lorentz, acting on the assumption that all matter consists of electrons, tried to explain this "Lorentz–Fitzgerald contraction" within a theory of the electron. Similar ideas were elaborated by the polymath Henri Poincaré[5] (1854–1912). But it was Albert Einstein in 1905 who put his finger on the solution.

4.2 Einsteinian Relativity

Physicists in the first years of the twentieth century were in a strange bind. Newton's equations (4.1.2) of matter and gravitation are invariant under the

[2] A. Michelson and E. W. Morley, Am. J. Sci. **34**, 333 (1887).
[3] T. S. Jaseja, A. Javan, J. Murray, and C. H. Townes, Phys. Rev. **133**, A1221 (1964).
[4] H. A. Lorentz, Versl. Kon. Akad. Wetensch. Amsterdam **I**, 74 (1892).
[5] H. Poincaré, Rendiconti del Circolo Matematico di Palermo **21**, 129 (1906).

Galilean transformation (4.1.1), while Maxwell's equations are not. That in itself was not so bad – it was possible to believe, as Maxwell did believe, that his equations only describe electromagnetism in one frame of reference, supposed to be the one at rest in the aether. But as we saw in the previous section, it was not possible to detect any effect of motion relative to the aether on the speed of light.

Postulate of Invariance

Einstein's solution to this conundrum was presented in 1905 in an article[6] "On the Electrodynamics of Moving Bodies." As suggested by the title, part of his motivation was a peculiar feature of electrodynamics. Consider a magnet moving past a conducting wire. To an observer at rest with respect to the wire, the changing magnetic field produces an electric field, which, as in an electric generator, drives a current in the wire. On the other hand, to an observer at rest with respect to the magnet, there is no electric field; instead the motion of the wire with velocity $\mathbf{v}$ through the magnetic field $\mathbf{B}$ produces a force per charge $\mathbf{v} \times \mathbf{B}/c$ that drives a current in the wire. Somehow the current is the same, although the two observers use different language to describe what is happening. So at least some electromagnetic phenomena are unaffected by the motion of the observer.

Einstein also mentioned in passing "unsuccessful attempts to detect a motion of the Earth" relative to what he called "the light medium," but did not give a reference to the Michelson–Morley experiment. In his 1905 paper he rejected the idea of Lorentz and Fitzgerald that the change in the speed of light due to the transformation (4.1.1) is somehow hidden from us by changes in the measuring apparatus due to motion. Instead, he insisted that Maxwell's equations are unaffected by uniform motion – only the change in coordinates due to uniform motion is not (4.1.1), but something else.

What was truly new and remarkable in Einstein's paper was that in working out this change of coordinates he supposed that the time coordinate, as well as the space coordinates, is affected by the motion of an observer. In writing the Galilean transformation in Eq. (4.1.1) I was careful to include the specification $t' = t$. That was an anachronism – no one before Einstein would have bothered to specify that the time coordinate is unaffected by the motion of an observer. It was then universally supposed that the flow of time is unaffected by motion or anything else. Now Einstein was contemplating the possibility that time as well as distance is affected by an observer's motion.

Einstein calculated the effect of motion on space and time coordinates by a variety of thought experiments, under the assumption that times and distances would be measured using light rays. Though he did not put it in this way, he

[6] A. Einstein, Ann. Phys. **17**, 891 (1905).

was in effect working out what coordinate transformations leave Maxwell's equations, and in particular the speed of light, unchanged. Of course, we can redefine spacetime coordinates any way we like. There is no physics content to a prescription of how to transform coordinates. As we shall see, what was new about the physics introduced by Einstein was not in his change of the coordinate transformation, to keep the speed of light constant, but his hypothesis that these new transformations leave the equations of mechanics as well as electrodynamics invariant. This was not true of Newton's equations, so Einstein had to change these equations, with profound consequences for physics.

This work of Einstein started what became one of the continuing preoccupations of modern physics: the study of hypothetical principles of invariance and their physical implications. Instead of working through Einstein's thought experiments, the discussion below adopts a more modern spirit. In this section we learn what transformations of space and time coordinates, known as Lorentz transformations, leave the speed of light invariant; in the following section we work out the consequences of the assumption that the laws that govern the rigidity of rulers and the ticking of clocks – whatever they are – are invariant under Lorentz transformations; in Section 4.4 we calculate the implications of the assumption that all the laws of mechanics are invariant under these transformations; in Section 4.5 we find the consequences of Lorentz invariance for the properties of photons; and in Section 4.6 we check that not only the speed of light but Maxwell's whole theory of electrodynamics is invariant under Lorentz transformation. In this work we shall make use of a compact spacetime notation introduced in 1907 by Herman Minkowski[7] (1864–1909).

Lorentz Transformations

Let us first consider what sort of spacetime transformation preserves the speed of light. If a light wave front shifts its position by a vector $\Delta\mathbf{x}$ in a time interval Δt, then if light travels at a speed c we have $|\Delta\mathbf{x}| = c\Delta t$, or in other words

$$0 = \Delta\mathbf{x}^2 - c^2(\Delta t)^2 \,. \tag{4.2.1}$$

So, what sort of transformation leaves invariant the quantity $\Delta\mathbf{x}^2 - c^2(\Delta t)^2$?

Before answering this question, it may be mentioned that there is a larger group of transformations that leave $\Delta\mathbf{x}^2 - c^2(\Delta t)^2$ invariant only when it vanishes. These are known as *conformal transformations*. One simple example is a rescaling $\mathbf{x} \to \lambda\mathbf{x}$, $t \to \lambda t$, with λ an arbitrary constant. Invariance of the laws of nature under conformal transformations would be enough to keep the speed of light the same for all observers, but it would apparently make it impossible to deal with non-zero masses. Nevertheless, conformal symmetry has been revived

---

[7] H. Minkowski, lecture delivered to the Math. Ges. Göttingen, November 5, 1907, published in Ann. Phys. **47**, 927 (1915).

again and again up to the present as a possible property of physical law at the most fundamental level, hidden from us through dynamical effects of one sort or another. Here we shall content ourselves with asking about the more limited class of transformations that leave $\Delta\mathbf{x}^2 - c^2(\Delta t)^2$ invariant, whether or not it vanishes.

As mentioned above, it will be very convenient to adopt the spacetime notation due to Minkowski, with a fourth coordinate $x^0 \equiv ct$. We use letters from the middle of the Greek alphabet to label the coordinates of events in spacetime, as x^μ, x^ν, etc. Then the right-hand side of Eq. (4.2.1) may be written

$$(\Delta\mathbf{x})^2 - c^2(\Delta t)^2 = \eta_{\mu\nu}\Delta x^\mu \Delta x^\nu ,$$

it being understood that repeated indices are summed over the values 1, 2, 3, 0. Here $\eta_{\mu\nu}$ is the matrix

$$\eta_{\mu\nu} = \begin{cases} 1 & \mu = \nu = 1, 2, 3 \\ -1 & \mu = \nu = 0 \\ 0 & \mu \neq \nu . \end{cases} \tag{4.2.2}$$

In this notation, the condition we impose on coordinate transformations $x^\mu \to x'^\mu$ may be written

$$\eta_{\mu\nu}\Delta x'^\mu \Delta x'^\nu = \eta_{\mu\nu}\Delta x^\mu \Delta x^\nu . \tag{4.2.3}$$

It can be shown[8] that the most general transformation of the spacetime coordinates that satisfies this condition is linear:

$$\Delta x'^\mu = \Lambda^\mu{}_\rho \Delta x^\rho , \tag{4.2.4}$$

with $\Lambda^\mu{}_\rho$ some set of constants. (We are excluding translations here, under which x^μ would change by a constant term a^μ, because Δx^μ is a difference of spacetime coordinates and hence unaffected by translations.) Recall that the repetition here of the index ρ indicates that this index is to be summed over the values 1, 2, 3, 0. Condition (4.2.3) now reads

$$\eta_{\mu\nu}\Lambda^\mu{}_\rho \, \Lambda^\nu{}_\sigma \, \Delta x^\rho \Delta x^\sigma = \eta_{\rho\sigma}\Delta x^\rho \Delta x^\sigma .$$

In order for this to be valid for any Δx^ρ, the coefficients of $\Delta x^\rho \Delta x^\sigma$ on both sides must be equal:

$$\eta_{\mu\nu}\Lambda^\mu{}_\rho \, \Lambda^\nu{}_\sigma = \eta_{\rho\sigma} , \tag{4.2.5}$$

for all values of the spacetime coordinate indices ρ and σ. Transformations (4.2.4) with $\Lambda^\mu{}_\nu$ satisfying (4.2.5) are known as *Lorentz transformations*.

---

[8] For a proof, see S. Weinberg, *Gravitation and Cosmology* (Wiley, New York, 1972), Section 2.1.

It will be instructive to consider the special class of coordinate transformations that act only on the $\mu = 3$ and $\mu = 0$ components of Δx^μ, which as we shall see is the case for transformations to a frame of reference moving along the 3-axis. For any such linear transformation, the only non-zero components of $\Lambda^\mu{}_\nu$ are

$$\Lambda^1{}_1 = \Lambda^2{}_2 = 1\ ,\quad \Lambda^3{}_3 = A\ ,\quad \Lambda^0{}_0 = B\ ,\quad \Lambda^3{}_0 = C\ ,\quad \Lambda^0{}_3 = D\ ,$$

with real constants A, B, C, and D that are constrained by the condition that this is a Lorentz transformation. In matrix notation, with $\Lambda^\mu{}_\nu$ given by the element of the matrix in row μ and column ν:

$$\Lambda^\mu{}_\nu = \begin{pmatrix} 1 & 0 & 0 & 0 \\ 0 & 1 & 0 & 0 \\ 0 & 0 & A & C \\ 0 & 0 & D & B \end{pmatrix} ,$$

the rows and columns being labeled in the order 1, 2, 3, 0.

Inserting the formulas for the components of $\Lambda^\mu{}_\nu$ into Eq. (4.2.5) gives nothing new if ρ or σ equals 1 or 2, while for $\rho = \sigma = 3$, $\rho = \sigma = 0$, and $\rho = 3, \sigma = 0$ (or $\rho = 0, \sigma = 3$), we get respectively

$$A^2 - D^2 = 1\ ,\quad C^2 - B^2 = -1\ ,\quad AC - DB = 0\ .$$

With three conditions on four parameters, there will be one free parameter left when all conditions are satisfied. We will take this parameter as

$$\beta \equiv C/B = D/A\ .$$

From $A^2 - D^2 = 1$ we then have

$$A^2 = \frac{1}{1-\beta^2}\ ,\quad D^2 = \frac{\beta^2}{1-\beta^2}\ ,$$

while from $C^2 - B^2 = -1$ we have

$$B^2 = \frac{1}{1-\beta^2}\ ,\quad C^2 = \frac{\beta^2}{1-\beta^2}\ .$$

To find the signs of A and B we impose an additional limitation on the transformations we are considering, that they can be obtained by a smooth change of parameters such as velocities and angles from a Lorentz transformation that does nothing. In our case, of transformations only of x^3 and x^0, a Lorentz transformation that does nothing has $A = B = 1$ and $C = D = 0$. Neither A nor B can vanish for any β, so if signs do not suddenly change as we change the parameters of the Lorentz transformation, we must have A and B positive for all Lorentz transformations of this form. Since $AC = DB$, this tells us also that C and D have the same sign, which by definition is the sign of β. So our conclusion is that the non-vanishing components of $\Lambda^\mu{}_\nu$ are

$$\Lambda^3{}_3 = \Lambda^0{}_0 = \gamma\,, \quad \Lambda^3{}_0 = \Lambda^0{}_3 = \beta\gamma\,,$$
$$\Lambda^1{}_1 = \Lambda^2{}_2 = 1\,, \tag{4.2.6}$$

where γ is the positive quantity

$$\gamma = +\frac{1}{\sqrt{1-\beta^2}}\,. \tag{4.2.7}$$

That is, in matrix notation,

$$\Lambda^\mu{}_\nu = \begin{pmatrix} 1 & 0 & 0 & 0 \\ 0 & 1 & 0 & 0 \\ 0 & 0 & \gamma & \beta\gamma \\ 0 & 0 & \beta\gamma & \gamma \end{pmatrix}\,. \tag{4.2.8}$$

The free parameter β can have any sign, but $|\beta| < 1$. We will see in Section 4.6 that not only the speed of light but also the complete set of Maxwell's equations are invariant under these transformations.

I'll pause to mention that there are other Lorentz transformations that cannot be obtained by a gradual variation of parameters from a Lorentz transformation that does nothing. These include the space inversion $x^3 \to -x^3$ with x^μ unchanged for $\mu \neq 3$, and the time reversal $x^0 \to -x^0$ with x^μ unchanged for $\mu \neq 0$. (Space inversion is often described as a change of sign of all three Cartesian coordinates, but this transformation can be produced by the reversal of any one coordinate, followed by an ordinary rotation of 180° around that coordinate direction.) As will be discussed in Section 6.5, experiments in the 1950s showed that invariance under space inversion is only a good approximation, being violated by the very weak forces that lead to the decay of some radioactive nuclei and elementary particles, and in Section 2.4 we have already mentioned that the same is true of invariance under time reversal. We will be concerned here only with transformations that *can* be obtained by a gradual variation of parameters from a Lorentz transformation that does nothing. (These are known as *proper orthochronous Lorentz transformations* – proper, meaning that the determinant of the matrix $\Lambda^\mu{}_\nu$ is unity, and orthochronous, meaning that $\Lambda^0{}_0 > 0$. In this book, I will refer to proper orthochronous Lorentz transformations simply as "proper.")

Now let us consider the physical meaning of β. Consider a tiny body at rest in the frame of reference with coordinates x^μ. At two different times, separated by a time difference Δt, the body is at the same position, so the separation of positions is $\Delta x^i = 0$ with $i = 1, 2, 3$. Now suppose we look at the same body in the frame of reference with coordinates x'^μ, given by the Lorentz transformation (4.2.6). The 1- and 2-coordinates will be unaffected, but the 3-coordinates and the times in the new frame of reference will be separated by

$$\Delta x'^3 = \Lambda^3{}_0 \Delta x^0 = \beta\gamma\,\Delta x^0\,, \quad \Delta t' = \Delta x'^0/c = \Lambda^0{}_0 \Delta x^0/c = \gamma\,\Delta x^0/c\,. \tag{4.2.9}$$

So in this frame the body has velocity

$$v = \Delta x'^3 / \Delta t' = c\beta \ . \tag{4.2.10}$$

Therefore $c\beta$ is the velocity in the 3-direction given to a body at rest by the Lorentz transformation (4.2.6).

The Galilean Limit

For velocities that are much less than the speed of light, $|\beta| = |v|/c$ is much less than one, and Eq. (4.2.7) then gives γ very close to one. In this case, setting $\beta = v/c$, $\gamma = 1$, and $x^0 = ct$, the transformation (4.2.9) becomes

$$\Delta x'^3 = v\Delta t \ , \qquad \Delta t' = \Delta t \ .$$

This is the same as the Galilean transformation (4.1.1) used for instance in working out the form of the Navier–Stokes equation in Section 2.5.

Maximum Speed

From Eqs. (4.2.7) and (4.2.10), we see that it is not possible for a finite Lorentz transformation to take a body from rest to a velocity greater than or even as large as c. (In Section 4.7 we will see that causality, the principle that effects cannot precede causes, rules out any signal traveling faster than light.) This may be surprising, because we can perform a pair of Lorentz transformations, each of which gives a body at rest a velocity in the 3-direction greater than $c/2$, which if these were Galilean transformations of the form (4.1.1) when combined would give a Galilean transformation from rest to a velocity greater than c. But velocities add differently in Einsteinian relativity.

Suppose we perform a Lorentz transformation $x^\mu \to x'^\mu = \Lambda_1{}^\mu{}_\nu x^\nu$ that gives a particle initially at rest a velocity $c\beta_1$ in the 3-direction and then perform a Lorentz transformation $x'^\mu \to x''^\mu = \Lambda_2{}^\mu{}_\nu x'^\nu$ that gives the particle that was initially at rest a velocity $c\beta_2$ in the same direction. The combined effect is a linear transformation

$$x^\mu \to x''^\mu = \Lambda_2{}^\mu{}_\rho \Lambda_1{}^\rho{}_\nu x^\nu = \Lambda_{21}{}^\mu{}_\nu x^\nu \ ,$$

where

$$\Lambda_{21}{}^\mu{}_\nu \equiv \Lambda_2{}^\mu{}_\rho \Lambda_1{}^\rho{}_\nu \ .$$

In matrix notation, this means that

$$\Lambda_{21}{}^\mu{}_\nu = \begin{pmatrix} 1 & 0 & 0 & 0 \\ 0 & 1 & 0 & 0 \\ 0 & 0 & \gamma_2 & \beta_2\gamma_2 \\ 0 & 0 & \beta_2\gamma_2 & \gamma_2 \end{pmatrix} \begin{pmatrix} 1 & 0 & 0 & 0 \\ 0 & 1 & 0 & 0 \\ 0 & 0 & \gamma_1 & \beta_1\gamma_1 \\ 0 & 0 & \beta_1\gamma_1 & \gamma_1 \end{pmatrix}$$

where, according to the general rules of matrix multiplication, the element in row μ and column ν of the product is the sum over ρ of the products of the terms in row μ and column ρ of the first matrix times the terms in row ρ and column ν in the second matrix. It is straightforward to calculate that

$$\Lambda_{21}{}^{\mu}{}_{\nu} = \begin{pmatrix} 1 & 0 & 0 & 0 \\ 0 & 1 & 0 & 0 \\ 0 & 0 & \gamma_{21} & \beta_{21}\gamma_{21} \\ 0 & 0 & \beta_{21}\gamma_{21} & \gamma_{21} \end{pmatrix}$$

where

$$\gamma_{21} = \gamma_1\gamma_2(1 + \beta_1\beta_2) \,, \quad \beta_{21}\gamma_{21} = \gamma_1\gamma_2(\beta_1 + \beta_2) \,,$$

and therefore

$$\beta_{21} = \frac{\beta_1 + \beta_2}{1 + \beta_1\beta_2} \,, \qquad \gamma_{21} = \frac{1}{\sqrt{1 - \beta_{21}^2}} \,.$$

Thus the relativistic rule for combining velocities is that a Lorentz transformation with velocity $v_1 = c\beta_1$ followed by a Lorentz transformation with velocity $v_2 = c\beta_2$ in the same direction gives a Lorentz transformation with velocity

$$v_{21} = c\beta_{21} = \frac{v_1 + v_2}{1 + v_1 v_2/c^2} \,. \tag{4.2.11}$$

Even if v_1 and v_2 both approach c, the combined velocity v_{21} approaches c, not $2c$.

General Directions

Of course there is nothing special about the 3-direction. Whatever velocity vector $\mathbf{v}$ is given to a body at rest by a given Lorentz transformation, we can always rotate our coordinate axes so that the 3-direction is in the direction of $\mathbf{v}$. The Lorentz transformation consequently will have the form (4.2.6), but with $\beta = |\mathbf{v}|/c$. If we rotate our coordinate axes back to their original direction, we find

$$\begin{aligned} \Lambda^i{}_j &= \delta_{ij} + (\gamma - 1)\hat{v}_i\hat{v}_j \,, \\ \Lambda^i{}_0 &= \Lambda^0{}_i = \gamma v_i/c \,, \quad \Lambda^0{}_0 = \gamma \,, \end{aligned} \tag{4.2.12}$$

where i and j run over the spatial coordinate indices 1, 2, 3; δ_{ij} is the unit matrix,

$$\delta_{ij} = \begin{cases} 1 & i = j \\ 0 & i \neq j \,, \end{cases}$$

and

$$\gamma = 1/\sqrt{1 - \mathbf{v}^2/c^2} \,. \tag{4.2.13}$$

Here $\hat{v}$ is the unit vector $\mathbf{v}/|\mathbf{v}|$. (To check, note that for $\mathbf{v}$ in the 3-direction, Eq. (4.2.12) gives $\Lambda^1{}_1 = \Lambda^2{}_2 = 1$, $\Lambda^3{}_3 = 1 + (\gamma - 1) = \gamma$, and so on for the other components.) Performing a Lorentz transformation with velocity $\mathbf{v}_1$ followed by a Lorentz transformation with velocity $\mathbf{v}_2$ in general does *not* give a Lorentz transformation of the form (4.2.12), unless $\mathbf{v}_1$ and $\mathbf{v}_2$ happen to be in the same direction. In general we get a rotation, followed by a Lorentz transformation of the form (4.2.12). This is not a contradiction, because rotations satisfy the condition (4.2.5) and therefore can be considered as belonging to a subgroup of the group of Lorentz transformations. Lorentz transformations of the special form (4.2.12) are often distinguished from more general Lorentz transformations by calling them *boosts*.

Special and General Relativity

A decade after presenting the special theory of relativity, Einstein gave us the general theory of relativity.[9] As its name implies, this theory is based on a more general principle of invariance than for special relativity: the laws of nature preserve their form under any possible change of spacetime coordinates, not just under Lorentz transformations.

But it should not be thought that special relativity is in any way superseded by general relativity. In general relativity it is still true that in certain *inertial frames* of reference, in free fall around any local matter and otherwise more or less at rest or in a state of uniform motion with respect to the average matter of the universe, the laws of physics are those of special relativity. For example, in inertial frames the separation Δx^μ of spacetime coordinates along a wave front of light satisfies $\eta_{\mu\nu}\Delta x^\mu \Delta x^\nu = 0$; we will see in the next section that the separation Δx^μ of the spacetime coordinates of two ticks of a moving clock whose ticks are T seconds apart at rest satisfies $\eta_{\mu\nu}\Delta x^\mu \Delta x^\nu = -T^2c^2$; and so on. If we make a coordinate transformation other than a Lorentz transformation, for instance to a frame of reference that accelerates or spins relative to the inertial frames, then the laws take a more general form, in which $\eta_{\mu\nu}$ is replaced with a field $g_{\mu\nu}(x)$. This field describes gravitation and satisfies differential equations that generalize and correct Newton's formula for gravitational attraction. In contrast, there is no field or any other physical quantity in special relativity that keeps track of the velocity of the coordinate system. So invariance plays a different role in general and special relativity. General relativity is a theory of the gravitational field, a quantity that keeps track of departures from inertial frames. Special relativity is a theory of invariance under Lorentz transformations from one inertial frame to another.

---

[9] A. Einstein, Ann. Phys. **49**, 769 (1916).

4.3 Clocks, Rulers, Light Waves

As a first application of Einsteinian relativity we now apply the assumption that the laws that govern the rigidity of rulers and the operation of clocks are invariant under the Lorentz transformations described in the previous section, and we use this assumption to calculate the effects of motion on observed distances and times. We shall also anticipate the demonstration in Section 4.6 that Maxwell's equations are Lorentz invariant, using this invariance to work out the effect of motion on the frequency and wave vectors of electromagnetic waves.

Clocks

Consider two ticks separated by a time interval T of a small clock at rest in the frame of reference with coordinates x^μ. The spacetime coordinates of these ticks are separated by $\Delta x^i = 0$, $\Delta x^0 = cT$, as usual with $i = 1, 2, 3$. Now perform a Lorentz transformation (4.2.6) that gives the clock a velocity $v = \beta c$ in the 3-direction. The 1- and 2- coordinates of the clock will be unaffected, while in the new reference frame the 3- and 0- coordinates of the clock at these two ticks will be separated by

$$\Delta x'^3 = \Lambda^3{}_0\, \Delta x^0 = \gamma\beta c T = \gamma v T \ , \tag{4.3.1}$$

$$\Delta x'^0 = \Lambda^0{}_0\, \Delta x^0 = \gamma c T \ , \tag{4.3.2}$$

where as before $\gamma \equiv (1 - v^2/c^2)^{-1/2}$. From Eq. (4.3.2) we see that the time interval between ticks of the moving clock is lengthened to $T' = \Delta x'^0/c = \gamma T$. This is what is seen by the observer who sees the clock moving with velocity $\mathbf{v}$; an observer who travels with the clock sees its ticks separated by T, just as if it were at rest.

There is another way of getting this result without ever looking at a specific Lorentz transformation. If the time interval between ticks of a clock at rest is T, then the spacetime separation between ticks at rest has components $\Delta x^i = 0$, $\Delta x^0 = cT$, which satisfy $\eta_{\mu\nu}\Delta x^\mu \Delta x^\nu = -c^2T^2$, where $\eta_{\mu\nu}$ is again the diagonal matrix (4.2.2) with elements 1, 1, 1, -1 on the diagonal, and the summation convention is again in force. If an observer sees the clock moving with velocity $\mathbf{v}$ in any direction and measures a time T' between ticks, then in the coordinates x'^μ used by this observer the spacetime separation between ticks has components $\Delta \mathbf{x}' = \mathbf{v}T'$, $\Delta x'^0 = cT'$, which satisfy $\eta_{\mu\nu}\Delta x'^\mu \Delta x'^\nu = (\mathbf{v}^2 - c^2)T'^2$. But, as discussed in the previous section, Lorentz transformations are designed to keep this quantity invariant, in the sense that $\eta_{\mu\nu}\Delta x'^\mu \Delta x'^\nu = \eta_{\mu\nu}\Delta x^\mu \Delta x^\nu$. Therefore $(\mathbf{v}^2 - c^2)T'^2 = -c^2T^2$, so as before $T' = T/\sqrt{1 - \mathbf{v}^2/c^2}$.

This lengthening of course applies to any kind of time interval, not just ticks of a clock. It is vividly displayed in the decay of unstable particles in cosmic rays. The collision of atomic nuclei in primary cosmic rays with atoms in the upper atmosphere produces particles known as muons, resembling electrons but about 210 times heavier. At rest, muons are observed to decay with a mean lifetime 2.2 microseconds, but although they are typically produced at an altitude of about 15 km, a good fraction of these muons reach the ground before decaying, so even traveling near the speed of light they must have survived for a time (as measured on the Earth's surface) at least 15 km/300 000 km/sec = 50 microseconds, and more if they reach the ground at a slant. If there were no relativistic time dilation, then the probability of a particle with a mean lifetime 2.2 microseconds surviving as long as 50 microseconds would be $\exp(-50/2.2) = 1.2 \times 10^{-10}$. Evidently the life of these muons is extended by their motion by a factor γ at least of order 10, which requires their velocity to be within a fraction of a percent of the speed of light.

Rulers

Next consider a ruler of length L at rest, lying along the 3-direction in a frame of reference with coordinates x^μ. At any fixed time its ends are separated in this frame by $\Delta x^3 = L$, $\Delta x^0 = 0$, and $\Delta x^1 = \Delta x^2 = 0$. Now perform a Lorentz transformation (4.2.6) that gives the ruler a velocity v (positive or negative) in the 3-direction. The spacetime coordinates x'^μ in the new reference frame will be separated by

$$\Delta x'^3 = \Lambda^3{}_3\, L = \gamma L\ , \quad \Delta x'^1 = \Delta x'^2 = 0 \tag{4.3.3}$$

$$\Delta x'^0 = \Lambda^0{}_3\, L = \gamma v L/c\ . \tag{4.3.4}$$

But Eq. (4.3.4) shows that in this frame the two ends of the ruler have been traveling for times that differ by an amount $\Delta t' = \gamma v L/c^2$, so to find the difference in the space coordinates at the *same* time t', we have to subtract $v\Delta t'$ from $\Delta x'^3$. The spatial separation of the ends of the ruler at the same time t' is then

$$\Delta x'^3 - v\Delta t' = \gamma L - \gamma L v^2/c^2 = L/\gamma\ . \tag{4.3.5}$$

This contraction of lengths in the direction of motion is similar to what Fitzgerald and Lorentz had proposed as the cause of the failure to measure the velocity of the Earth through the aether.

Light Waves

We saw in Eq. (3.1.3) that each component of the electromagnetic fields in a light wave in empty space can be written as a sum of terms proportional to $e^{\pm i\phi}$, where ϕ is the phase:

$$\phi = \mathbf{k}\cdot\mathbf{x} - \omega t\ . \tag{4.3.6}$$

We could always add a spacetime-independent term to ϕ by adjusting the phase of the coefficients **e** and **b** of $e^{i\phi}$ in the fields (3.1.3), so it is only the difference $\Delta\phi$ in ϕ between spacetime points that has physical significance. We expect such phase differences to be Lorentz invariant, because, as we shall see in Section 4.6, Lorentz transformations subject electromagnetic fields to real linear transformations. So we need to give **k** and ω Lorentz transformation properties that ensure the Lorentz invariance of the phase differences:

$$\Delta\phi = \mathbf{k}\cdot\Delta\mathbf{x} - \omega\Delta t\ . \tag{4.3.7}$$

To see how to manage this, we once again introduce a four-dimensional notation, taking $k^0 = \omega/c$, so that Eq. (4.3.7) reads

$$\Delta\phi = \mathbf{k}\cdot\Delta\mathbf{x} - k^0\Delta x^0 = \eta_{\mu\nu}k^\mu\Delta x^\nu\ , \tag{4.3.8}$$

where $\eta_{\mu\nu}$ is again the diagonal matrix (4.2.2) with elements 1, 1, 1, -1 on the diagonal, and the summation convention is again in force. It is obvious then that $\Delta\phi$ will be Lorentz invariant if we ascribe to k^μ the same Lorentz transformation as Δx^μ – that is, if under a Lorentz transformation $\Delta x^\mu \to \Lambda^\mu{}_\nu\Delta x^\nu$ we have

$$k^\mu \to \Lambda^\mu{}_\nu k^\nu\ , \tag{4.3.9}$$

where again $\Lambda^\mu{}_\nu$ satisfies the condition (4.2.5) for a Lorentz transformation, $\eta_{\mu\nu}\Lambda^\mu{}_\rho\Lambda^\nu{}_\sigma = \eta_{\rho\sigma}$. Note that the transformation (4.3.9) also preserves the condition

$$0 = c^2|\mathbf{k}|^2 - \omega^2 = c^2\eta_{\mu\nu}k^\mu k^\nu\ , \tag{4.3.10}$$

which says that the wave $\propto e^{i\phi}$ travels at the speed of light.

For example, consider a light wave traveling in the +3-direction, which has wave vector with $k^1 = k^2 = 0$ and $k^3 = \omega/c$ and frequency $\nu = \omega/2\pi$ in a reference frame with spacetime coordinates x^μ. Suppose we perform a Lorentz transformation $x^\mu \to x'^\mu = \Lambda^\mu{}_\nu x^\nu$ that gives bodies at rest in the first reference frame a velocity v (positive or negative) in the 3-direction in the new reference frame. With $\Lambda^\mu{}_\nu$ given by Eq. (4.2.6), the frequency in the new reference frame is given by

$$\begin{aligned}\nu' \equiv \omega'/2\pi = ck'^0/2\pi &= c(\Lambda^0{}_3k^3 + \Lambda^0{}_0k^0)/2\pi \\ &= (\Lambda^0{}_3 + \Lambda^0{}_0)\nu = \gamma(1 + v/c)\nu\ .\end{aligned} \tag{4.3.11}$$

Using $\gamma^2 = 1/(1 - v^2/c^2)$, we can rewrite this in a more revealing form:

$$\nu' = \gamma^{-1}(1 - v/c)^{-1}\nu\ . \tag{4.3.12}$$

The factor $1/\gamma$ is the relativistic time dilation discussed above for moving clocks: if time intervals are lengthened by a factor γ, then frequencies are decreased by a factor $1/\gamma$.

The factor $(1 - v/c)^{-1}$ is the usual Doppler shift, which applies in both non-relativistic and relativistic contexts, and indeed was first observed in sound waves. If the source of the light wave is at rest in the reference frame with coordinates x^μ, then the Lorentz transformation $\Lambda^\mu{}_\nu$ gives the source a velocity v in the 3-direction, which for v positive is along the direction of the light wave and hence toward whoever is observing the wave. If the time interval between wave crests emitted by the source at rest is $1/\nu$, then, apart from relativistic effects, the observer will see these crests arrive at a time interval less by a factor $1 - v/c$, since the distance that each crest has to travel is less than that for the previous crest by a factor $1 - v/c$, and hence the observed frequency, the rate at which wave crests arrive at the observer, is increased (apart from relativistic time dilation) by a factor $1/(1 - v/c)$. For negative v the source is moving away from the observer, and the factor $(1 - v/c)^{-1}$ gives a decrease in frequency, as seen in the redshift of light from receding galaxies at great distances.

4.4 Mass, Energy, Momentum, Force

Einstein published two papers on relativity theory in 1905. Shortly after the first paper, which is cited in Section 4.2, he published in the same journal another paper[10] with the title "Does the inertia of a body depend on its energy content?" This is often referred to as "the $E = mc^2$ paper," but as can be gathered from the title, it would be better called "the $m = E/c^2$ paper." In this paper, Einstein showed that the mass of a body decreases by an amount E/c^2 when the body emits radiation with energy E. Here "mass" was defined as inertial mass, by the prescription that, as in Newtonian mechanics, the kinetic energy of a particle of mass m with velocity $v \ll c$ is $mv^2/2$.

Einstein's Thought Experiment

Here is the proof of Einstein's result. Consider a particle such as an atomic nucleus, at rest in a reference frame with coordinates x^μ, in an excited state A. Suppose that it decays into a state B of lower energy, emitting two "back-to-back" photons of equal energy traveling in opposite directions along the 3-axis. The symmetry of the problem rules out any recoil of the particle in its final state B, so there is no kinetic energy in the initial or final states and hence each photon must carry energy $(E_A - E_B)/2$ and therefore have frequency $\nu = (E_A - E_B)/2h$.

Now consider the same process as observed in a reference frame with coordinates $x'^\mu = \Lambda^\mu{}_\nu x^\nu$, with $\Lambda^\mu{}_\nu$ the Lorentz transformation (4.2.6). In this frame

---

[10] A. Einstein, Ann. Phys. **18**, 639 (1905).

the decaying particle is traveling with velocity v in the $+3$-direction both before and after the decay. Suppose that $v \ll c$. Before it decays, the total energy of the particle is its internal energy E_A plus its kinetic energy:

$$E_{\text{before}} = E_A + \frac{1}{2} m_A v^2 \; .$$

According to Eq. (4.3.11), in this reference frame the frequencies of the photons that travel in the $+3$- and -3-directions are respectively

$$\nu_{\pm} = \frac{(1 \pm v/c)\nu}{\sqrt{1 - v^2/c^2}} = \frac{(1 \pm v/c)}{\sqrt{1 - v^2/c^2}} (E_A - E_B)/2h$$

so the total energy of the final state is

$$\begin{aligned} E_{\text{after}} &= E_B + \frac{1}{2} m_B v^2 + h\nu_+ + h\nu_- \\ &= E_B + \frac{1}{2} m_B v^2 + (E_A - E_B)/\sqrt{1 - v^2/c^2} \; , \end{aligned}$$

or, since we are assuming that $v \ll c$,

$$E_{\text{after}} = E_B + \frac{1}{2} m_B v^2 + (E_A - E_B)(1 + v^2/2c^2) \; .$$

The conservation of energy requires that $0 = E_{\text{before}} - E_{\text{after}}$, so

$$0 = E_A - E_B + \frac{1}{2}(m_A - m_B) v^2 - (E_A - E_B)(1 + v^2/2c^2) \; .$$

In order for this to be possible with velocity-independent internal energies and masses, we must have

$$m_A - m_B = (E_A - E_B)/c^2 \tag{4.4.1}$$

as was to be proved.

Despite our use of the approximation $v \ll c$, Eq. (4.4.1) is not an approximate result. No one can stop us from making a Lorentz transformation with an arbitrarily small velocity, so we can reduce any error we have made along the way in deriving Eq. (4.4.1) to be as small as we like, simply by making v/c sufficiently small.

Equation (4.4.1) is not yet the famous $E = mc^2$. As long as we are dealing only with a single body changing its state, as in the above Einstein thought experiment, it is only changes in its energy that matter for the conservation of energy, not the energy itself, and we might as well *define* the energy of any one state, say the lowest state, as mc^2. But $E = mc^2$ goes beyond Einstein's result (4.4.1) when we consider a reaction involving a number of bodies, coming into and going out of existence, and exchanging energy with each other.

General Formulas for Energy and Momentum

The question of the energy of a massive particle at rest is part of a larger question: what are the energy and momentum of a particle moving with arbitrary velocity? It is largely up to us what we want to call energy and momentum. Historically, as we saw in Section 2.1, physicists gave these names to certain quantities that they had found to be conserved. A three-vector at first called "quantity of motion" by Newton was found to be conserved as a consequence of the equality of action and reaction, and later became known as momentum. A rotationally invariant quantity at first called *vis viva* was found by Huygens to be conserved when bodies come into contact, and was later called kinetic energy. The concept of energy then had to be broadened to preserve the conservation of energy in more general processes, as for instance by including potential energy. It is the conservation of energy and momentum that makes these concepts useful, whether we want to calculate how much fuel to use to boil a given mass of water or how fast an alpha particle must be traveling to give a certain velocity to a gold nucleus that it strikes.

We are not in a position in this chapter to prove the conservation of whatever we call energy and momentum. As we shall see in Section 5.7 of the chapter on quantum mechanics, these conservation laws follow from the invariance of the laws of nature under translations in time and space. But we can here learn a lot from the requirement that the conservation of the total energy and momentum of a number of colliding particles must be Lorentz invariant.

Hence, in order to express the momentum and energy of a body as functions of its velocity, we impose two conditions on these functions:

- The conservation of energy and momentum is Lorentz invariant. That is, if one observer sees these quantities conserved, then so must any other observer related to the first by a Lorentz transformation.
- For velocities much less than c, the momentum and (up to a constant term) the energy must be given by the same formulas as in Newtonian mechanics.

To accomplish this, we shall assume that the momentum $\mathbf{p}$ and energy E of a particle can be assembled into a four-component quantity p^μ with p^0 proportional to the energy E, which transform just like the components of Δx^μ. That is, in changing our spacetime coordinates from x^μ to $x'^\mu = \Lambda^\mu{}_\nu x^\nu$, the energy–momentum four-vector p^μ of any particle is changed to

$$p'^\mu = \Lambda^\mu{}_\nu p^\nu \,. \tag{4.4.2}$$

If the observer who uses the coordinates x^μ sees that in a collision the momentum four-vectors p_n^μ of the various colliding particles satisfy the condition of total energy and momentum conservation,

$$\sum_{n,\text{before}} p_n^\mu - \sum_{n,\text{after}} p_n^\mu = 0,$$

then an observer who uses coordinates $x'^\mu = \Lambda^\mu{}_\nu x^\nu$ will see that

$$\sum_{n,\text{before}} p'^\mu_n - \sum_{n,\text{after}} p'^\mu_n = \Lambda^\mu{}_\nu \left[\sum_{n,\text{before}} p^\nu_n - \sum_{n,\text{after}} p^\nu_n \right] = 0$$

and so will see energy and momentum again conserved.

The transformation property (4.4.2) allows us to calculate the energy and momentum of a particle with an arbitrary velocity if we know its energy and momentum when it is at rest. At rest the spatial components p^i must vanish (which way would this vector point?) and we can take p^0 to be some number that we shall temporarily call N, characterizing the type of particle. It follows that the momentum four-vector of a particle with velocity $\mathbf{v}$ is given by

$$p^\mu(\mathbf{v}) = \Lambda^\mu{}_0(\mathbf{v})N$$

where $\Lambda(\mathbf{v})$ is the Lorentz transformation (the "boost") that takes the particle at rest to velocity $\mathbf{v}$. In particular, for $\mathbf{v}$ in the 3-direction, $\Lambda(\mathbf{v})$ is the Lorentz transformation (4.2.6), so $\mathbf{p}$ is in the 3-direction, with value

$$p^3(v) = \Lambda^3{}_0(v)N = \gamma(v/c)N\ , \tag{4.4.3}$$

and

$$p^0(v) = \Lambda^0{}_0(v)N = \gamma N \tag{4.4.4}$$

where again $\gamma \equiv 1/\sqrt{1 - v^2/c^2}$.

To implement the second condition above, we next consider the limit $v \ll c$. Here Eq. (4.4.3) gives

$$p^3(v) = N[v/c + O(v^3/c^3)]\ .$$

In order for this to give the Newtonian result $p^3(v) = mv$ for $v \ll c$, we must take $N = mc$, so that

$$\mathbf{p}(v) = m\gamma\mathbf{v} = m\mathbf{v}[1 + \mathbf{v}^2/2c^2 + \cdots]\ . \tag{4.4.5}$$

Also, for $v \ll c$, Eq. (4.4.4) now gives

$$p^0 = mc[1 + v^2/2c^2 + O(v^4/c^4)]\ .$$

In order for this to give the Newtonian result $mv^2/2$ for the kinetic energy, we must choose the constant of proportionality between p^0 and E so that $E = cp^0$, and hence

$$E(\mathbf{v}) = mc^2\gamma = mc^2 + m\mathbf{v}^2/2 + m\mathbf{v}^4/6c^2 + \cdots\ . \tag{4.4.6}$$

Note that we cannot leave out the term mc^2 in the energy (4.4.6), or change it to any other constant term. If we did, then p^μ would not satisfy the condition

(4.4.2) for a four-vector, and the conservation of energy and momentum would not be Lorentz invariant.

We can eliminate the velocity $\mathbf{v}$ from Eqs. (4.4.5) and (4.4.6) to derive a relation between energy and momentum. Since $\gamma^2(1 - \mathbf{v}^2/c^2) = 1$, we have $E^2 - \mathbf{p}^2c^2 = m^2c^4$, or in other words,

$$E = \sqrt{\mathbf{p}^2c^2 + m^2c^4}\ . \tag{4.4.7}$$

This can also be derived directly by noting that $E^2 - c^2\mathbf{p}^2 = -c^2\eta_{\mu\nu}p^\mu p^\nu$ takes the value m^2c^4 in the reference frame in which the body is at rest, and is Lorentz invariant, so it takes the same value m^2c^4 in all reference frames.

$E = mc^2$

Einstein suggested in his 1905 paper that the reduction of mass accompanying the emission of energy might be detected by the study of radioactive salts. This proved difficult, because it is not easy to measure accurately the atomic weights of different states of a radioactive isotope. In the early 1930s it became possible to verify Einstein's relation between energy and mass by studying reactions among stable isotopes, such as ${}^1\mathrm{H} + {}^7\mathrm{Li} \to 2\ {}^4\mathrm{He}$. The masses of the atoms of ${}^1\mathrm{H}$, ${}^7\mathrm{Li}$, and ${}^4\mathrm{He}$ are respectively $1.007825\,m_1$, $7.016003\,m_1$, and $4.002603\,m_1$, where m_1 is the mass of unit atomic weight, defined today as $1/12$ the mass of the carbon isotope ${}^{12}\mathrm{C}$. The mass lost in this reaction is thus $\Delta m = 0.018622\,m_1 = 3.09 \times 10^{-26}\,\mathrm{g} = 17.3\ \mathrm{MeV}/c^2$. Thus it is expected that the kinetic energies of the two ${}^4\mathrm{He}$ nuclei in the final state should exceed the kinetic energies of the ${}^1\mathrm{H}$ and ${}^7\mathrm{Li}$ nuclei in the initial state by $\Delta mc^2 = 17.3$ MeV, and this is observed, verifying $E = mc^2$ and not just Eq. (4.4.1).

Force

Because of the presence of the factor γ in Eqs. (4.4.5) and (4.4.6), the quantity $m\gamma$ is sometimes called the relativistic mass. I will not use this terminology, because it suggests that we can calculate the acceleration produced by any force just by replacing m in Newton's $F = ma$ with $m\gamma$, which is not the case. To find how bodies respond to forces in special relativity, we need to formulate a general Lorentz-invariant version of Newton's second law.

Though the time coordinate is Galilean invariant it is not Lorentz invariant, so neither is the time derivative d/dt. To replace the time derivative in Newton's second law, we note that $d\tau$ is Lorentz invariant, where

$$d\tau \equiv \sqrt{-\eta_{\mu\nu}dx^\mu dx^\nu/c^2} = \sqrt{dt^2 - \mathbf{dx}^2/c^2} = \sqrt{dt^2 - dt^2\mathbf{v}^2/c^2} = dt/\gamma\ . \tag{4.4.8}$$

So, in place of the Newtonian formula $d\mathbf{p}/dt = \mathbf{F}$, the requirement of Lorentz invariance suggests that

$$\frac{dp^\mu}{d\tau} = F^\mu \,, \tag{4.4.9}$$

where F^μ is a four-vector with the same Lorentz-transformation properties as Δx^μ or k^μ or p^μ. The space components of Eq. (4.4.9) give

$$\gamma \frac{d\mathbf{p}}{dt} = \mathbf{F} \tag{4.4.10}$$

but $\mathbf{p}$ is not just $m\mathbf{v}$, and the factor γ in Eq. (4.4.10) is outside the time derivative.

Incidentally, we do not need a special determination of the time component F^0. We have already noted that $\eta_{\mu\nu} p^\mu p^\nu = -m^2c^2$, so

$$0 = \frac{d}{d\tau}(\eta_{\mu\nu} p^\mu \, p^\nu) = 2\eta_{\mu\nu} \frac{dp^\mu}{d\tau} \, p^\nu \,.$$

Hence

$$0 = \eta_{\mu\nu} F^\mu \, p^\nu = \mathbf{F} \cdot \mathbf{p} - F^0 E/c \tag{4.4.11}$$

and therefore

$$F^0 = c\mathbf{F} \cdot \mathbf{p}/E = \mathbf{F} \cdot \mathbf{v}/c \,. \tag{4.4.12}$$

We will see in Section 4.6 how to construct the four-vector F^μ for the forces exerted by electric and magnetic fields on a moving charged particle.

4.5 Photons as Particles

As we saw in Section 3.2, Einstein in 1905 proposed that the energy of radiation of a given frequency ν is always an integer multiple of $h\nu$. This led to the further conjecture that the radiation consists of particles, later called photons, each with energy $h\nu$. A state with energy $nh\nu$ would then be interpreted as consisting of n photons.

Photon Momentum

If we suppose that photons are real particles, then we need to work out the relation between their energy and the magnitude of their momentum. In order for the conservation of energy and momentum to be Lorentz invariant when photons interact with other particles, the photon energy E and momentum $\mathbf{p}$ must form a four-vector p^μ, with $p^0 = E/c$, just as for other particles. That is, in changing coordinates from x^μ to $x'^\mu = \Lambda^\mu{}_\nu x^\nu$, the photon momentum four-vector is changed to $p'^\mu = \Lambda^\mu{}_\nu p^\nu$. But we cannot work out formulas for the components of p^μ in the way we did for other particles, by expressing p^μ as a Lorentz transformation acting on the four-momentum of a particle at rest, because photons never can be at rest.

Instead, we return to the starting point, that the energy of a quantum of radiation is proportional to the frequency. This implies that the time component of p^μ is proportional to the time component of another four-vector, the wave vector k^μ discussed in Section 4.3. Specifically, using the result $k^0 = \omega/c$ given there, we have

$$p^0 \equiv E/c = h\nu/c = \hbar\omega/c = \hbar k^0 \ .$$

Then in all Lorentz frames

$$0 = p^0 - \hbar k^0 \ . \tag{4.5.1}$$

It is a general rule that if the time component a^0 of a four-vector a^μ vanishes in all coordinate systems, then the whole four-vector a^μ vanishes. For if for any arbitrary Lorentz transformation Λ we have $a^0 = a'^0 = 0$ where $a'^\mu = \Lambda^\mu{}_\nu a^\nu$, then

$$0 = a'^0 = \Lambda^0{}_i a^i \ ,$$

which implies that a^i vanishes. (If $\mathbf{a} \neq 0$ we can rotate our coordinate axes so that the 3-axis is in the direction of $\mathbf{a}$, and take $\Lambda^\mu{}_\nu$ to be a Lorentz transformation (4.2.6) along this direction, in which case $0 = \beta\gamma|\mathbf{a}|$, so $\mathbf{a} = 0$.) The whole four-vector a^μ thus vanishes, as was to be proved. Taking $a^\mu = p^\mu - \hbar k^\mu$, we conclude then from Eq. (4.5.1) that the photon four-momentum is

$$p^\mu = \hbar k^\mu \tag{4.5.2}$$

and in particular

$$|\mathbf{p}| = \hbar|\mathbf{k}| = \hbar\omega/c = E/c \ . \tag{4.5.3}$$

This is just the relation between energy and momentum that we would expect from Eq. (4.4.7) if we treat the photon as a particle of zero mass.

Compton Scattering

If photons carry momentum, then when a photon is scattered by an electron at rest the electron should recoil. Suppose the incoming and outgoing photons have wave vectors $\mathbf{k}$ and $\mathbf{k}'$, respectively. According to Eq. (4.4.7), the energy of an electron of momentum $\mathbf{p}_e$ is given by

$$E_e = \sqrt{\mathbf{p}_e^2 c^2 + m_e^2 c^4} \ . \tag{4.5.4}$$

The conservation of energy in the scattering of a photon by an electron at rest requires that

$$c\hbar|\mathbf{k}| + m_e c^2 = c\hbar|\mathbf{k}'| + \sqrt{\mathbf{p}_e^2 c^2 + m_e^2 c^4}\ ,$$

where $\mathbf{p}_e$ is the momentum of the recoiling electron. According to Eq. (4.5.2), the conservation of momentum gives

$$\mathbf{p}_e = \hbar\mathbf{k} - \hbar\mathbf{k}'\ ,$$

so the conservation of energy becomes

$$c\hbar|\mathbf{k}| + m_e c^2 = c\hbar|\mathbf{k}'| + \sqrt{c^2\hbar^2\big(|\mathbf{k}|^2 + |\mathbf{k}'|^2 - 2\cos\theta|\mathbf{k}||\mathbf{k}'|\big) + m_e^2 c^4}\ ,$$

where θ is the angle between the initial and final photon wave vectors. Subtracting $c\hbar|\mathbf{k}'|$ from both sides and squaring, we have

$$\begin{aligned} &c^2\hbar^2(\mathbf{k}^2 - 2|\mathbf{k}||\mathbf{k}'| + \mathbf{k}'^2) + 2c^3\hbar m_e(|\mathbf{k}| - |\mathbf{k}'|) + m_e^2 c^4 \\ &= c^2\hbar^2(\mathbf{k}^2 + \mathbf{k}'^2 - 2\cos\theta|\mathbf{k}||\mathbf{k}'|) + m_e^2 c^4\ . \end{aligned}$$

Cancelling the terms $c^2\hbar^2\mathbf{k}^2$. $c^2\hbar^2\mathbf{k}'^2$, and $m_e^2 c^4$ on both sides leaves us with

$$|\mathbf{k}| - |\mathbf{k}'| = |\mathbf{k}||\mathbf{k}'|(1 - \cos\theta)\hbar/m_e c\ .$$

It is conventional to write this in terms of the wavelengths $\lambda = 2\pi/|\mathbf{k}|$ and $\lambda' = 2\pi/|\mathbf{k}'|$, and $h = 2\pi\hbar$:

$$\lambda' - \lambda = (1 - \cos\theta)h/m_e c\ . \qquad (4.5.5)$$

The quantity $h/m_e c$ equals 2.425×10^{-10} cm, and gives the increase in wavelength for a photon scattered at right angles to its original direction. This is known as the *Compton wavelength of the electron*, in honor of Arthur Holly Compton (1892–1962).

Compton at Washington University studied the scattering of monochromatic X-ray photons, with energy 17 keV. These photons were created by X-ray fluorescence: atoms of high atomic number, such as platinum, were exposed to a beam of high-energy electrons in a tube something like the cathode ray tubes used by Thomson (with whom Compton had worked at Cambridge). The beam of high-energy electrons knocked electrons out of these atoms, some from inner orbits. Then other electrons of nearly zero energy fell into these orbits, emitting monochromatic radiation, which, as we saw in our discussion of atomic number in Section 3.4, is at X-ray wavelengths for atoms with $Z \gg 1$. In Compton's experiment these photons were directed at a graphite target, where they were scattered by an outer electron of the carbon atom. These outer electrons have energies of the order of an eV, or at most tens of eV, negligible compared with the 17 keV energy of the incoming X-ray photon, so they scattered the X-ray photons just as if they were at rest. The wavelength of the scattered photon was measured by diffraction scattering, using a single crystal as a diffraction grating. Compton's experiment verified Eq. (4.5.5) in 1923, giving a significant boost to the acceptance of the quantum of light as a particle of zero mass. It was the chemist G. N. Lewis (1875–1946) who a few years later gave this particle the name “photon.”

There are other types of particle with zero mass. One is the *graviton*, the quantum of gravitational radiation. This radiation has been observed, but there is unfortunately no prospect of observing its quantum nature in the foreseeable future. There are also eight types of *gluons*, massless particles that in our present Standard Model are supposed to mediate strong nuclear forces. They interact so strongly when pulled away from other strongly interacting particles that they cannot even in principle be observed in isolation, but there is plenty of indirect evidence of their existence.

4.6 Electromagnetic Fields and Forces

Recall that Maxwell’s equations take the form

$$\nabla \times \mathbf{B} - \frac{1}{c}\frac{\partial \mathbf{E}}{\partial t} = \frac{4\pi}{c}\mathbf{J}\,, \qquad \nabla \cdot \mathbf{E} = 4\pi\rho\,, \tag{4.6.1}$$

$$\nabla \times \mathbf{E} + \frac{1}{c}\frac{\partial \mathbf{B}}{\partial t} = 0\,, \qquad \nabla \cdot \mathbf{B} = 0\,, \tag{4.6.2}$$

where $\mathbf{E}$ and $\mathbf{B}$ are the electric and magnetic fields, while ρ and $\mathbf{J}$ are the densities of electric charge and electric current. Are these equations Lorentz invariant, as required by Einsteinian relativity?

That is not quite the right question. We have no *a priori* knowledge of the Lorentz-transformation properties of the electric and magnetic fields. The real question that confronts us here is: what Lorentz-transformation properties can be supposed for the fields and densities in these equations that will make the equations Lorentz invariant? In the course of answering this question, we will encounter some algebraic devices that are useful in judging the Lorentz invariance of all sorts of field theories.

Density and Current

Let’s start by considering the charge density $\rho(\mathbf{x}, t)$ and current density $\mathbf{J}(\mathbf{x}, t)$ appearing on the right-hand sides of the Maxwell equations (4.6.1). Following the same arguments as in Section 2.5, because electric charge is conserved these satisfy a continuity equation like Eq. (2.5.2):

$$\frac{\partial}{\partial t}\rho(\mathbf{x}, t) + \nabla \cdot \mathbf{J}(\mathbf{x}, t) = 0\,. \tag{4.6.3}$$

This can be derived directly from the inhomogeneous Maxwell equations (4.6.1); just add c times the divergence of the first equation to the time derivative of the second equation. So how should $\rho(\mathbf{x}, t)$ and $\mathbf{J}(\mathbf{x}, t)$ behave under Lorentz transformations in order for Eq. (4.6.3) to be Lorentz invariant?

It helps to put the continuity equation in a revealing four-dimensional form. Define a four-component quantity $J^\mu(x)$ with $J^0(x) = c\rho(x)$. Then, recalling that $x^0 \equiv ct$, Eq. (4.6.3) reads

$$\frac{\partial}{\partial x^\mu} J^\mu(x) = 0 \,, \tag{4.6.4}$$

with repeated indices summed as usual over the values 1, 2, 3, 0. Now, how does the partial derivative $\partial/\partial x^\mu$ transform if we perform a Lorentz transformation $x^\mu \to x'^\mu = \Lambda^\mu{}_\nu x^\nu$? The chain rule of partial differentiation tells us that

$$\frac{\partial}{\partial x^\nu} = \frac{\partial x'^\mu}{\partial x^\nu} \frac{\partial}{\partial x'^\mu}$$

so in our case

$$\Lambda^\mu{}_\nu \frac{\partial}{\partial x'^\mu} = \frac{\partial}{\partial x^\nu} \,. \tag{4.6.5}$$

Therefore, if we suppose that $J^\mu(x)$ transforms as a four-vector under the Lorentz transformation $x^\mu \to x'^\mu = \Lambda^\mu{}_\nu x^\nu$, in the sense that the current $J'^\mu(x')$ measured by an observer who uses spacetime coordinates x'^μ is

$$J'^\mu(x') = \Lambda^\mu{}_\nu J^\nu(x) \,, \tag{4.6.6}$$

then

$$\frac{\partial}{\partial x'^\mu} J'^\mu(x') = \frac{\partial}{\partial x'^\mu} \Lambda^\mu{}_\nu J^\nu(x) = \frac{\partial}{\partial x^\nu} J^\nu(x) \,.$$

This is the Lorentz transformation of what is called a *scalar*. The quantity $\partial J^\mu/\partial x^\mu$ is seen by different observers to have the same value at the same point in spacetime, although these observers use different spacetime coordinate systems to label that point.

So, if an observer who uses spacetime coordinates x^μ sees $\partial J^\mu/\partial x^\mu$ to vanish at some particular value x^μ_I of these coordinates, then an observer who uses spacetime coordinates $x'^\mu = \Lambda^\mu{}_\nu x^\nu$ will see $\partial J'^\mu/\partial x'^\mu$ vanish at the corresponding coordinates $x'^\mu_I = \Lambda^\mu{}_\nu x^\nu_I$. In particular, if the first observer sees $\partial J^\mu/\partial x^\mu$ vanish everywhere, then so will any other observer whose coordinates are related to those of the first observer by a Lorentz transformation. So the Lorentz transformation (4.6.6) does make the conservation condition (4.6.4) Lorentz invariant.

The Inhomogeneous Maxwell Equations

We next consider how to rewrite the inhomogeneous Maxwell equations (4.6.1). The Lorentz invariance of these equations requires that we give $\mathbf{E}$ and $\mathbf{B}$

Lorentz-transformation properties such that their first derivatives with respect to space and time coordinates can be assembled into a four-component field that transforms as a four-vector, in the same sense (4.6.6) as J^μ. We cannot assemble a four-vector from the six components of $\mathbf{E}$ and $\mathbf{B}$ themselves, but we can assemble them into a different sort of quantity, an antisymmetric array $F^{\mu\nu}(x) = -F^{\nu\mu}(x)$ with *two* vector indices. We take

$$E_1 = F^{01} = -F^{10}, \quad E_2 = F^{02} = -F^{20}, \quad E_3 = F^{03} = -F^{30} \,, \tag{4.6.7}$$

$$B_1 = F^{23} = -F^{32}, \quad B_2 = F^{31} = -F^{13}, \quad B_3 = F^{12} = -F^{21} \,, \tag{4.6.8}$$

and $F^{\mu\nu} = 0$ if $\mu = \nu$. In this notation the 3-component of the first of the inhomogeneous equations (4.6.1) reads

$$\frac{4\pi}{c} J^3 = \frac{\partial B_2}{\partial x^1} - \frac{\partial B_1}{\partial x^2} - \frac{\partial E_3}{\partial x^0} = \frac{\partial F^{3\nu}}{\partial x^\nu}$$

with the understanding that in accordance with the summation convention the repeated index ν is summed over the values 1, 2, 3, 0, with the $\nu = 3$ term here vanishing because $F^{33} = 0$. The same applies to the 1-component and 2-component of the first of equations (4.6.1). Further, in this notation the second of equations (4.6.1) reads

$$\frac{4\pi}{c} J^0 = \nabla \cdot \mathbf{E} = \frac{\partial}{\partial x^1} F^{01} + \frac{\partial}{\partial x^2} F^{02} + \frac{\partial}{\partial x^3} F^{03} = \frac{\partial F^{0\nu}}{\partial x^\nu} \,.$$

So, in this notation all of the inhomogeneous Maxwell equations (4.6.1) can be summarized in the single four-component equation

$$\frac{\partial}{\partial x^\nu} F^{\mu\nu}(x) = \frac{4\pi}{c} J^\mu(x) \,. \tag{4.6.9}$$

It is now almost obvious how to make the inhomogeneous Maxwell equations Lorentz invariant. We suppose that under a Lorentz transformation $x^\mu \to x'^\mu = \Lambda^\mu{}_\nu x^\nu$ the field $F^{\mu\nu}(x)$ transforms like $J^\mu(x)$, but with a *pair* of four-valued indices. That is, the observer who uses coordinates x'^μ measures electric and magnetic fields with

$$F'^{\mu\nu}(x') = \Lambda^\mu{}_\rho \, \Lambda^\nu{}_\sigma F^{\rho\sigma}(x) \,. \tag{4.6.10}$$

Fields with this sort of transformation property are known as *tensors*.

To see that this makes Eq. (4.6.9) Lorentz invariant, consider a general Lorentz transformation $x^\mu \to x'^\mu = \Lambda^\mu{}_\nu x^\nu$. Multiplying Eq. (4.6.9) with $\Lambda^\rho{}_\mu$ and using Eq. (4.6.5) again to set $\partial/\partial x^\nu = \Lambda^\sigma{}_\nu \partial/\partial x'^\sigma$ gives

$$\Lambda^\rho{}_\mu \Lambda^\sigma{}_\nu \frac{\partial}{\partial x'^\sigma} F^{\mu\nu}(x) = \Lambda^\rho{}_\mu \frac{\partial}{\partial x^\nu} F^{\mu\nu}(x) = \frac{4\pi}{c} \Lambda^\rho{}_\mu J^\mu(x) \,.$$

Using the transformation properties (4.6.6) and (4.6.10), this becomes

$$\frac{\partial}{\partial x'^{\sigma}} F'^{\rho\sigma}(x') = \frac{4\pi}{c} J'^{\rho}(x') \, .$$

Thus Eq. (4.6.9) holds in the frame of reference with coordinates x'^{μ} if it holds in the frame of reference with coordinates x^{μ}, which is what we mean when we say it is Lorentz invariant. This then is a partial answer to our question: the inhomogeneous Maxwell equations (4.6.1) are Lorentz invariant if the electric and magnetic fields transform as components (4.6.7) and (4.6.8) of an antisymmetric tensor field.

This represents a unification of electricity and magnetism beyond anything of which Oersted, Ampére, Faraday, or even Maxwell could have dreamed. Not only are electric and magnetic fields coupled in the field equations – putting an observer into motion can change electric or magnetic fields into combinations of both electric and magnetic fields. For example, suppose an observer using coordinates x^{μ} finds a uniform electric field E_1 in the 1-direction, and no magnetic field, so that the only non-vanishing component of $F^{\mu\nu}$ is $F^{01} = -F^{10} = E_1$. Suppose a second observer uses coordinates $x'^{\mu} = \Lambda^{\mu}{}_{\nu}(v)x^{\nu}$, where $\Lambda^{\mu}{}_{\nu}(v)$ is the Lorentz transformation (4.2.6) that gives a body at rest a velocity v in the 3-direction, whose non-vanishing components are:

$$\Lambda^{3}{}_{3} = \Lambda^{0}{}_{0} = \gamma \, , \quad \Lambda^{3}{}_{0} = \Lambda^{0}{}_{3} = \beta\gamma \, ,$$
$$\Lambda^{1}{}_{1} = \Lambda^{2}{}_{2} = 1 \, ,$$

where $\beta = v/c$, and γ is again the positive quantity $\gamma = +1/\sqrt{1-\beta^2}$. The second observer sees an electromagnetic field

$$F'^{\mu\nu} = \Lambda^{\mu}{}_{\rho}(v)\, \Lambda^{\nu}{}_{\sigma}(v)\, F^{\rho\sigma} = \big(\Lambda^{\mu}{}_{0}(v)\, \Lambda^{\nu}{}_{1}(v) - \Lambda^{\mu}{}_{1}(v)\, \Lambda^{\nu}{}_{0}(v)\big) E_1 \, .$$

Its only non-vanishing components in this case are

$$E'_1 = F'^{01} = -F'^{10} = \Lambda^{0}{}_{0} E_1 = \gamma E_1 \, ,$$
$$B'_2 = F'^{31} = -F'^{13} = \Lambda^{3}{}_{0} E_1 = \beta\gamma E_1 \, .$$

Not only is the electric field increased; a magnetic field appears where before there was none. This is the sort of thing that had led Einstein to his 1905 paper.

Upstairs, Downstairs

We still have to verify that, with electromagnetic fields obeying the transformation rule (4.6.10) that makes the inhomogeneous Maxwell equations (4.6.1) Lorentz invariant, the homogeneous Maxwell equations (4.6.2) are also Lorentz invariant. To check this, we need to widen our ideas about vectors and tensors.

In general, we define a four-vector field $V^\mu(x)$ as a quantity that has the same Lorentz-transformation property as Δx^μ or p^μ or J^μ:

$$V^\mu(x) \to V'^\mu(x') = \Lambda^\mu{}_\nu V^\nu(x) .$$

There is a different kind of four-vector field, conventionally written with a lower index, that transforms according to

$$U_\mu(x) \to U'_\mu(x') = \Lambda_\mu{}^\nu U_\nu(x), \tag{4.6.11}$$

where $\Lambda_\mu{}^\nu$ is the transposed inverse of the matrix $\Lambda^\mu{}_\nu$ in the sense that

$$\Lambda_\mu{}^\rho \Lambda^\mu{}_\sigma = \Lambda^\rho{}_\mu \Lambda_\sigma{}^\mu = \delta^\rho_\sigma \equiv \begin{cases} 1 & \rho = \sigma \\ 0 & \rho \neq \sigma . \end{cases} \tag{4.6.12}$$

The classic example of a vector that is naturally defined with a lower index is the partial derivative. If we multiply Eq. (4.6.5) with $\Lambda_\rho{}^\nu$, sum over the repeated index ν, and use Eq. (4.6.12), we find

$$\frac{\partial}{\partial x'^\rho} = \Lambda_\rho{}^\nu \frac{\partial}{\partial x^\nu} . \tag{4.6.13}$$

It is trivial to calculate the transposed inverse $\Lambda_\mu{}^\nu$ of any given Lorentz transformation $\Lambda^\mu{}_\nu$. To see this, recall the defining characteristic (4.2.5) of Lorentz transformations:

$$\eta_{\mu\nu} \Lambda^\mu{}_\rho \, \Lambda^\nu{}_\sigma = \eta_{\rho\sigma} .$$

Multiplying with $\Lambda_\kappa{}^\rho$, summing over ρ, and using Eq. (4.6.12) gives

$$\eta_{\kappa\nu} \Lambda^\nu{}_\sigma = \eta_{\rho\sigma} \Lambda_\kappa{}^\rho . \tag{4.6.14}$$

That is, for i and j each running over 1, 2, 3:

$$\Lambda_i{}^j = \Lambda^i{}_j , \quad \Lambda_0{}^j = -\Lambda^0{}_j , \quad \Lambda_i{}^0 = -\Lambda^i{}_0 , \quad \Lambda_0{}^0 = \Lambda^0{}_0 .$$

In general, a tensor can have both upper and lower indices, and transforms with a Λ or its transposed inverse for each. For instance, a tensor $t^{\mu\nu}{}_\rho$ has the transformation property

$$t^{\mu\nu}{}_\rho \to \Lambda^\mu{}_\lambda \Lambda^\nu{}_\kappa \Lambda_\rho{}^\sigma t^{\lambda\kappa}{}_\sigma .$$

If we set an upper index equal to a lower index and (following the summation convention) sum over this index, we get another tensor with one less upper index and one less lower index. For instance, in the above example, if we set $\nu = \rho$ and sum, we obtain a quantity $v^\mu \equiv t^{\mu\nu}{}_\nu$, with the transformation property of a tensor with one index – that is, a vector:

$$v^\mu \to \Lambda^\mu{}_\lambda \Lambda^\nu{}_\kappa \Lambda_\nu{}^\sigma t^{\lambda\kappa}{}_\sigma = \Lambda^\mu{}_\lambda \, \delta^\sigma_\kappa \, t^{\lambda\kappa}{}_\sigma = \Lambda^\mu{}_\lambda \, t^{\lambda\sigma}{}_\sigma = \Lambda^\mu{}_\lambda \, v^\lambda ,$$

as required for a vector. One case has been already encountered: if we define a tensor $t^\mu_\nu \equiv \partial J^\mu / \partial x^\nu$ and set the upper and lower indices equal and sum, we obtain a quantity that we already know is a scalar: $t^\mu_\mu \equiv \partial J^\mu / \partial x^\mu$.

Although it is important not to confuse upper and lower indices, the difference between them is just a matter of the sign of the time components. We can use the matrix $\eta_{\mu\nu}$ to lower an index on any tensor, giving a new tensor. For instance, returning to our earlier example, if $t^{\mu\nu}{}_{\rho}$ is a tensor with transformation property

$$t^{\mu\nu}{}_{\rho} \to \Lambda^{\mu}{}_{\lambda}\Lambda^{\nu}{}_{\kappa}\Lambda_{\rho}{}^{\xi}t^{\lambda\kappa}{}_{\xi} \ ,$$

we can lower the index ν, defining a new tensor:

$$u^{\mu}{}_{\sigma\rho} \equiv \eta_{\nu\sigma}t^{\mu\nu}{}_{\rho} \ .$$

Using Eq. (4.6.14), we see that this has the transformation property

$$\begin{aligned} u^{\mu}{}_{\sigma\rho} &\to \eta_{\nu\sigma}\Lambda^{\mu}{}_{\lambda}\,\Lambda^{\nu}{}_{\kappa}\Lambda_{\rho}{}^{\xi}t^{\lambda\kappa}{}_{\xi} \\ &= \Lambda^{\mu}{}_{\lambda}\,\Lambda_{\sigma}{}^{\tau}\eta_{\tau\kappa}\Lambda_{\rho}{}^{\xi}t^{\lambda\kappa}{}_{\xi} = \Lambda^{\mu}{}_{\lambda}\,\Lambda_{\sigma}{}^{\tau}\Lambda_{\rho}{}^{\xi}u^{\lambda}{}_{\tau\xi} \end{aligned}$$

as is appropriate for a tensor with one upper index and two lower indices. (It is also possible to raise any lower indices on tensors, but we won't need to do this here.)

The Homogeneous Maxwell Equations

With our new-found power to lower indices, let introduce a new tensor:

$$H_{\mu\nu\lambda} \equiv \frac{\partial F_{\mu\nu}}{\partial x^{\lambda}} + \frac{\partial F_{\nu\lambda}}{\partial x^{\mu}} + \frac{\partial F_{\lambda\mu}}{\partial x^{\nu}} \ . \tag{4.6.15}$$

It is easy to see that $H_{\mu\nu\lambda}$ is totally antisymmetric. For instance, interchanging μ and λ gives

$$H_{\lambda\nu\mu} = \frac{\partial F_{\lambda\nu}}{\partial x^{\mu}} + \frac{\partial F_{\nu\mu}}{\partial x^{\lambda}} + \frac{\partial F_{\mu\lambda}}{\partial x^{\nu}} = -\frac{\partial F_{\nu\lambda}}{\partial x^{\mu}} - \frac{\partial F_{\mu\nu}}{\partial x^{\lambda}} - \frac{\partial F_{\lambda\mu}}{\partial x^{\nu}} = -H_{\mu\nu\lambda} \ .$$

Therefore $H_{\mu\nu\lambda}$ vanishes unless all three indices are unequal. In four spacetime dimensions, this means that $H_{\mu\nu\lambda}$ has only four independent components. Lowering the indices in Eqs. (4.6.7) and (4.6.8), we have

$$\begin{aligned} &E_1 = -F_{01}, \quad E_2 = -F_{02}, \quad E_3 = -F_{03} \ , \\ &B_1 = F_{23}, \quad B_2 = F_{31}, \quad B_3 = F_{12} \ , \end{aligned} \tag{4.6.16}$$

so

$$\begin{aligned} H_{123} &\equiv \frac{\partial F_{12}}{\partial x^3} + \frac{\partial F_{23}}{\partial x^1} + \frac{\partial F_{31}}{\partial x^2} = \frac{\partial B_3}{\partial x^3} + \frac{\partial B_1}{\partial x^1} + \frac{\partial B_2}{\partial x^2} = \nabla \cdot \mathbf{B} \ , \\ H_{120} &\equiv \frac{\partial F_{12}}{\partial x^0} + \frac{\partial F_{20}}{\partial x^1} + \frac{\partial F_{01}}{\partial x^2} = \frac{1}{c}\frac{\partial B_3}{\partial t} + \frac{\partial E_2}{\partial x^1} - \frac{\partial E_1}{\partial x^2} \\ &= \left[\frac{1}{c}\frac{\partial \mathbf{B}}{\partial t} + \nabla \times \mathbf{E}\right]_3 \ . \end{aligned}$$

Likewise

$$H_{230} = \left[\frac{1}{c}\frac{\partial \mathbf{B}}{\partial t} + \boldsymbol{\nabla} \times \mathbf{E}\right]_1 , \quad H_{310} = \left[\frac{1}{c}\frac{\partial \mathbf{B}}{\partial t} + \boldsymbol{\nabla} \times \mathbf{E}\right]_2 .$$

Hence the homogeneous Maxwell equations are the same as the requirement that, for all μ, ν, and λ,

$$H_{\mu\nu\lambda} = 0 . \tag{4.6.17}$$

This is a manifestly Lorentz-invariant condition; if $H_{\mu\nu\lambda}(x)$ vanishes, then so does $H'_{\rho\sigma\kappa}(x') = \Lambda_\rho{}^\mu \Lambda_\sigma{}^\nu \Lambda_\kappa{}^\lambda H_{\mu\nu\lambda}(x)$.

(We will not here use the formalism of differential forms, but for anyone interested in this subject, I mention in passing that a completely antisymmetric tensor with p lower indices is known as a p-form. Thus $F_{\mu\nu}$ is a 2-form, and $H_{\mu\nu\lambda}$ is a 3-form. Given a p-form, we can form a $p + 1$-form, known as the exterior derivative, by taking the spacetime derivative and antisymmetrizing. Thus, $H_{\mu\nu\lambda}$ is the exterior derivative of $F_{\mu\nu}$. A p-form whose exterior derivative vanishes is said to be *closed*; a p-form that can be written as the exterior derivative of a $p - 1$-form is said to be *exact*. Thus $H_{\mu\nu\lambda}$ is exact, and the homogeneous Maxwell equations (4.6.2) tell us that $F_{\mu\nu}$ is closed. It is easy to see that, because partial derivatives commute, any exact p-form is closed, and a profound theorem due to Poincaré tells us that in simply connected spaces any closed p-form is exact but that this is not necessarily true in spaces with more complicated topology.[11] In electrodynamics, since $F_{\mu\nu}$ is closed, we can conclude that in ordinary spacetime it is exact, so it can be written as the exterior derivative of a 1-form A_μ known as the four-vector potential; that is, $F_{\mu\nu} = \partial A_\mu/\partial x^\nu - \partial A_\nu/\partial x^\mu$. Maxwell originally wrote his equations as differential equations for $\mathbf{A}$ and A^0, not $\mathbf{E}$ and $\mathbf{B}$.)

Electric and Magnetic Forces

We saw in Section 4.4 that in special relativity Newton's $\mathbf{F} = m\mathbf{a}$ is replaced with the Lorentz-invariant formula (4.4.9)

$$\frac{dp^\mu}{d\tau} = F^\mu \tag{4.6.18}$$

where p^μ is the four-vector of energy and momentum, $c\,d\tau \equiv [-\eta_{\mu\nu}dx^\mu dx^\nu]^{1/2}$, and F^μ is a four-vector subject to the constraint

$$\eta_{\mu\nu} p^\mu F^\nu = 0 . \tag{4.6.19}$$

[11] For a more thorough treatment, see e.g. H. Flanders, *Differential Forms* (Academic Press, New York, 1963).

So, what should we take for F^μ in the case of a particle of charge q in a space that is empty except for being pervaded by electric and magnetic fields? Just as for the momentum four-vector of massive particles, the force four-vector is uniquely determined by the condition that it takes the known form for a particle at rest, and is a four-vector, so that it is given by a Lorentz transformation for a particle of any velocity.

There is an obvious four-vector that is linear in the electric and magnetic fields and (because $F^{\mu\rho} = -F^{\rho\mu}$) satisfies Eq. (4.6.19):

$$f^\mu \equiv \eta_{\rho\sigma} F^{\mu\rho} p^\sigma$$

so we can guess that $F^\mu \propto f^\mu$. To check that this gives the right answer for a particle at rest and to find the coefficient of proportionality, let us evaluate f^μ for a particle at rest, for which $\mathbf{p} = 0$ and $p^0 = mc$. In this limit

$$f^i \to -mcF^{i0} = mcE_i \ , \quad f^0 \to -mcF^{00} = 0$$

with $i = 1, 2, 3$. Therefore to have agreement with the familiar formula $d\mathbf{p}/dt = q\mathbf{E}$ for the acceleration of a particle of charge q and zero velocity by an electric field, we take

$$F^\mu = \frac{q}{mc} f^\mu = \frac{q}{mc} \eta_{\rho\sigma} F^{\mu\rho} p^\sigma \ . \tag{4.6.20}$$

That is, for a general velocity,

$$F^i = \frac{q}{mc} [F^{ij} p^j - F^{i0} p^0] \ , \tag{4.6.21}$$

and in three-vector notation, recalling that $p^0 = mc\gamma$, $\mathbf{p} = m\gamma\mathbf{v}$,

$$\mathbf{F} = \frac{q}{mc} [p^0 \mathbf{E} + \mathbf{p} \times \mathbf{B}] = q\gamma [\mathbf{E} + \mathbf{v} \times \mathbf{B}/c] \ . \tag{4.6.22}$$

Since $d\tau = dt/\gamma$, this gives

$$m \frac{d}{dt} [\gamma \mathbf{v}] = q [\mathbf{E} + \mathbf{v} \times \mathbf{B}/c] \ . \tag{4.6.23}$$

Given the existence of the force exerted by electric fields, the force exerted by magnetic fields is an inevitable consequence of Lorentz invariance. It is a special feature of electromagnetic forces that the only change in the equation of motion introduced by special relativity is the replacement of the mass m in the momentum with $m\gamma$, which in this one case allows us to treat $m\gamma$ as a relativistic mass.

4.7 Causality

We saw in Section 4.2 that no Lorentz transformation acting on a body at rest could give it a speed greater than c, the speed of light. We can derive a stronger result, that no influence whatever can travel faster than light. This is not just a confession of technological inadequacy, but a consequence of an assumption of causality, that effects always come after causes.

Invariance of Temporal Order

Suppose that in some coordinate frame the difference between the spacetime coordinates of an event and the event that cause it is Δx^μ:

$$x^\mu_{\text{effect}} - x^\mu_{\text{cause}} \equiv \Delta x^\mu \ .$$

According to the principle of causality, we must have $\Delta t = \Delta x^0/c > 0$. Now suppose we perform a Lorentz transformation $\Lambda^\mu{}_\nu$ that would give a body at rest a velocity $\mathbf{v}$ in the direction opposite to the spatial separation $\Delta\mathbf{x}$. Without loss of generality, we can rotate our coordinate system so that $\Delta\mathbf{x}$ and thus $-\mathbf{v}$ are in the 3-direction. Then $\Lambda^\mu{}_\nu$ takes the form (4.2.6) with $\beta = -|\mathbf{v}|/c$, and in the new coordinate frame the difference between the times of effect and cause is

$$\Delta t' = \Lambda^0{}_\mu \Delta x^\mu/c = \gamma[\Delta t - v|\Delta\mathbf{x}|/c^2] \tag{4.7.1}$$

where $v = |\mathbf{v}|$ and $\gamma = 1/\sqrt{1 - v^2/c^2}$. Now, v can be anything, except that it must be less than c, so if $|\Delta\mathbf{x}|/c$ is greater than Δt we could make $\Delta t'$ *negative* by taking v in the range $1 > v/c > c\Delta t/|\Delta\mathbf{x}|$. So the observer using coordinates x'^μ would see the effect precede the cause.

To rule this out, we must assume that the difference Δx^μ between the spacetime coordinates of an event and the event that causes it satisfies the inequality

$$|\Delta\mathbf{x}|/c \leq \Delta t \ . \tag{4.7.2}$$

Whatever physical influence is exerted by the cause to produce the effect travels at a speed $|\Delta\mathbf{x}|/\Delta t$; the inequality (4.7.2) says that the speed of this influence must be no greater than c.

Fortunately, if the bound (4.7.2) is seen to be satisfied by one observer then it is satisfied for any other observer related to the first by the sort of proper Lorentz transformation discussed in this chapter. The inequality (4.7.2) is equivalent to the inequality

$$-\eta_{\mu\nu}\Delta x^\mu \, \Delta x^\nu = c^2(\Delta t)^2 - |\Delta\mathbf{x}|^2 \geq 0 \ . \tag{4.7.3}$$

This quantity is Lorentz invariant, so if Eq. (4.7.3) is satisfied for one observer using coordinates x^μ, the corresponding inequality must be satisfied for coordinates $x'^\mu = \Lambda^\mu{}_\nu x^\nu$, so we must also have

$$c^2(\Delta t')^2 \geq |\Delta\mathbf{x}'|^2 \ . \tag{4.7.4}$$

Since this gives a non-zero lower bound on $|\Delta t'|$, in order for $\Delta t'$ to have an opposite sign from Δt the Lorentz transformation would have to produce a discontinuous jump in the coordinates. This is not possible for the sort of "proper" Lorentz transformation that concerns us in this chapter, which as discussed in Section 4.2 can be produced from the identity transformation $x \to x$ by a smooth change of parameters. So if one observer sees $\Delta t > 0$ and $|\Delta \mathbf{x}|/c \leq \Delta t$, then any observer related to the first by a proper Lorentz transformation will see $\Delta t' > 0$ and $|\Delta \mathbf{x}'|/c \leq \Delta t'$.

Light Cone

These conclusions are well illustrated by introducing the *light cone*, the spacetime surface with $\eta_{\mu\nu}\Delta x^\mu \, \Delta x^\nu = 0$. Points outside the light cone fall on hyperboloids with $\eta_{\mu\nu}\Delta x^\mu \, \Delta x^\nu = a > 0$. Any point on one of these hyperboloids can be taken to any other point on the same hyperboloid (that is, with the same value of a) by a proper Lorentz transformation, even if this entails a change of sign of Δx^0. On these hyperboloids $|\Delta \, \mathbf{x}| > c|\Delta t|$, so it is not possible for any influence traveling at less than the speed of light to traverse a spacetime interval Δx^μ outside the light cone. Thus as long as we assume that physical influences never travel faster than light, the circumstance that proper Lorentz transformations can change the sign of Δt outside the light cone presents no challenge to causality.

Points inside the light cone fall on hyperboloids with $\eta_{\mu\nu}\Delta x^\mu \, \Delta x^\nu = b < 0$. For each value of b there are two disconnected hyperboloids, one inside the future light cone, with $\Delta x^0 > 0$, and one inside the past light cone, with $\Delta x^0 < 0$. Any point on one of these connected hyperboloids can be taken to any other point on the same hyperboloid by a proper Lorentz transformation, but proper Lorentz transformations cannot take us from inside the future light cone to inside the past light cone. Causality requires that the difference Δx^μ in the coordinates of an effect and its cause be on or within the future light cone, and if one observer sees this to be the case then so will all other observers related to the first by a proper Lorentz transformation.

5
Quantum Mechanics

Our modern understanding of atoms, molecules, solids, atomic nuclei, and elementary particles is largely based on quantum mechanics. Quantum mechanics grew in the mid-1920s out of two independent developments: the 1925 matrix mechanics of Werner Heisenberg[1] (1901–1976), and the 1926 wave mechanics of Erwin Schrödinger[2] (1887–1961). For the most part in this chapter we will follow the path of wave mechanics, which is far more convenient for all but the simplest calculations. After a look at the historical inspiration for wave mechanics in Section 5.1 the Schrödinger equation will be introduced in Section 5.2 and used to derive not only the hydrogen energy levels found by Bohr but also their degeneracy. The general principles of the wave mechanical formulation of quantum mechanics are laid out in Section 5.3 and provide a basis for the discussion of spin in Section 5.4, identical particles in Section 5.5, and scattering processes in Section 5.6. In Section 5.7 the general principles are supplemented with the canonical formalism, which is used in Section 5.8 to work out the Schrödinger equation for charged particles in a general electromagnetic field. This will provide us with examples of the application of a widely useful approximation scheme, perturbation theory, which is outlined in general terms in Section 5.9.

The two approaches of wave and matrix mechanics were unified by Paul Dirac (1902–1984) in a more abstract formalism, which he called transformation theory.[3] This has evolved into a modern approach in which physical states are represented by vectors in an abstract space known as Hilbert space, with wave functions arising as components of these vectors in a suitable basis. The Hilbert space approach is briefly described in Section 5.10.

---

1 W. Heisenberg, Zeit. Phys. **33**, 879 (1925). This article is reprinted in English in Van der Waerden, *Sources of Quantum Mechanics*, listed in the bibliography.

2 E. Schrödinger, Ann. Physik **79**, 361, 409 (1926). These articles are reprinted in English in Shearer, *Collected Papers on Wave Mechanics*, listed in the bibliography.

3 This approach is described in Dirac, *The Principles of Quantum Mechanics*, listed in the bibliography.

5.1 De Broglie Waves

Free-Particle Wave Functions

Wave mechanics can be traced to the 1923 Paris Ph.D. thesis of Louis de Broglie[4] (1892–1987). De Broglie was inspired by the quantum interpretation of electromagnetic radiation. If an electromagnetic wave can somehow be interpreted as a stream of particles, photons, then might not electrons, which are undoubtedly particles, be described somehow as waves? As we saw in Section 4.5, the momentum $\mathbf{p}$ and energy E of the photons making up an electromagnetic wave that is proportional to $\exp(i\mathbf{k}\cdot\mathbf{x} - i\omega t)$ are given by $\mathbf{p} = \hbar\mathbf{k}$ and $E = \hbar\omega$, so the wave has the spacetime dependence

$$\mathbf{E} \text{ and } \mathbf{B} \propto \exp\left[i\mathbf{p}\cdot\mathbf{x}/\hbar - iEt/\hbar\right] , \tag{5.1.1}$$

plus the complex conjugates. De Broglie in his thesis suggested that an electron of momentum $\mathbf{p}$ is associated with a complex wave function of similar form

$$\psi_{\mathbf{p}}(\mathbf{x}, t) \propto \exp\left[i\mathbf{p}\cdot\mathbf{x}/\hbar - iE(\mathbf{p})t/\hbar\right] , \tag{5.1.2}$$

where now the energy is not $c|\mathbf{p}|$, as for a photon, but rather is given by the formula (4.5.4):

$$E(\mathbf{p}) = \sqrt{m_e^2c^4 + \mathbf{p}^2c^2} ,$$

with m_e the electron mass.

Group Velocity

The association of the wave (5.1.2) with a moving electron gained plausibility from the remark that a localized packet of these waves travels with the velocity of the electron. Consider a packet of these waves:

$$\psi(\mathbf{x}, t) = \int d^3p\; g(\mathbf{p}) \exp\left[i\mathbf{p}\cdot\mathbf{x}/\hbar - iE(\mathbf{p})t/\hbar\right] \tag{5.1.3}$$

where $g(\mathbf{p})$ is a smooth function of momentum that is peaked at some value $\mathbf{P}$. Suppose also that $g(\mathbf{p})$ is chosen so that at $t = 0$ the integral is peaked at $\mathbf{x} = 0$. (This will be the case if $g(\mathbf{p})$ varies little over some range around $\mathbf{P}$ that is large enough that if $\mathbf{x}$ is not near zero then the factor $\exp[i\mathbf{p}\cdot\mathbf{x}/\hbar)]$ in Eq. (5.1.3) at $t = 0$ will undergo many oscillations over the range of the integral, which makes the integral exponentially small except near $\mathbf{x} = 0$.) Then, by expanding the argument of the exponential around $\mathbf{P}$, we have

$$E(\mathbf{p}) \simeq E(\mathbf{P}) + \mathbf{V}\cdot(\mathbf{p} - \mathbf{P}) + \cdots$$

[4] L. de Broglie, Comptes Rendus Acad. Sci. **177**, 507, 548, 630 (1923).

where

$$V_i = \left.\frac{\partial E(\mathbf{p})}{\partial p_i}\right]_{\mathbf{p}=\mathbf{P}} . \tag{5.1.4}$$

This gives the wave function for $t \neq 0$:

$$\psi(\mathbf{x},t) \simeq \exp\left[i\mathbf{P}\cdot\mathbf{x}/\hbar - i[E(\mathbf{P}) - \mathbf{V}\cdot\mathbf{P}]t/\hbar\right] \\ \times \int d^3p\ g(\mathbf{p})\ \exp(i\mathbf{p}\cdot[\mathbf{x}-\mathbf{V}t]) . \tag{5.1.5}$$

Because of the way we have constructed the packet function $g(\mathbf{p})$, the magnitude of (5.1.5) is peaked at $\mathbf{x} = \mathbf{V}t$, which shows that the packet moves at velocity $\mathbf{V}$, known as its *group velocity*. But $V_i = \partial E(\mathbf{p})/\partial p_i = c^2 p_i/E(\mathbf{p})$, which as shown in Eqs. (4.4.5) and (4.4.6) is indeed the velocity of a particle of momentum $\mathbf{p}$.

Application to Hydrogen

De Broglie's hypothesis met with just one initial success. The electron in a hydrogen atom is not free, but moves under the influence of the proton's attraction. Nevertheless, de Broglie supposed that the electron is described by the free-particle wave function (5.1.2), but with the waves traveling in a circle around the proton like sound waves in a toroidal organ pipe. To avoid a discontinuity in ψ, it is necessary that a whole number n of wavelengths λ should fit around the circle, so the radius of the circle is constrained by the condition that $2\pi r = n\lambda$, with $n = 1, 2, \ldots$ According to Eq. (5.1.2), $\lambda = 2\pi\hbar/p$, where $p \equiv |\mathbf{p}|$, so de Broglie's condition was

$$pr = n\hbar \tag{5.1.6}$$

which for non-relativistic electrons with $p = m_e v$ is the same as Bohr's condition (3.4.2), but now with no need of the correspondence principle to infer that $\hbar = h/2\pi$. De Broglie could then repeat Bohr's calculation, using the non-relativistic formula $E = m_e v^2/2 - e^2/r$ for energy and the formula $m_e v^2/r = e^2/r$ for centripetal acceleration, and thereby obtain Bohr's formula $E = -e^4 m_e/2\hbar^2 n^2$ for the hydrogen energy levels. Nothing new had been learned about hydrogen, but de Broglie's derivation at least gave a hint at an explanation of Bohr's quantization condition.

Davisson–Germer Experiment

There is a story that in his oral Ph.D. examination, de Broglie was asked if there was some direct way of observing the wave nature of electrons, and he answered that it might be possible to observe the diffraction of electron waves by a crystal lattice, like the well-known diffraction of X-rays used for instance in measuring the increase of wavelength in Compton scattering. Whether or not this story is true, the idea was a good one. According to Eq. (5.1.2), the wavelength of a non-relativistic electron with kinetic energy $E_e \ll m_e c^2$ is given by

$$\lambda = 2\pi\hbar/p_e = 2\pi\hbar/\sqrt{2m_e E_e} = 12.26 \times 10^{-8}\,\text{cm}\,[E_e(\text{eV})]^{-1/2} . \quad (5.1.7)$$

Hence we only need electrons with energy a bit larger than 10 eV to get wavelengths nearly as small as a typical lattice spacing, about 10^{-8} cm. This is no coincidence. In de Broglie's interpretation of the Bohr quantization assumption, the wavelength of an electron with an energy of a few eV, which is typical of atomic binding energies, must fit a few times around an atomic orbit, and therefore must be similar to the size of the atom, which is similar to the spacing of atoms in crystals.

Several physicists tried and failed to observe the diffraction of electron waves, until it was finally measured in 1927 by Clinton Davisson (1881–1958) and Lester Germer (1896–1971) at the old Bell Telephone Laboratories building on West Street in Manhattan.[5] (It was also measured at about the same time at the University of Aberdeen by George Paget Thomson (1892–1975), a son of J. J. Thomson.) They used a beam of electrons with kinetic energy 54 eV, incident on a single crystal of nickel with a spacing of lattice planes $d = 0.91 \times 10^{-8}$ cm (already known from measurements using X-ray diffraction). Electrons are reflected not only from the surface of the crystal, but from numerous planes within the nickel. At certain angles θ between the incident and reflected waves all these reflected waves go off with the same phase and therefore add constructively, leading to enhanced reflection at these angles. According to a 1913 formula (derived in the appendix to this section) of William Henry Bragg (1862–1942) and his son Lawrence Bragg (1890–1971), for any sort of wave the angles θ_n between incident and reflected waves at which reflection is enhanced in this way satisfy the Bragg formula:

$$n\lambda = 2d\cos(\theta_n/2) , \quad (5.1.8)$$

where $n = 1, 2, 3, \ldots$ Davisson and Germer found an enhanced $n = 1$ reflection at $\theta_1 = 50^\circ$, giving a wavelength

$$\lambda = 2 \times 0.91 \times 10^{-8}\,\text{cm} \times \cos(25^\circ) = 1.68 \times 10^{-8}\,\text{cm} ,$$

in satisfactory agreement with the wavelength 1.67×10^{-8} cm expected from Eq. (5.1.7) for a kinetic energy of 54 eV.

The wave nature of the electron allowed the development of a new instrument, the electron microscope. Recall that a photon of energy E has wavelength

[5] C. Davisson and L. Germer, Phys. Rev. **30**, 707 (1927).

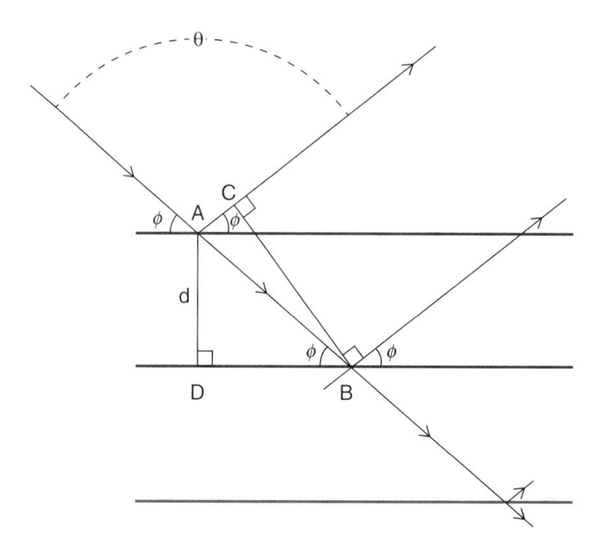

Figure 5.1 Derivation of the Bragg formula. The bold lines represent the planes of the crystal lattice, seen edge on. Arrows indicate the direction of the light rays.

$\lambda_\gamma = 2\pi\hbar c/E$, so the ratio of the wavelength (5.1.7) of an electron of energy E to the wavelength of a photon of the same energy is

$$\frac{\lambda_e}{\lambda_\gamma} = \sqrt{\frac{E}{2m_e c^2}} \; .$$

For energies in the range of 10 eV to 10 keV this is very much less than one, giving electron microscopes much better resolution than microscopes using photons of the same energy.

Appendix: Derivation of the Bragg Formula

Suppose that a wave of some sort is incident on a crystal lattice, with a ray striking one plane of the lattice at point A, where it makes an angle ϕ between the ray and the plane. (See Figure 5.1.) Part of the wave is reflected, with the reflected ray making the same angle ϕ with the plane. Another part of the wave continues in its original direction to the next plane, with the ray striking this plane at point B, again at angle ϕ. Part of this ray is reflected at B, again at angle ϕ, while another part continues to deeper planes. Draw a line from B in a direction normal to the first reflected ray, intersecting this ray at a point C. The purpose of this construction is that the two parallel reflected rays travel the same distance from B and C to any distant detector, so the difference in the total distance that each ray travels to the detector is AB − AC. The two rays interfere constructively if this difference is a whole number n of wavelengths λ:

$$AB - AC = n\lambda\ , \qquad n = 1, 2, 3, \ldots$$

If this is satisfied, then the difference in the distance traveled by rays reflected from the second and third planes will also be $n\lambda$, and so on down into deeper and deeper planes, so all these reflected waves will interfere constructively. The same is true of any rays that strike the crystal along parallel directions, whether they are reflected from the first, second, or any other crystal plane. We then have a very strong enhancement of the reflection. So we have to ask, how do AB and AC depend on the lattice spacing and on the angle ϕ between the rays and the crystal planes?

Draw a line from A to the second plane, which intersects it at a right angle at a point D. The length of the line AD is the spacing d of lattice planes. Looking at the right triangle ADB (whose hypotenuse is AB) we see that

$$AB = d/\sin\phi\ .$$

To calculate AC, note that the angle at B between BA and BC is $180^\circ - 2\phi - 90^\circ = 90^\circ - 2\phi$, so looking at the right triangle BAC (whose hypotenuse is AB), we see that

$$AC = AB\sin(90^\circ - 2\phi) = AB\cos(2\phi) = AB[1 - 2\sin^2\phi]$$

so

$$AB - AC = 2AB\sin^2\phi = 2d\sin\phi$$

and the condition for constructive interference is therefore

$$n\lambda = 2d\sin\phi\ .$$

It is common to describe the reflection in terms of the angle θ between the incident and reflected rays, $\theta = 180^\circ - 2\phi$, so $\phi = 90^\circ - \theta/2$, and the condition for constructive interference is then

$$n\lambda = 2d\cos(\theta/2)\ ,$$

as was to be shown.

5.2 The Schrödinger Equation

Wave Equation in a Potential

De Broglie in 1923 had described the wave function associated with a free electron and had scored some success in applying this to the electron in a hydrogen atom, imagining a free electron wave running around the electron orbit. But of course electrons in atoms are not free. Starting in 1925, Erwin Schrödinger struggled to extend the idea of the wave function to an electron moving in a potential.[6]

Schrödinger's starting point was de Broglie's wave theory. Equation (5.1.2) gives the wave function of a free electron of momentum **p** as

$$\psi_{\mathbf{p}}(\mathbf{x}, t) \propto \exp\left[i\mathbf{p}\cdot\mathbf{x}/\hbar - iE(\mathbf{p})t/\hbar\right] .$$

The content of this explicit formula can be expressed as a pair of differential equations

$$-i\hbar\nabla\psi_{\mathbf{p}}(\mathbf{x}, t) = \mathbf{p}\psi_{\mathbf{p}}(\mathbf{x}, t) , \tag{5.2.1}$$

$$i\hbar\frac{\partial}{\partial t}\psi_{\mathbf{p}}(\mathbf{x}, t) = E(\mathbf{p})\psi_{\mathbf{p}}(\mathbf{x}, t) , \tag{5.2.2}$$

where, now in a non-relativistic approximation,

$$E(\mathbf{p}) \simeq m_e c^2 + \frac{1}{2m_e}\mathbf{p}^2 . \tag{5.2.3}$$

An electron that is bound in an atom cannot have a definite momentum – classically, it goes round and round its orbit – so we would not expect the bound electron wave function to satisfy an equation like (5.2.1). On the other hand, we can try to use something like Eq. (5.2.2) to find the wave function of a bound electron, with Eq. (5.2.1) used only to interpret **p** as $-i\hbar\nabla$ in $E(\mathbf{p})$. Schrödinger thus took the equation for a bound electron as

$$i\hbar\frac{\partial}{\partial t}\psi(\mathbf{x}, t) = E(-i\hbar\nabla,\ \mathbf{x})\psi(\mathbf{x}, t) , \tag{5.2.4}$$

where now E is given a dependence on **x** to account for the presence of potential energy. For a non-relativistic electron in a potential $V(\mathbf{x})$ the energy is $E(\mathbf{p}, \mathbf{x}) = m_e c^2 + \mathbf{p}^2/2m_e + V(\mathbf{x})$, and Eq. (5.2.4) reads

$$i\hbar\frac{\partial}{\partial t}\psi(\mathbf{x}, t) = \left[m_e c^2 - \frac{\hbar^2}{2m_e}\nabla^2 + V(\mathbf{x})\right]\psi(\mathbf{x}, t) . \tag{5.2.5}$$

This is known as the *time-dependent Schrödinger equation.*

Because the potential is assumed time-independent, Eq. (5.2.5) has solutions of the form

$$\psi(\mathbf{x}, t) = \exp\left[-i(m_e c^2 + E)t/\hbar\right]\psi(\mathbf{x}) , \tag{5.2.6}$$

where

$$\left[-\frac{\hbar^2}{2m_e}\nabla^2 + V(\mathbf{x})\right]\psi(\mathbf{x}) = E\psi(\mathbf{x}) . \tag{5.2.7}$$

[6] E. Schrödinger, Ann. Phys. **79**, 361, 409 (1926).

This is known as the *time-independent Schrödinger equation*. It is interpreted as the condition for $\psi(\mathbf{x})$ to represent a state with definite energy E, relative to the rest-mass energy $m_e c^2$.

Boundary Conditions

This all began as just guesswork. Schrödinger and other physicists at first imagined that $\psi(\mathbf{x}, t)$ gives an indication of how much of the electron is near $\mathbf{x}$ at time t. As we will see when we come to scattering in Section 5.6, it was only a few years later that Max Born correctly interpreted $|\psi(\mathbf{x}, t)|^2$ as a probability density – that is, $|\psi(\mathbf{x}, t)|^2 \, d^3x$ is the probability that the electron is in a small volume d^3x at position $\mathbf{x}$ and time t. For the present, all we need to know is that the relevant solutions of Eq. (5.2.7) are those with

$$\int |\psi(\mathbf{x})|^2 \, d^3x < \infty \tag{5.2.8}$$

so that by dividing $\psi(\mathbf{x})$ by the square root of this integral we obtain a normalized wave function, corresponding to a probability density for which the total probability of the electron being somewhere is 100%.

If we assume (as is generally the case in practice) that $V(\mathbf{x})$ vanishes at large distances $|\mathbf{x}|$, then at large $|\mathbf{x}|$ Eq. (5.2.7) becomes

$$E\psi(\mathbf{x}) \to -\frac{\hbar^2}{2m_e}\nabla^2 \psi(\mathbf{x}) \, . \tag{5.2.9}$$

A bound electron must have $E < 0$ (since otherwise it would be energetically possible for the electron to escape to infinite distance) so Eq. (5.2.9) has solutions that at large $|\mathbf{x}|$ behave as

$$\psi(\mathbf{x}) \to P(\mathbf{x}) \, \exp(\pm\kappa|\mathbf{x}|) \tag{5.2.10}$$

where κ is the positive square root,

$$\kappa = +\sqrt{2m_e|E|/\hbar^2} \, , \tag{5.2.11}$$

and $P(\mathbf{x})$ is some function such as a polynomial that varies much more slowly than an exponential for large $|\mathbf{x}|$. (A derivative $\partial/\partial x_i$ acting on the exponential in Eq. (5.2.10) yields a constant factor $\pm\kappa$, while a gradient acting on a function $P(\mathbf{x})$ in Eq. (5.2.10) that grows as a power of $|\mathbf{x}|$ gives a factor for $|\mathbf{x}| \to \infty$ proportional to $1/|\mathbf{x}|$.) Solutions of the time-independent Schrödinger equation thus come in pairs, one of which (the one with a minus sign in the exponential in Eq. (5.2.10)) satisfies the condition (5.2.8) at least as far as convergence at large $|\mathbf{x}|$ is concerned, while the other does not.

We shall see that there is also a smoothness condition on $\psi(\mathbf{x})$ at $\mathbf{x} \to 0$ that must be imposed on the wave function. We can always find solutions of the Schrödinger equation (5.2.7) that satisfy either this condition at $\mathbf{x} \to 0$ or the condition that $\psi(\mathbf{x}) \propto \exp(-\kappa|\mathbf{x}|)$ at $\mathbf{x} \to \infty$, but we cannot impose both conditions except for certain discrete values of E. These are the allowed energy levels of the bound electron. Schrödinger was justly proud that the existence of discrete energy levels, and hence the existence of atomic spectra discovered over a century earlier, were now explained as a mathematical consequence of boundary conditions imposed on a wave equation, rather than Bohr's *ad hoc* assumption of angular momentum quantization.

Spherical Symmetry

We now specialize to the case of spherical symmetry, which applies in one-electron atoms (and approximately for each electron in atoms with many electrons), for which the potential is only a function of $r = |\mathbf{x}|$. There is a mathematical identity that is useful for a wide variety of problems with spherical symmetry in various branches of mathematical physics:

$$\nabla^2 f(\mathbf{x}) = \frac{1}{r^2}\frac{\partial}{\partial r}\left[r^2 \frac{\partial f(\mathbf{x})}{\partial r}\right] + \frac{1}{r^2}(\mathbf{x} \times \nabla)^2 f(\mathbf{x}) \tag{5.2.12}$$

where $f(\mathbf{x})$ is an arbitrary differentiable function of position. (This can be derived in the same way that in ordinary vector algebra we derive the familiar identity $(\mathbf{a} \times \mathbf{b})^2 = \mathbf{a}^2\mathbf{b}^2 - (\mathbf{a} \cdot \mathbf{b})^2$, but here keeping track of the order of the position variable and derivatives that act on it.) As already mentioned, Schrödinger assumed that $-i\hbar\nabla$ should be interpreted as the operator representing the momentum, so the operator representing the orbital angular momentum is

$$\mathbf{L} \equiv -i\hbar\, \mathbf{x} \times \nabla\,, \tag{5.2.13}$$

and we can write Eq. (5.2.12) as the identity

$$\nabla^2 f(\mathbf{x}) = \frac{1}{r^2}\frac{\partial}{\partial r}\left[r^2 \frac{\partial f(\mathbf{x})}{\partial r}\right] - \frac{1}{\hbar^2 r^2}\mathbf{L}^2 f(\mathbf{x})\,. \tag{5.2.14}$$

The time-independent Schrödinger equation (5.2.7) thus takes the form

$$-\frac{\hbar^2}{2m_e}\frac{1}{r^2}\frac{\partial}{\partial r}\left[r^2 \frac{\partial \psi(\mathbf{x})}{\partial r}\right] + \frac{1}{2m_e r^2}\mathbf{L}^2\psi(\mathbf{x}) + V(r)\psi(\mathbf{x}) = E\psi(\mathbf{x})\,. \tag{5.2.15}$$

Radial and Angular Wave Functions

In order for the gradient operator in the Schrödinger equation to be well-defined, we need the wave function to be analytic in $\mathbf{x}$, by which is meant that it can be expanded in a power series about any point, and in particular about the origin $\mathbf{x} = 0$. As we shall see, this condition can be imposed on the wave function unless the potential is very singular.

Suppose that for some particular wave function, the smallest power of $\mathbf{x}$ in the expansion of the wave function around the origin is some integer $\ell = 0, 1, 2, \ldots$ Then for $\mathbf{x} \to 0$ the wave function is dominated by a homogeneous polynomial of order ℓ in the coordinates x_i – that is, a sum of terms each of which has ℓ factors of the coordinates x_i. For instance, a homogeneous polynomial in $\mathbf{x}$ of order zero is a constant, a homogeneous polynomial in $\mathbf{x}$ of order one is a linear combination of x_1, x_2, and x_3, and a homogeneous polynomial in $\mathbf{x}$ of order two is a linear combination of

$$x_1^2, \quad x_2^2, \quad x_3^2, \quad x_1 x_2, \quad x_2 x_3, \quad x_3 x_1 \;.$$

Defining a radial coordinate $r \equiv |\mathbf{x}|$ and a unit vector $\hat{x} \equiv \mathbf{x}/r$, the wave function for $r \to 0$ can now be written

$$\psi(\mathbf{x}) \to r^\ell Y_\ell(\hat{x}) \;, \tag{5.2.16}$$

where Y_ℓ is some homogeneous polynomial of order ℓ in the unit vector $\hat{x}$. (As we shall see, for $\ell \geq 1$ there is more than one such polynomial, which will later have to be distinguished by attaching an additional label to Y_ℓ.)

Just knowing the value of ℓ is enough to tell us how $\mathbf{L}^2$ acts on the wave function. Note that $\mathbf{L}$ does not act on functions of r, because $\mathbf{L} f(r) = -i\hbar(\mathbf{x} \times \hat{x} f'(r)) = 0$, so $\mathbf{L}^2$ acts only on the direction $\hat{x}$. For a wave function that goes as (5.2.16) for $r \to 0$, the first two terms on the left of Eq. (5.2.15) go as

$$-\frac{\hbar^2}{2m_e}\frac{1}{r^2}\frac{\partial}{\partial r}\left[r^2\frac{\partial\psi(\mathbf{x})}{\partial r}\right] \to -\frac{\hbar^2\ell(\ell+1)}{2m_e} r^{\ell-2} Y_\ell(\hat{x}) \;,$$

$$\frac{1}{2m_e r^2}\mathbf{L}^2\psi(\mathbf{x}) \to \frac{1}{2m_e} r^{\ell-2}\mathbf{L}^2 Y_\ell(\hat{x}) \;,$$

while $E\psi$ and (as long as the potential does not blow up as fast as $1/r^2$ for $r \to 0$) also $V(r)\psi$ are negligible for $r \to 0$ compared with $r^{\ell-2}$. Hence the time-independent Schrödinger equation (5.2.15) requires that

$$\mathbf{L}^2 Y_\ell(\hat{x}) = \hbar^2\ell(\ell+1) Y_\ell(\hat{x}) \;. \tag{5.2.17}$$

We can therefore find solutions of the Schrödinger equation (5.2.15) of the form

$$\psi(\mathbf{x}) = R(r) Y_\ell(\hat{x}) \;, \tag{5.2.18}$$

where $R(r)$ satisfies the radial wave equation

$$-\frac{\hbar^2}{2m_e}\frac{1}{r^2}\frac{d}{dr}\left[r^2\frac{dR(r)}{dr}\right] + \frac{\hbar^2\ell(\ell+1)}{2m_e r^2}R(r) + V(r)R(r) = ER(r) \;, \tag{5.2.19}$$

with boundary conditions

$$R(r) \propto r^\ell \;\text{ for }\; r \to 0 \qquad R(r) \propto P(r)\exp(-\kappa r) \;\text{ for }\; r \to \infty \;.$$

Here the term proportional to $\ell(\ell+1)$ acts as a positive and hence repulsive potential arising from the centrifugal force acting on an electron with non-zero angular momentum. The function $R(r)$ will turn out to depend on an index in addition to ℓ, on which the energy also depends.

Angular Multiplicity

As we will see when we come to the periodic table of elements, it is important to know the number of independent solutions of Eq. (5.2.17) for a given ℓ. For this purpose, it is convenient first to recast Eq. (5.2.17) as a condition on $r^\ell Y_\ell(\hat{x})$, a homogeneous polynomial of order ℓ in the three-vector $\mathbf{x}$. From Eq. (5.2.14), we see that Eq. (5.2.17) is equivalent to the condition

$$\nabla^2\big(r^\ell Y_\ell(\hat{x})\big) = 0 \,. \tag{5.2.20}$$

To distinguish among the solutions of Eq. (5.2.20), it is convenient to consider the action of the operator $L_3 \equiv -i\hbar(x_1\partial/\partial x_2 - x_2\partial/\partial x_1)$. Note that

$$L_3(x_1 \pm i x_2) = -i\hbar(-x_2 \pm i x_1) = \pm\hbar(x_1 \pm i x_2)\,, \quad L_3 x_3 = 0\,.$$

We can take a complete set of independent homogeneous polynomials in $\mathbf{x}$ of order ℓ as the products of $\nu_\pm$ factors of $x_1 \pm i x_2$ and $\ell - \nu_+ - \nu_-$ factors of x_3 for various non-negative integers $\nu_\pm$. The action of L_3 on these products is

$$\begin{aligned} &L_3\Big((x_1+ix_2)^{\nu_+}(x_1-ix_2)^{\nu_-}x_3^{\ell-\nu_+-\nu_-}\Big) \\ &\quad = \hbar(\nu_+ - \nu_-)\big[(x_1+ix_2)^{\nu_+}(x_1-ix_2)^{\nu_-}x_3^{\ell-\nu_+-\nu_-}\big]\,. \end{aligned}$$

Now, for an arbitrary function $f(\mathbf{x})$,

$$\begin{aligned} \frac{\partial^2}{\partial x_1^2}(L_3 f) &= -2i\hbar\frac{\partial^2 f}{\partial x_1 \partial x_2} + L_3\frac{\partial^2 f}{\partial x_1^2}\,, \\ \frac{\partial^2}{\partial x_2^2}(L_3 f) &= +2i\hbar\frac{\partial^2 f}{\partial x_2 \partial x_1} + L_3\frac{\partial^2 f}{\partial x_2^2}\,, \\ \frac{\partial^2}{\partial x_3^2}(L_3 f) &= L_3\frac{\partial^2 f}{\partial x_3^2}\,, \end{aligned}$$

so

$$\nabla^2(L_3 f) = L_3\nabla^2 f\,.$$

(We shall see in Section 5.4 that this is just a consequence of the rotational invariance of the Laplacian ∇^2.) It follows that when ∇^2 acts on a sum of terms of the form $(x_1+ix_2)^{\nu_+}(x_1-ix_2)^{\nu_-}x_3^{\ell-\nu_+-\nu_-}$, all with the same value of $\nu_+ - \nu_-$, it gives a sum of terms that again all have that value of $\nu_+ - \nu_-$. We can therefore find solutions of Eq. (5.2.20) that are sums of products of coordinates all with the same value of $m \equiv \nu_+ - \nu_-$, and label the solutions as $r^\ell Y_\ell^m(\hat{x})$, where

$$L_3 Y_\ell^m(\hat{x}) = \hbar m Y_\ell^m(\hat{x}) \ .$$

How many solutions of Eq. (5.2.20) are there for a given ℓ? First let's ask how many independent homogeneous polynomials in $\mathbf{x}$ of order ℓ of the form $(x_1 + ix_2)^{\nu_+}(x_1 - ix_2)^{\nu_-}x_3^{\ell-\nu_+-\nu_-}$ there are for a given ℓ. The exponent ν_+ can be any integer from 0 to ℓ and, for a given ν_+, the exponent ν_- can be any integer from 0 to $\ell - \nu_+$, so the number of these independent homogeneous polynomials of order ℓ in $\mathbf{x}$ is therefore

$$N_\ell = \sum_{\nu_+=0}^{\ell} \sum_{\nu_-=0}^{\ell-\nu_+} 1 = \sum_{\nu_+=0}^{\ell} (\ell - \nu_+ + 1) = (\ell+1)^2 - \frac{\ell(\ell+1)}{2}$$
$$= \frac{(\ell+1)(\ell+2)}{2} \ .$$

We also have to impose the condition (5.2.20). The function $\nabla^2(r^\ell Y_\ell(\hat{x}))$ is itself a homogeneous polynomial of order $\ell - 2$ in the three-vector $\mathbf{x}$, so setting this function equal to zero imposes $N_{\ell-2}$ conditions on $r^\ell Y_\ell(\hat{x})$, and the number of independent Y_ℓ subject to these conditions is thus

$$N_\ell - N_{\ell-2} = \frac{(\ell+1)(\ell+2)}{2} - \frac{(\ell-1)\ell}{2} = 2\ell + 1 \ .$$

But this is the same as the number of possible values of $m \equiv \nu_+ - \nu_-$, ranging from $m = -\ell$ to $m = +\ell$, so there must be just one function $Y_\ell^m(\hat{x})$ for each ℓ and m.

The index m does not appear in Eq. (5.2.19), so the $2\ell + 1$ states that differ only in the value of m all have the same energy as long as the spherical symmetry of the atom is maintained. The degeneracy of these states can be lifted by exposing the atom to an external perturbation which marks out a preferred direction, in which case the energies of the different states will be split from one another. Where the external perturbation is a magnetic field this is known as the *Zeeman effect*, after Pieter Zeeman (1865–1943), who first reported it in 1897. It was not possible to understand the details of the splitting of energy levels in the Zeeman effect until the discovery of electron spin, to be discussed in Section 5.4. We will calculate the Zeeman effect in Section 5.9, using the methods of perturbation theory, as an application of the quantum theory of the interaction of electrons with electromagnetic fields, to be described in Section 5.8.

Spherical Harmonics

Explicit formulas for spherical harmonics are needed in some applications of quantum mechanics but they are not needed in the calculations of energy levels in one-electron atoms. Nevertheless, to make this discussion of angular dependence concrete, we give here all the spherical harmonics for $\ell = 0$, $\ell = 1$, and $\ell = 2$:

$$Y_0^0 = \sqrt{\frac{1}{4\pi}} ,$$

$$Y_1^1 = -\sqrt{\frac{3}{8\pi}}(\hat{x}_1 + i\hat{x}_2) = -\sqrt{\frac{3}{8\pi}} \sin\theta \, e^{i\phi} ,$$

$$Y_1^0 = \sqrt{\frac{3}{4\pi}}\hat{x}_3 = \sqrt{\frac{3}{4\pi}} \cos\theta ,$$

$$Y_1^{-1} = \sqrt{\frac{3}{8\pi}}(\hat{x}_1 - i\hat{x}_2) = \sqrt{\frac{3}{8\pi}} \sin\theta \, e^{-i\phi} ,$$

$$Y_2^2 = \sqrt{\frac{15}{32\pi}}(\hat{x}_1 + i\hat{x}_2)^2 = \sqrt{\frac{15}{32\pi}}(\sin\theta)^2 \, e^{2i\phi} ,$$

$$Y_2^1 = -\sqrt{\frac{15}{8\pi}}(\hat{x}_1 + i\hat{x}_2)\hat{x}_3 = -\sqrt{\frac{15}{8\pi}} \sin\theta \, \cos\theta \, e^{i\phi} ,$$

$$Y_2^0 = \sqrt{\frac{5}{16\pi}}(2\hat{x}_3^2 - \hat{x}_1^2 - \hat{x}_2^2) = \sqrt{\frac{5}{16\pi}}(3(\cos\theta)^2 - 1) ,$$

$$Y_2^{-1} = \sqrt{\frac{15}{8\pi}}(\hat{x}_1 - i\hat{x}_2)\hat{x}_3 = \sqrt{\frac{15}{8\pi}} \sin\theta \, \cos\theta \, e^{-i\phi} ,$$

$$Y_2^{-2} = \sqrt{\frac{15}{32\pi}}(\hat{x}_1 - i\hat{x}_2)^2 = \sqrt{\frac{15}{32\pi}}(\sin\theta)^2 \, e^{-2i\phi} .$$

They are written in terms of the angles appearing in spherical polar coordinates:

$$x_1 = r \sin\theta \, \cos\phi , \quad x_2 = r \sin\theta \, \sin\phi , \quad x_3 = r \cos\theta . \tag{5.2.21}$$

The numerical factors have been chosen to make the spherical harmonics orthonormal, in the sense that

$$\int_0^{2\pi} d\phi \int_0^{\pi} \sin\theta \, d\theta \, Y_{\ell'}^{m'*}(\theta, \phi) \, Y_\ell^m(\theta, \phi) = \delta_{\ell\ell'}\delta_{mm'}. \tag{5.2.22}$$

Hydrogenic Energy Levels

Let's now specialize further to the case of a one-electron atom, such as neutral hydrogen, singly ionized helium, etc., with nuclear charge Ze. The electrostatic potential felt by the electron is then the Coulomb potential $V(r) = -Ze^2/r$. We seek a solution of the form (5.2.18), $\psi(\mathbf{x}) = R(r)Y_\ell^m(\hat{x})$, where $R(r)$ is some function only of r. Then the radial wave equation (5.2.19) reads

$$-\frac{\hbar^2}{2m_e}\frac{1}{r^2}\frac{\partial}{\partial r}\left[r^2\frac{\partial R(r)}{\partial r}\right]+\frac{\ell(\ell+1)\hbar^2}{2m_e r^2}R(r)-\frac{Ze^2R(r)}{r}=ER(r)\ . \quad (5.2.23)$$

We can easily find a family of exact solutions of this equation. Recall that, according to the definition of ℓ, for $r \to 0$ we have $R(r) \propto r^\ell$, while as shown above, for $r \to \infty$ the function $R(r)$ goes as $\exp(-\kappa r)$ times some function of r that grows more slowly than $\exp(\kappa r)$. So let us try for a solution of the form

$$R(r) = r^\ell \exp(-\kappa r)\ . \quad (5.2.24)$$

The first term in Eq. (5.2.23) then contains a contribution in which both derivatives act on the exponential, which takes the form $-(\hbar^2\kappa^2/2m_e)R(r)$, which according to Eq. (5.2.11) just matches $ER(r)$ on the right-hand side. It also contains a contribution in which both derivatives act on powers of r; this gives a contribution

$$-\frac{\hbar^2\exp(-\kappa r)}{2m_e}\frac{1}{r^2}\frac{\partial}{\partial r}\left[r^2\frac{\partial r^\ell}{\partial r}\right]=-\frac{\hbar^2\ell(\ell+1)R(r)}{2m_e r^2}\ ,$$

which cancels the second term on the left-hand side of Eq. (5.2.23). The first term in Eq. (5.2.23) also contains contributions in which one derivative acts on $\exp(-\kappa r)$, giving a factor $-\kappa$, while the other derivative acts on a power of r, giving a factor $[(\ell+2)+\ell]/r = 2(\ell+1)/r$, so these contribution add up to

$$+\frac{2(\ell+1)\hbar^2\kappa}{2m_e r}R(r)\ .$$

This remaining contribution must cancel the Coulomb term $-Ze^2R(r)/r$, so the necessary and sufficient condition for a solution of the form (5.2.24) is

$$\frac{(\ell+1)\hbar^2\kappa}{m_e} = Ze^2\ .$$

We conclude that these solutions have

$$E = -\frac{\hbar^2\kappa^2}{2m_e} = -\frac{Z^2e^4m_e}{2\hbar^2(\ell+1)^2}\ , \quad (5.2.25)$$

which agrees with the Bohr formula (3.4.6) if we identify n with $\ell+1$.

This is not the only class of solutions. It is straightforward though tedious to show that, in addition to solutions of the form (5.2.24), there are more general solutions of the form

$$\psi(r) = r^\ell P_{\ell,\nu}(r)\exp(-\kappa r)\ , \quad (5.2.26)$$

where $P_{\ell,\nu}(r)$ is a polynomial of order ν. Without actually constructing these polynomials, we can relate the energy to ℓ and ν by considering Eq. (5.2.23) in the limit $r \to \infty$. In this limit, Eq. (5.2.26) gives

$$R(r) \propto r^{\ell+\nu} \exp(-\kappa r) \; .$$

By repeating the arguments previously applied to the solution (5.2.24), but now only for $r \to \infty$, we see that the first term in Eq. (5.2.23) contains a contribution in which both derivatives act on the exponential, which matches the term ER on the right-hand side, and another contribution in which both derivatives act on powers of r, which is negligible for $r \to \infty$, as is the centrifugal potential and terms in which one or two derivatives act on sub-leading terms in $P_{\ell,\nu}(r)$. This leaves the potential term and the part of the first term in Eq. (5.2.23) in which one derivative acts on the exponential and the other on the leading power of r in $P_{\ell,\nu}(r)$, which gives a contribution that cancels the potential term if $(\ell+\nu+1)\hbar^2\kappa/m_e = Ze^2$, or in other words if E is given by the Bohr formula

$$E = -\frac{\hbar^2\kappa^2}{2m_e} = -\frac{Z^2e^4m_e}{2\hbar^2n^2} \; , \tag{5.2.27}$$

where now

$$n = \nu + \ell + 1 \; , \tag{5.2.28}$$

with ν the order of the polynomial in Eq. (5.2.26), and hence a non-negative integer.

The positive-definite integer $n \geq \ell + 1$ defined by Eq. (5.2.28), on which the energy solely depends, is known as the *principal quantum number.* Spectroscopists have developed a terminology, in which the letters s, p, d, f and so on stand for $\ell = 0,\ 1,\ 2,\ 3$, etc. A state is labeled first with n, and then with the letter indicating ℓ, so in hydrogen the states are $1s$, $2s$, $2p$, $3s$, $3p$, $3d$, and so on.

We can now work out the degeneracy of these energy levels. Since the energy depends only on n, according to Eq. (5.2.28) for each energy we can have ℓ equal to anything between $\ell = 0$ and $\ell = n-1$ (for which respectively $\nu = n-1$ and $\nu = 0$). We have seen that for each ℓ there are $2\ell+1$ states distinguished by different values of m. So, according to this reasoning, the total number of states for a given energy and hence a given value of n is

$$\#_n = \sum_{\ell=0}^{n-1}(2\ell+1) = 2\frac{(n-1)n}{2} + n = n^2 \; . \tag{5.2.29}$$

As we will see in Section 5.4, because electron spin has been left out in this calculation the degeneracy (5.2.29) is too small by a factor 2.

5.3 General Principles of Quantum Mechanics

As we have seen in the story so far, quantum mechanics began with guesswork: Einstein's guess that the energy and momentum of light waves comes in particles; Bohr's guess that if the energy and momentum of radiation are quantized then so are other things, such as the angular momentum of electrons in atomic orbits; de Broglie's guess that if electromagnetic waves consist of particles then particles such as electrons behave like waves; and Schrödinger's guess that the differential equations for de Broglie's waves could be modified for atomic electrons by inserting a potential. It is time that we move on from this guesswork and describe the general principles of quantum mechanics as they emerged in the formalism of wave mechanics soon after 1925. Then in following sections we shall go on to applications of quantum mechanics in contexts more general than those considered so far.

States and Wave Functions

The first general principle of wave mechanics is that physical states are represented by wave functions, functions $\psi(\mathbf{x}_1, \mathbf{x}_2, \ldots)$, with one coordinate argument for each particle in the system (and, as we shall see in the next section, with the wave function depending also on the 3-component of each particle's spin angular momentum). As anticipated in Section 5.2 for the case of single-particle wave functions, the probability in a state represented by wave function ψ that one particle is in a small volume d^3x_1 around $\mathbf{x}_1$, another particle is in a small volume d^3x_2 around $\mathbf{x}_2$, and so on, is

$$dP = |\psi(\mathbf{x}_1, \mathbf{x}_2, \ldots)|^2 \, d^3x_1 \, d^3x_2 \cdots . \tag{5.3.1}$$

Since with 100% probability the particles have to be somewhere, this requires the wave function to satisfy the normalization condition

$$\int d^3x_1 \, d^3x_2 \cdots |\psi(\mathbf{x}_1, \mathbf{x}_2, \ldots)|^2 = 1 . \tag{5.3.2}$$

Two wave functions that differ only by a constant phase factor of absolute value unity represent the same state. In solving differential equations for the wave function, the important thing is that the integral (5.3.2) should be finite – in that case we can always find a ψ that satisfies Eq. (5.3.2) by dividing the wave function by the square root of this integral.

Observables and Operators

The second general principle of wave mechanics is that observable physical quantities are represented by linear operators on these wave functions, here generally distinguished by upper case letters. By an operator A being "linear" is meant that, for any pair of wave functions ψ_1 and ψ_2 and numbers a_1 and a_2, we have

$$A[a_1\psi_1 + a_2\psi_2] = a_1 A\psi_1 + a_2 A\psi_2 . \tag{5.3.3}$$

As part of this principle, a state represented by a wave function ψ has a definite value α for the observable represented by an operator A if and only if

$$A\psi = \alpha\psi \ . \tag{5.3.4}$$

In this case we say that ψ is an *eigenfunction* of A with *eigenvalue a*.

For instance, the operator P_{nj} that represents the jth component (with $j = 1, 2, 3$) of the momentum of the nth particle acts as $-i\hbar$ times the partial derivative $\partial/\partial x_{nj}$ with respect to the jth component of the coordinate of the nth particle, acting on whatever function is to the right. This is clearly linear in the sense of Eq. (5.3.3). In order for a one-particle state to have a definite value $\mathbf{p}$ for the momentum of the particle, it is necessary that the wave function should satisfy an equation of the form (5.3.4), which in this case reads

$$-i\hbar\nabla\psi(\mathbf{x}) = \mathbf{p}\psi(\mathbf{x}) \ ,$$

which has as a solution the de Broglie wave function

$$\psi(\mathbf{x}) \propto \exp(i\mathbf{p}\cdot\mathbf{x}/\hbar) \ .$$

This raises a problem, which is endemic to values of observables that like momentum lie in a continuous spectrum of possible values: the integral (5.3.2) is infinite for a wave function of this form and therefore cannot be normalized. But we can find a wave function that is arbitrarily close to this form for an arbitrarily large range of position:

$$\psi(\mathbf{x}) = (\pi L)^{-3/2}\exp(-\mathbf{x}^2/2L^2)\exp(i\mathbf{p}\cdot\mathbf{x}/\hbar) \ ,$$

in which the constant factor $(\pi L)^{-3/2}$ is chosen so that this ψ satisfies the normalization condition (5.3.2). The constant L can be chosen as some very large length, in which case the particle is almost certainly in a very large volume L^3, where it almost certainly has the momentum $\mathbf{p}$.

The operator X_{nj} that represents the jth component of the position vector of the nth particle, acting on any function of position to its right such as a wave function or the derivative of a wave function, simply multiplies that function by the argument x_{nj} and is obviously linear. Here again we have the problem that its eigenfunctions cannot be normalized. In the one-particle case, a wave function $\psi(\mathbf{x})$ that represents a state with a definite position $\mathbf{a}$ would have $\mathbf{X}\psi(\mathbf{x}) \equiv \mathbf{x}\psi(\mathbf{x})$ equal to $\mathbf{a}\psi(\mathbf{x})$ for all $\mathbf{x}$, so that it would have to vanish for all $\mathbf{x} \neq \mathbf{a}$, and the integral (5.3.2) would vanish. But we can find a normalized wave function that represents a state in which the particle is almost certainly very close to position $\mathbf{a}$:

$$\psi(\mathbf{x}) = (\pi d)^{-3/2}\exp\big(-(\mathbf{x}-\mathbf{a})^2/2d^2\big) \ ,$$

where d is here some very small length.

From operators we can construct other operators, which may or may not represent physical quantities. Linear combinations provide a trivial example: if A and B are operators while a and b are ordinary complex numbers, then $aA + bB$ is an operator for which

$$[aA + bB]\psi = aA\psi + bB\psi \ .$$

The product AB of any two operators A and B is defined by associativity: it is an operator that, acting on any function f to its right, gives the same result as acting first with B and then acting to the right on Bf with A:

$$(AB)f \equiv A(Bf) \ . \tag{5.3.5}$$

The Hamiltonian

One linear operator formed in this way is the *Hamiltonian*, which represents the energy. For instance, for a single non-relativistic particle moving in a potential V the Hamiltonian is

$$H = \frac{1}{2m}\mathbf{P}^2 + V(\mathbf{X}) \ . \tag{5.3.6}$$

The time-independent Schrödinger equation (5.2.7) is just the statement $H\psi = E\psi$, which tells us that ψ represents a state with energy E. The eigenfunctions of this Hamiltonian with negative eigenvalues are normalizable, a condition we imposed in finding the bound state energy values in Section 5.2, but there are also eigenfunctions with positive eigenvalues, representing unbound states, which can only be normalized in the same approximate sense as the eigenfunctions of position and momentum.

Adjoints

There is another process for producing new operators from other operators, analogous to taking the complex conjugate of a number. For any operator A, we define the *adjoint* $A^\dagger$ as the operator for which

$$\int [A\psi_1]^*\psi_2 = \int \psi_1^*[A^\dagger\psi_2] \tag{5.3.7}$$

where ψ_1 and ψ_2 are any two wave functions. Here and below we use the abbreviation

$$\int \psi_1^*\psi_2 \equiv \int d^3x_1 \int d^3x_2 \cdots \psi_1^*(\mathbf{x}_1, \mathbf{x}_2, \ldots)\psi_2(\mathbf{x}_1, \mathbf{x}_2, \ldots) \ . \tag{5.3.8}$$

It is easy to see that the adjoint of a product is the product of the adjoints in the opposite order:

$$[AB]^\dagger = B^\dagger A^\dagger \tag{5.3.9}$$

because

$$\int [AB\psi_1]^* \psi_2 = \int [B\psi_1]^* [A^\dagger \psi_2] = \int \psi_1^* [B^\dagger A^\dagger \psi_2] \,.$$

It is also obvious that the adjoint of a linear combination of operators is the same linear combination of the adjoints, but with complex conjugate coefficients

$$[aA + bB]^\dagger = a^* A^\dagger + b^* B^\dagger \,, \tag{5.3.10}$$

and the adjoint of an adjoint gives back the original operator:

$$[A^\dagger]^\dagger = A \,. \tag{5.3.11}$$

There is an important class of linear operators that are their own adjoints

$$A^\dagger = A \,. \tag{5.3.12}$$

A physical quantity represented by such an operator can only have real values in any state, for if $A\psi = \alpha\psi$ for some wave function ψ, then

$$\alpha \int \psi^* \psi = \int \psi^* [A\psi] = \int [A\psi]^* \psi = \alpha^* \int \psi^* \psi$$

so $\alpha = \alpha^*$. Such operators are called self-adjoint, or *Hermitian*. The coordinate operator $\mathbf{X}_n$ is obviously self-adjoint, and the momentum operator $\mathbf{P}_{ni}$ is too, because the minus sign produced by taking the complex conjugate of $-i$ is cancelled by the minus sign produced by integration by parts:

$$\int [P_{ni}\psi_1]^* \psi_2 = \int [-i\hbar \boldsymbol{\nabla}_{ni} \psi_1]^* \psi_2 = +i\hbar \int [\boldsymbol{\nabla}_{ni} \psi_1]^* \psi_2$$
$$= -i\hbar \int \psi_1^* [\boldsymbol{\nabla}_{ni} \psi_2] = \int \psi_1^* [P_{ni} \psi_2] \,.$$

Assuming the potential to be real, the Hamiltonian (5.3.6) is also self-adjoint, so the allowed energy values are all real. Also, all components of the angular momentum operator $\mathbf{L} = \mathbf{X} \times \mathbf{P}$ for a single particle are self-adjoint. For instance

$$L_3^\dagger = (X_1 P_2 - X_2 P_1)^\dagger = P_2 X_1 - P_1 X_2 = X_1 P_2 - X_2 P_1$$

the last step being valid because P_2 does not act on x_1 and P_1 does not act on x_2; both only act on whatever function of $\mathbf{x}$ that L_3 is acting on. Likewise of course for L_1 and L_2.

Self-adjoint operators have another important property. If A is self-adjoint and $A\psi_1 = \alpha_1 \psi_1$ and $A\psi_2 = \alpha_2 \psi_2$, then

$$\alpha_1 \int \psi_2^* \psi_1 = \int \psi_2^* [A\psi_1] = \int [A\psi_2]^* \psi_1 = \alpha_2 \int \psi_2^* \psi_1 \,,$$

so if $\alpha_1 \neq \alpha_2$ then $\int \psi_2^* \psi_1 = 0$. Such wave functions are said to be *orthogonal*. For instance any two different spherical harmonics such as those listed in

Section 5.2 are orthogonal, because they are eigenfunctions of the self-adjoint operators $\mathbf{L}^2$ and L_3 with different eigenvalues $\hbar^2\ell(\ell+1)$ and/or $\hbar m$.

Expectation Values

The interpretation given by Eq. (5.3.1) of $|\psi|^2$ as a probability density tells us that if we measure any function $f(\mathbf{x}_1, \mathbf{x}_2, \dots)$ of positions many times in the state represented by wave function ψ, the mean value of the measured values will be

$$\langle f\rangle_\psi = \int f(\mathbf{x}_1, \mathbf{x}_2, \dots)\, |\psi(\mathbf{x}_1, \mathbf{x}_2, \dots)|^2 d^3x_1\, d^3x_2 \cdots ,$$

provided that ψ is normalized so that $\int |\psi(\mathbf{x}_1, \mathbf{x}_2, \dots)|^2 d^3x_1\, d^3x_2 \cdots = 1$. Since $\psi(\mathbf{x}_1, \mathbf{x}_2, \dots)$ is an eigenfunction of the operator $f(\mathbf{X}_1, \mathbf{X}_2, \dots)$ with eigenvalue $f(\mathbf{x}_1, \mathbf{x}_2, \dots)$, this can be written

$$\langle f\rangle_\psi = \int \psi^*(\mathbf{x}_1, \mathbf{x}_2, \dots)\big[f(\mathbf{X}_1, \mathbf{X}_2, \dots)\psi](\mathbf{x}_1, \mathbf{x}_2, \dots)\big] ,$$

or, in our abbreviated notation,

$$\langle f\rangle_\psi = \int \psi^*[F\psi] ,$$

where F is the operator $f(\mathbf{X}_1, \mathbf{X}_2, \dots)$. It is only a short step from this to a third postulate of quantum mechanics, which states that when any physical quantity represented by an operator A is measured many times, each time in the state represented by normalized wave function ψ, then the average value found for this quantity is

$$\langle A\rangle_\psi = \int \psi^*[A\psi] \tag{5.3.13}$$

or, if the wave function is not normalized,

$$\langle A\rangle_\psi = \frac{\int \psi^*[A\psi]}{\int \psi^*\psi} .$$

This is called the *expectation value* of A for the wave function ψ. For a self-adjoint operator

$$\int \psi^*[A\psi] = \int [A\psi]^*\psi = \left(\int \psi^*[A\psi]\right)^* ,$$

so the expectation value of a self-adjoint operator is real for any wave function.

It is obvious that if $A\psi = \alpha\psi$ then the expectation value $\langle A\rangle_\psi$ of A for the wave function ψ is just α, but expectation values give useful information even for wave functions that do not represent states with a definite value for the observable. For instance, the mean square spread of values of an observable represented by A around its mean value is

$$(\Delta A)^2 \equiv \left\langle (A - \langle A \rangle)^2 \right\rangle . \tag{5.3.14}$$

Probabilities

Suppose a physical system is in a state represented by a normalized wave function ψ, and we measure an observable represented by a Hermitian operator A that (to start with the simplest case) has only discrete non-degenerate eigenvalues α_n with eigenfunctions φ_n. Even though ψ will not in general be one of these eigenfunctions, it can generally be expanded as a series of terms proportional to the eigenfunctions

$$\psi = \sum_n c_n \varphi_n ,$$

where c_n are some numerical coefficients. (The proof of the possibility of such an expansion depends on detailed properties of the operator A.) As we have seen, such eigenfunctions are orthogonal and, if properly normalized, can be taken as *orthonormal* in the sense that

$$\int \varphi_m^* \varphi_n = \delta_{nm} \equiv \begin{cases} 1 & n = m \\ 0 & n \neq m . \end{cases}$$

We can find the coefficients c_m by multiplying the expansion with φ_m^* and integrating over all values of the arguments, which gives

$$\int \varphi_m^* \psi = \sum_n c_n \int \varphi_m^* \varphi_n = c_m .$$

The expectation value of the observable A is then

$$\langle A \rangle_\psi = \int \psi^* A \psi = \sum_{nm} c_n^* c_m \int \varphi_n^* A \varphi_m$$

or, since $A\varphi_m = \alpha_m \varphi_m$,

$$\langle A \rangle_\psi = \sum_{nm} \alpha_m c_n^* c_m \int \varphi_n^* \varphi_m = \sum \alpha_m |c_m|^2 = \sum_m \alpha_m \left| \int \varphi_m^* \psi \right|^2 .$$

Since a corresponding result is true for any function of A, the inevitable interpretation is that when the observable represented by A is measured in a state represented by the normalized wave function ψ, the probability of finding the result α_m is

$$P_m(\psi) = \left| \int \varphi_m^* \psi \right|^2 . \tag{5.3.15}$$

This is known as the *Born rule* and can be taken instead of Eq. (5.3.13) as the third postulate of quantum mechanics.

Continuum Limit

We can also calculate probability densities for an observable that takes a continuum of values, by taking the limit of the case in which the observable takes a very large number of very close discrete values. If the number of values of the index n for which the eigenvalue α_n is in a range from α to $\alpha + d\alpha$ is $\mathcal{N}(\alpha)d\alpha$, then in the state represented by normalized wave function ψ the probability of finding the observable in this range is

$$dP(\alpha) = \mathcal{N}(\alpha)d\alpha \times \left| \int \varphi_\alpha^* \psi \right|^2 , \tag{5.3.16}$$

where φ_α is any normalized eigenfunction of A with eigenvalue in this narrow range. For any such observable, instead of working with the conventional wave functions $\psi(\mathbf{x}_1, \mathbf{x}_2, \dots)$ we can use wave functions

$$\psi(\alpha) \equiv \sqrt{\mathcal{N}(\alpha)} \int \varphi_\alpha^* \psi \tag{5.3.17}$$

for which Eq. (5.3.16) gives the probability of finding the observable in the range from α to $\alpha + d\alpha$:

$$dP(\alpha) = |\psi(\alpha)|^2 \, d\alpha \, . \tag{5.3.18}$$

The classic example of such continuum operators and alternative wave functions is provided by momentum.

Momentum Space

Consider for instance a particle in a cubical box of edge L. The normalized wave function representing a state with definite momentum $\mathbf{p}$ is

$$\varphi_{\mathbf{p}}(\mathbf{x}) = L^{-3/2} \exp(i\mathbf{p} \cdot \mathbf{x}/\hbar) \, .$$

Pretty much as we saw for photons in Section 3.2, the allowed momenta take the form $\mathbf{p} = 2\pi\mathbf{n}\hbar/L$, where $\mathbf{n}$ is a vector with integer components, so that this wave function should have the same values on opposite sides of the box. In a state represented by a normalized wave function $\psi(\mathbf{x})$, the probability of finding the momentum to have value $\mathbf{p}$ is

$$P_{\mathbf{p}} = \left| \int_{L^3} \varphi_{\mathbf{p}}^*(\mathbf{x}) \psi(\mathbf{x}) \, d^3x \right|^2 ,$$

the integral here taken over the interior of the box. We can pass to the continuum limit by taking the box to be very large, so that the allowed momentum values are very close together. Since the allowed vectors $\mathbf{n}$ form a lattice of cubes, each of volume unity, the number dN of these allowed momenta in a small volume of momentum space d^3p around $\mathbf{p}$ equals the corresponding volume in the space of vectors $\mathbf{n}$:

$$dN = d^3\mathbf{n} = \left(\frac{L}{2\pi\hbar}\right)^3 d^3p$$

so the probability of finding the momentum in this range is

$$\left(\frac{L}{2\pi\hbar}\right)^3 d^3p \times \left|\int_{L^3} \varphi^*_{\mathbf{p}}(\mathbf{x})\psi(\mathbf{x})\, d^3x\right|^2 = |\psi^{\P}(\mathbf{p})|^2 d^3p\,, \tag{5.3.19}$$

where

$$\begin{aligned}\psi^{\P}(\mathbf{p}) &\equiv \left(\frac{L}{2\pi\hbar}\right)^{3/2} \int_{L^3} \varphi^*_{\mathbf{p}}(\mathbf{x})\psi(\mathbf{x})\, d^3x \\ &\to (2\pi\hbar)^{-3/2} \int \exp(-i\mathbf{p}\cdot\mathbf{x}/\hbar)\psi(\mathbf{x})\, d^3x\,,\end{aligned} \tag{5.3.20}$$

with the last integral taken over all space. We can just as well say that the state of the system is represented by the momentum-space wave function $\psi^{\P}(\mathbf{p})$ as by the coordinate-space wave function $\psi(\mathbf{x})$. Indeed, as we will see in Section 5.10, both the coordinate-space wave function ψ and the momentum-space wave function $\psi^{\P}$ are nothing but the components in different bases of a vector in an abstract space, known as Hilbert space.

Commutation Relations

The commutator of two operators A and B, written $[A, B]$, is defined by

$$[A, B] \equiv AB - BA\,. \tag{5.3.21}$$

In order for the physical quantities represented by operators A and B to have definite numerical values α and β in a state represented by wave function ψ it is necessary that the commutator $[A, B]$ acting on ψ should vanish, because

$$[A, B]\psi = \beta A\psi - \alpha B\psi = \beta\alpha\psi - \alpha\beta\psi = 0\,.$$

In particular, it is never possible for any state to have definite values for both of two quantities represented by operators whose commutator is simply a non-zero number, because such a commutator can never give zero when acting on any ψ.

It is helpful in evaluating commutators to note that commutation acts like differentiation. For instance,

$$\begin{aligned}[A, BC] = ABC - BCA &= ABC - BAC + BAC - BCA \\ &= [A, B]C + B[A, C]\,.\end{aligned}$$

Thus, since all components of momentum commute with one another, they also commute with any function only of momenta, such as the total kinetic energy operator $\sum_n \mathbf{P}_n^2/2m_n$.

Uncertainty Principle

Note that the commutator of X_{ni} and P_{mj} acting on any multi-particle wave function ψ is

$$[X_{ni}, P_{mj}]\psi = -i\hbar x_{ni}\frac{\partial\psi}{\partial x_{mj}} + i\hbar\frac{\partial}{\partial x_{mj}}(x_{ni}\psi)$$
$$= -i\hbar x_{ni}\frac{\partial\psi}{\partial x_{mj}} + i\hbar\delta_{ij}\delta_{nm}\psi + i\hbar x_{ni}\frac{\partial\psi}{\partial x_{mj}} = +i\hbar\delta_{ij}\delta_{nm}\psi\ ,$$

a result we write as a commutation relation

$$[X_{ni}, P_{mj}] = i\hbar\delta_{ij}\delta_{nm}\ . \tag{5.3.22}$$

This shows in particular that there can be no state in which a component of some particle's position and the same component of the same particle's momentum both have definite values.

Indeed, using this commutation relation, it is possible to set a lower bound on the product of the root mean square spread of values of position and momentum:

$$\Delta X_{ni}\Delta P_{ni} \geq \hbar/2 \tag{5.3.23}$$

a result known as the *Heisenberg uncertainty principle*.[7]

We can see this in a simple example. The normalized wave function for a particle confined to a distance d around some position $\mathbf{a}$ can be written as a superposition of wave functions with definite momentum:

$$(\pi d)^{-3/2}\exp\big(-(\mathbf{x}-\mathbf{a})^2/2d^2\big)$$
$$= \left(\frac{d}{2\pi^2\hbar^2}\right)^{3/2}\int d^3p\,\exp\big(i\mathbf{p}\cdot[\mathbf{x}-\mathbf{a}]/\hbar\big)\exp(-d^2\mathbf{p}^2/2\hbar^2)\ . \tag{5.3.24}$$

We see that if the spread in values of $\mathbf{x}$ is of order d, then the spread in values of $\mathbf{p}$ is of order $\hbar/d$, and the product of the spreads is of order $\hbar$, in accordance with the uncertainty principle.

---

[7] W. Heisenberg, Zeit. Phys. **43**, 172 (1927). For a textbook proof, see Weinberg, *Lectures on Quantum Mechanics*, listed in the bibliography.

Time Dependence

In the earliest formulation of quantum mechanics, wave functions were given a time dependence governed by the time-dependent Schrödinger equation:

$$i\hbar\frac{\partial}{\partial t}\psi(\mathbf{x}_1,\mathbf{x}_2,\ldots;t) = H\psi(\mathbf{x}_1,\mathbf{x}_2,\ldots;t) \tag{5.3.25}$$

where H is the Hamiltonian operator, representing the energy of the system. The wave function of a state with a definite energy E thus has a trivial time-dependence, contained in a phase factor $\exp(-iEt/\hbar)$. The expectation value (5.3.13) of any operator for such a wave function is independent of time. More generally, assuming that the Hamiltonian is self-adjoint, the time dependence of the expectation value of an observable represented by an operator A in a state represented by a normalized wave function ψ that satisfies Eq. (5.3.25) is governed by the differential equation

$$\begin{aligned}\frac{d}{dt}\langle A\rangle &= \int \psi^*[(-i/\hbar)AH\psi] + \int [(-i/\hbar)H\psi]^*[A\psi] \\ &= (-i/\hbar)\left[\int \psi^*[AH\psi] - \int \psi^*[HA\psi]\right]\end{aligned}$$

and therefore

$$i\hbar\frac{d}{dt}\langle A\rangle = \langle [A,H]\rangle\ . \tag{5.3.26}$$

In particular, the normalization integral $\int \psi^*\psi$ is the expectation value (5.3.13) of the unit operator, which acting on any wave function just gives the same wave function. Since this operator commutes with the Hamiltonian (or anything else), the normalization integral is constant in time; once normalized, wave functions remain normalized.

For instance, in the case of a single particle moving in an external potential, with Hamiltonian (5.3.6),

$$H = \frac{1}{2m}\mathbf{P}^2 + V(\mathbf{X})\ ,$$

we have

$$[\mathbf{X},H] = \frac{1}{2m}[\mathbf{X},\mathbf{P}^2] = \frac{i\hbar}{m}\mathbf{P} \tag{5.3.27}$$

and

$$[\mathbf{P},H] = [\mathbf{P},V(\mathbf{X})] = -i\hbar\boldsymbol{\nabla}V(\mathbf{X})\ . \tag{5.3.28}$$

The equations of motion of the expectation values are then

$$\frac{d}{dt}\langle\mathbf{X}\rangle = \langle\mathbf{P}\rangle/m\ , \qquad \frac{d}{dt}\langle\mathbf{P}\rangle = -\langle\boldsymbol{\nabla}V(\mathbf{X})\rangle\ . \tag{5.3.29}$$

This is much the same as in classical physics, but note that $\langle \nabla V(\mathbf{X}) \rangle$ is not the same as $\nabla V(\langle \mathbf{X} \rangle)$, so this is not a closed set of equations.

Conservation Laws

Operators that commute with the Hamiltonian deserve special attention, in part because their expectation values (and the expectation values of any functions of them) are time-independent for any wave function. These represent what are called *conserved quantities*. Among these operators is of course H itself, so the mean energy $\langle H \rangle$ of any state is constant in time. The momenta of particles moving in an external potential is not conserved, but the total momentum of a number of particles is conserved if the potential depends only on the differences of their coordinates. For instance, for such a two-particle system,

$$\left[\mathbf{P}_1 + \mathbf{P}_2, V(\mathbf{X}_1 - \mathbf{X}_2)\right] = -i\hbar \nabla_1 V(\mathbf{X}_1 - \mathbf{X}_2) - i\hbar \nabla_2 V(\mathbf{X}_1 - \mathbf{X}_2) = 0 \, .$$

What about angular momentum? For simplicity, consider just a single particle, whose orbital angular momentum is $\mathbf{L} = \mathbf{X} \times \mathbf{P}$, in a potential that depends only on $R = \sqrt{\mathbf{X}^2}$. It is straightforward to work out that the commutators of a general linear combination $\mathbf{e} \cdot \mathbf{L}$ of the components of $\mathbf{L}$ with the position and momentum operators are

$$[\mathbf{e} \cdot \mathbf{L} \, , \, \mathbf{X}] = -i\hbar \, \mathbf{e} \times \mathbf{X} \, , \tag{5.3.30}$$

$$[\mathbf{e} \cdot \mathbf{L} \, , \, \mathbf{P}] = -i\hbar \, \mathbf{e} \times \mathbf{P} \, . \tag{5.3.31}$$

For instance,

$$\begin{aligned}
&[\mathbf{e} \cdot \mathbf{L}, X_1] \\
&= \left[\big(e_1(X_2 P_3 - X_3 P_2) + e_2(X_3 P_1 - X_1 P_3) + e_3(X_1 P_2 - X_2 P_1)\big), X_1\right] \\
&= -i\hbar(e_2 X_3 - e_3 X_2) = -i\hbar(\mathbf{e} \times \mathbf{X})_1 \, .
\end{aligned}$$

It follows that each component of $\mathbf{L}$ commutes with $\mathbf{P}^2$ and $\mathbf{X}^2$, and hence also with $V(\sqrt{\mathbf{X}^2})$, and so with the Hamiltonian.

Another reason for us to give special attention to operators that commute with the Hamiltonian is that states with a given energy can be classified according to the eigenvalues of these conserved quantities. For instance, for a Coulomb potential the states with a given principal quantum number n and hence with a given energy can be classified according to the eigenvalues of $\mathbf{L}^2$ and L_3, both of which commute with the Hamiltonian as well as with each other. Of course, L_1 and L_2 also commute with the Hamiltonian and with $\mathbf{L}^2$, but as we shall see in the next section they do not commute with each other, or with L_3, so the best we can do is to classify states according to the eigenvalues of $\mathbf{L}^2$ and L_3 as well as H.

Heisenberg and Schrödinger Pictures

The formalism described here, in which the wave function depends on time but operators are time-independent, is known as the *Schrödinger picture*. There is another formalism, known as the *Heisenberg picture*, in which wave functions are time-independent and operators depend on time. To make clear the relation between these pictures, just for the present we will use subscripts H and S to distinguish wave functions and operators in the Heisenberg and Schrödinger pictures.

The time-dependent Schrödinger equation (5.3.25) has a formal solution

$$\psi_{\rm S}(t) = e^{-iHt/\hbar}\psi_{\rm S}(0);$$

so, if we define the Heisenberg-picture wave function as the Schrödinger-picture wave function at zero time,

$$\psi_{\rm H} \equiv \psi_{\rm S}(0) \tag{5.3.32}$$

then wave functions in the two pictures are related by

$$\psi_{\rm S}(t) = e^{-iHt/\hbar}\psi_{\rm H}\ . \tag{5.3.33}$$

In the Heisenberg picture, in order to preserve Eq. (5.3.26) we must give operators a time dependence with

$$i\hbar\frac{d}{dt}A_{\rm H}(t) = [A_{\rm H}(t), H]\ . \tag{5.3.34}$$

The commutators of position and momentum with the Hamiltonian given in Eqs. (5.3.27) and (5.3.28) show that Eq. (5.3.34) gives these operators in the Heisenberg picture the same time dependence as the corresponding quantities in classical mechanics. To satisfy Eq. (5.3.34), we define the Heisenberg-picture operator $A_{\rm H}(t)$ in terms of the Schrödinger-picture operator $A_{\rm S}$ representing the same observable by

$$A_{\rm H}(t) = e^{iHt/\hbar}A_{\rm S}e^{-iHt/\hbar}\ . \tag{5.3.35}$$

We can go back and forth between the two pictures. For instance, for an arbitrary operator A and wave function ψ, the Schrödinger-picture wave function corresponding to the Heisenberg-picture operator $A_{\rm H}(t)$ acting on the Heisenberg-picture wave function $\psi_{\rm H}$ is

$$e^{-iHt/\hbar}A_{\rm H}(t)\psi_{\rm H} = A_{\rm S}e^{-iHt/\hbar}\psi_{\rm H} = A_{\rm S}\psi_{\rm S}(t)\ ,$$

just as if we had worked from the beginning in the Schrödinger picture.

The Heisenberg picture and Schrödinger picture are physically equivalent, but useful in different contexts. The Schrödinger picture is more naturally used in calculating bound state energies, and, as we shall see in Section 5.6, it can also be used for scattering processes. The Heisenberg picture is invaluable when we want to use known equations of motion for observables to motivate a choice of Hamiltonian, as we will do in Section 5.8. Also, in field theories where observables depend on position, in order to preserve the appearance of Lorentz invariance it is necessary to work in the Heisenberg picture, so that these observables will depend on time as well as on space coordinates.

5.4 Spin and Orbital Angular Momentum

Spin Discovered

The counting of states described in Section 5.2 was already known in 1925 to be in conflict with spectroscopic data. The problem emerged most clearly in the study of alkali metals. These are elements such as lithium, sodium, potassium, etc. that were known to readily lose a single electron.[8] In the contemporary atomic models of the time, this meant that an alkali metal atom has one loosely bound electron outside inner shells of more tightly bound electrons. The potential felt by this outer electron is spherically symmetric but it is not a Coulomb potential, which would be proportional to $1/r$; so because L_3 and $\mathbf{L}^2$ commute with each other and with H it was still expected that states of definite energy would also have definite values $\hbar^2\ell(\ell+1)$ for $\mathbf{L}^2$ and $2\ell+1$ states of equal energy for any given ℓ, distinguished by different eigenvalues $\hbar m$ of L_3, but no further degeneracy was expected. States could still be labeled with a principal quantum number n, defined so that the number of nodes of the wave function (values of r where the wave function vanishes) is $n-\ell-1$, as it is in hydrogen, but, unlike the case of hydrogen, here the energies depend on ℓ as well as on n.

There is a very well studied "D-line" in the spectrum of sodium vapor (which gives sodium vapor lamps their orange color) with wavelength about 5890 angstroms, interpreted as a $3p \to 3s$ transition between states of the outermost electron with $n = 3$. But even with moderate resolution, spectroscopists were able to see that this line was doubled, having two components with wavelengths 5896 angstroms and 5890 angstroms. Wolfgang Pauli (1900–1958) was led to suggest that, on the basis of this and other data, there is a fourth quantum number, besides n, ℓ, and m, which takes just two values in all states with $\ell \geq 1$. But the physical significance of this quantum number was at first mysterious.

Then in 1925 two young Dutch physicists, Samuel Goudsmit (1902–1978) and George Uhlenbeck (1900–1988), suggested[9] that the extra quantum number

[8] With the charge of the electron and the atomic weights of these elements known, it could be concluded from the ratio of the metal mass produced in electrolysis to the electric charge used that one electron is needed to convert one ion in a solution of the metal salt to an alkali metal atom, so the atom in becoming an ion had to lose just one electron. This was in contrast with metals like beryllium, magnesium, calcium, etc., which require two electrons to convert an ion to an atom.

[9] S. Goudsmit and G. Uhlenbeck, Naturwiss. **13**, 953 (1925).

was associated with an internal angular momentum, or *spin*, of the electron. At first this idea seemed absurd. If the spin $\mathbf{S}$ is anything like $\mathbf{L}$, then any one component of $\mathbf{S}$ should take $2s + 1$ values, running by unit steps from $-\hbar s$ up to $+\hbar s$, where s is given by $\mathbf{S}^2 = \hbar^2 s(s+1)$. But to have $2s + 1 = 2$ we need $s = 1/2$, while ℓ is always an integer.

The notion of spin $s = 1/2$ was not understood until physicists adopted a more mature view of the nature of angular momentum, that it is an operator whose existence and properties are dictated by the invariance of the laws of nature under rotations, rather than by experience with classical spinning bodies. This takes some explanation.

Rotations

In general, an infinitesimal rotation changes any vector $\mathbf{v}$ by an amount

$$\delta\mathbf{v} = \mathbf{e} \times \mathbf{v} \tag{5.4.1}$$

where $\mathbf{e}$ is an infinitesimal 3-vector characterizing the rotation. This is a rotation, because it leaves all scalar products unchanged:

$$\delta(\mathbf{v} \cdot \mathbf{v}') = \mathbf{v} \cdot (\mathbf{e} \times \mathbf{v}') + (\mathbf{e} \times \mathbf{v}) \cdot \mathbf{v}' = 0 \,. \tag{5.4.2}$$

It is in fact (though we don't need to know this here) a rotation by an infinitesimal angle of $|\mathbf{e}|$ radians counterclockwise around the direction of $\mathbf{e}$. For instance, if $\mathbf{e}$ is in the 3-direction then (5.4.1) gives

$$\delta v_1 = -|\mathbf{e}| v_2 \,, \quad \delta v_2 = +|\mathbf{e}| v_1 \,, \quad \delta v_3 = 0 \,.$$

Now, suppose that one observer sees that a physical system is in a state represented by a wave function ψ, and suppose a second observer views the same state using coordinate axes that have been subjected to a slight rotation, which changes any vector $\mathbf{v}$ by an infinitesimal amount $\mathbf{e} \times \mathbf{v}$. What does she see? For $\mathbf{e}$ infinitesimal the change in the wave function must be linear in $\mathbf{e}$, and can therefore be written

$$\delta\psi = (i/\hbar)\, \mathbf{e} \cdot \mathbf{J}\, \psi \tag{5.4.3}$$

where $\mathbf{J}$ is some triplet of operators and the factor $(i/\hbar)$ is inserted for future convenience. We would not want the rotation to change the total probability $\int \psi^* \psi = 1$ that the particles in the system are somewhere, so we require that

$$0 = \delta \int \psi^* \psi = \int [(i/\hbar)\mathbf{e} \cdot \mathbf{J}\psi]^* \psi + \int \psi^* (i/\hbar)\, \mathbf{e} \cdot \mathbf{J}\, \psi$$
$$= (i/\hbar) \int \psi^* \mathbf{e} \cdot (-\mathbf{J}^\dagger + \mathbf{J})\psi$$

and therefore $\mathbf{J}$ must be self-adjoint;

$$\mathbf{J}^\dagger = \mathbf{J} \,. \tag{5.4.4}$$

(We are here using the abbreviation (5.3.8):

$$\int \psi_a^* \psi_b \equiv \int d^3x_1 \int d^3x_2 \cdots \psi_a^*(\mathbf{x}_1, \mathbf{x}_2, \ldots)\psi_b(\mathbf{x}_1, \mathbf{x}_2, \ldots)$$

and the definition (5.3.7) of the adjoint $A^\dagger$ of any operator A, except that as we shall see we must include discrete variables along with coordinates.)

In order for the transformation of the wave function to correspond to a rotation, it is necessary that it should produce a rotation of expectation values. That is, if $\mathbf{V}$ is an operator representing an observable that transforms as a vector under the general infinitesimal rotation (5.4.1), we must have

$$\delta\langle \mathbf{V}\rangle = \mathbf{e} \times \langle \mathbf{V}\rangle \,. \tag{5.4.5}$$

From Eqs. (5.4.3) and (5.4.4) we see that

$$\begin{aligned}\delta \int \psi^* \mathbf{V}\psi &= \int [(-i/\hbar)\mathbf{e}\cdot\mathbf{J}\psi]^* \mathbf{V}\psi + \int \psi^* \mathbf{V}(-i/\hbar)\mathbf{e}\cdot\mathbf{J}\psi \\ &= (i/\hbar)\int \psi^*[\mathbf{e}\cdot\mathbf{J}, \mathbf{V}]\psi\end{aligned}$$

and therefore we require

$$[\mathbf{e}\cdot\mathbf{J}, \mathbf{V}] = -i\hbar\, \mathbf{e}\times\mathbf{V}. \tag{5.4.6}$$

The same reasoning shows that $\mathbf{J}$ commutes with any rotationally invariant operator. For instance, for any pair of vector operators $\mathbf{V}$ and $\mathbf{V}'$, it is a consequence of Eq. (5.4.6) that

$$[\mathbf{e}\cdot\mathbf{J}\,,\; \mathbf{V}\cdot\mathbf{V}'] = -i\hbar\big([\mathbf{e}\times\mathbf{V}]\cdot\mathbf{V}' + \mathbf{V}\cdot[\mathbf{e}\times\mathbf{V}']\big) = 0\,,$$

just as in Eq. (5.4.2). In particular, as long as the Hamiltonian is rotationally invariant it commutes with $\mathbf{J}$,

$$[\mathbf{J}, H] = 0\,, \tag{5.4.7}$$

so, according to Eq. (5.3.26) angular momentum is conserved, in the sense that any expectation value of $\mathbf{J}$ is time-independent.

The requirement that the product $\mathbf{e}\cdot\mathbf{J}$ in Eq. (5.4.3) should not depend on the orientation of the coordinate axes implies that the operator $\mathbf{J}$ is itself a vector and hence satisfies Eq. (5.4.6);

$$[\mathbf{e}\cdot\mathbf{J}, \mathbf{J}] = -i\hbar\, \mathbf{e}\times\mathbf{J} \tag{5.4.8}$$

for any $\mathbf{e}$. From the coefficients of the different components of $\mathbf{e}$ in this equation, we easily find the equivalent commutation relations:

$$[J_1, J_2] = i\hbar J_3\,, \quad [J_2, J_3] = i\hbar J_1\,, \quad [J_3, J_1] = i\hbar J_2\,. \tag{5.4.9}$$

(For instance, the 2-component of Eq. (5.4.8) is

$$[\mathbf{e}\cdot\mathbf{J}, J_2] = -i\hbar\, (\mathbf{e}\times\mathbf{J})_2 = -i\hbar\, [e_3 J_1 - e_1 J_3],$$

in which the coefficient of e_1 is $[J_1, J_2] = i\hbar J_3$.) Also, $\mathbf{J}^2$ like any other scalar commutes with $\mathbf{J}$:

$$[J_i, \mathbf{J}^2] = 0 \,. \tag{5.4.10}$$

As we shall see, it is the commutation relations (5.4.9) that determine the possible values of $\mathbf{J}^2$ and the possible values of J_3 for a given $\mathbf{J}^2$.

Spin and Orbital Angular Momenta

The discussion so far in this section may produce some sense of déjà vu. We saw in Eqs. (5.3.30) and (5.3.31) that the orbital angular momentum $\mathbf{L}$ has commutators just like (5.4.6) with coordinates and momenta and hence with any vector $\mathbf{V}$ formed from coordinates and momenta:

$$[\mathbf{e} \cdot \mathbf{L}, \mathbf{V}] = -i\hbar\, \mathbf{e} \times \mathbf{V} \,. \tag{5.4.11}$$

Since $\mathbf{L}$ is itself a vector formed from coordinates and momenta, this also applies with $\mathbf{V} = \mathbf{L}$, and hence

$$[L_1, L_2] = i\hbar L_3 \,, \quad [L_2, L_3] = i\hbar L_1 \,, \quad [L_3, L_1] = i\hbar L_2 \,, \tag{5.4.12}$$

just like the commutators of the components of $\mathbf{J}$. Of course, we can also calculate these commutators directly from the commutators of momentum and position operators. For instance,

$$\begin{aligned}
[L_1, L_2] &= [X_2 P_3 - X_3 P_2, X_3 P_1 - X_1 P_3] \\
&= X_2 P_1 [P_3, X_3] + P_2 X_1 [X_3, P_3] \\
&= i\hbar(-X_2 P_1 + P_2 X_1) = i\hbar L_3 \,.
\end{aligned}$$

But this does not mean that $\mathbf{J} = \mathbf{L}$. Instead, we can consider the possibility that

$$\mathbf{J} = \mathbf{L} + \mathbf{S} \,, \tag{5.4.13}$$

where $\mathbf{S}$, known as the *spin*, is some operator whose properties we will now work out.

First, because $\mathbf{J}$ satisfies Eq. (5.4.6) for any vector operator $\mathbf{V}$, and $\mathbf{L}$ satisfies Eq. (5.4.11) for any vector operator $\mathbf{V}$ formed from positions and momenta, the difference of these equations tells us that

$$[S_i, V_j] = 0 \tag{5.4.14}$$

for any vector operator $\mathbf{V}$ formed from positions and momenta. The spin operator has nothing to do with positions and momenta.

In particular, since $\mathbf{L}$ is a vector formed from positions and momenta,

$$[S_i, L_j] = 0 \,. \tag{5.4.15}$$

It follows that

$$[J_i, J_j] = [L_i, L_j] + [S_i, S_j]$$

so the S_i satisfy the same commutation relations with each other as in Eq. (5.4.9) for $\mathbf{J}$ and Eq. (5.4.12) for $\mathbf{L}$:

$$[S_1, S_2] = i\hbar S_3 \ , \quad [S_2, S_3] = i\hbar S_1 \ , \quad [S_3, S_1] = i\hbar S_2 \ . \tag{5.4.16}$$

Multiplets

We next show how to use the commutation relations (5.4.9) to find the allowed values of $\mathbf{J}^2$ and the range of allowed values of J_3 for a given $\mathbf{J}^2$. Though presented here for the total angular momentum $\mathbf{J}$, precisely the same reasoning and corresponding results apply to any angular momentum operators with corresponding commutation relations, such as the orbital angular momentum vector $\mathbf{L}$ that satisfies Eq. (5.4.12) and the spin angular momentum vector $\mathbf{S}$ that satisfies Eq. (5.4.16).

First, we note that

$$[J_3\,,\,(J_1 \pm i J_2)] = i\hbar J_2 \pm i\,(-i\hbar J_1) = \pm\hbar\,(J_1 \pm i J_2)\ . \tag{5.4.17}$$

Therefore $J_1 \pm i J_2$ act as *raising and lowering operators*: for a wave function ψ^m that satisfies the eigenvalue condition $J_3\psi^m = \hbar m \psi^m$ (with any m), we have

$$J_3\,(J_1 \pm i J_2)\psi^m = (m \pm 1)\hbar(J_1 \pm i J_2)\psi^m \ ,$$

so if $(J_1 \pm i J_2)\psi^m$ does not vanish then it is an eigenfunction of J_3 with eigenvalue $\hbar(m \pm 1)$. Since $\mathbf{J}^2$ commutes with J_3, we can choose ψ^m to be an eigenfunction of $\mathbf{J}^2$ as well as J_3, and, since $\mathbf{J}^2$ commutes with $(J_1 \pm i J_2)$, all the wave functions that are connected with each other by lowering and/or raising operators will have the same eigenvalue for $\mathbf{J}^2$. We say that such wave functions form an angular momentum *multiplet*.

Now, there must be a maximum and a minimum to the eigenvalues of J_3 that can be reached in this way, because the square of any eigenvalue of J_3 is necessarily *not more* than the eigenvalue of $\mathbf{J}^2$. The reason is that for any wave function ψ that has an eigenvalue a for J_3 and an eigenvalue b for $\mathbf{J}^2$, we have

$$b - a^2 = \langle(\mathbf{J}^2 - J_3^2)\rangle = \langle(J_1^2 + J_2^2)\rangle \geq 0\ .$$

It is conventional to define a quantity j as the maximum value of the eigenvalues of $J_3/\hbar$ for a particular multiplet of wave functions that are related by raising and lowering operators. We will also temporarily define j' as the minimum eigenvalue of $J_3/\hbar$ for these wave functions. The wave function ψ^j for which J_3 takes its maximum eigenvalue $\hbar j$ must satisfy

$$(J_1 + i J_2)\psi^j = 0\ , \tag{5.4.18}$$

since otherwise $(J_1 + iJ_2)\psi^j$ would be a wave function with a larger eigenvalue of J_3. Likewise, acting on the wave function ψ^j with $(J_1 - iJ_2)$ gives an eigenfunction of J_3 with eigenvalue $\hbar(j-1)$, unless of course this wave function vanishes. Continuing in this way, we must eventually get to a wave function $\psi^{j'}$ with the minimum eigenvalue $\hbar j'$ of J_3, which satisfies

$$(J_1 - iJ_2)\psi^{j'} = 0 , \tag{5.4.19}$$

since otherwise $(J_1 - iJ_2)\psi^{j'}$ would be a wave function with an even smaller eigenvalue of J_3. We get to $\psi^{j'}$ from ψ^j by applying the lowering operator $(J_1 - iJ_2)$ a whole number of times, so $j - j'$ must be a whole number.

To go further, we use the commutation relations of J_1 and J_2 to show that

$$(J_1 - iJ_2)(J_1 + iJ_2) = J_1^2 + J_2^2 + i[J_1, J_2] = \mathbf{J}^2 - J_3^2 - \hbar J_3 , \tag{5.4.20}$$

$$(J_1 + iJ_2)(J_1 - iJ_2) = J_1^2 + J_2^2 - i[J_1, J_2] = \mathbf{J}^2 - J_3^2 + \hbar J_3 . \tag{5.4.21}$$

According to Eq. (5.4.18), the operator (5.4.20) gives zero when acting on ψ^j, so

$$\mathbf{J}^2\psi^j = \hbar^2 j(j+1)\psi^j . \tag{5.4.22}$$

On the other hand, according to Eq. (5.4.19) the operator (5.4.21) gives zero when acting on $\psi^{j'}$, so

$$\mathbf{J}^2\psi^{j'} = \hbar^2 j'(j'-1)\psi^{j'} . \tag{5.4.23}$$

But all these wave functions are eigenfunctions of $\mathbf{J}^2$ with the same eigenvalue, so $j'(j'-1) = j(j+1)$. This quadratic equation for j' has two solutions, $j' = j+1$, and $j' = -j$. The first solution is impossible, because j' is the minimum eigenvalue of $J_3/\hbar$ and therefore cannot be greater than the maximum eigenvalue j. This leaves us with the other solution,

$$j' = -j . \tag{5.4.24}$$

But we saw that $j - j' = 2j$ must be a non-negative integer, so *j must be a non-negative integer or half integer.* The eigenvalues of J_3 range over the $2j+1$ values of $\hbar m$ with m running by unit steps from $-j$ to $+j$. The corresponding eigenfunctions will be denoted ψ_j^m, so that

$$J_3\psi_j^m = \hbar m\,\psi_j^m , \qquad m = -j, -j+1, \ldots, +j \tag{5.4.25}$$

$$\mathbf{J}^2\,\psi_j^m = \hbar^2 j(j+1)\,\psi_j^m . \tag{5.4.26}$$

These are the same eigenvalues as those we found in the previous section in the case of orbital angular momentum, with the one big difference that j and m may be half-integers rather than integers. This justifies the guess of Goudsmit and Uhlenbeck that electrons could have an intrinsic angular momentum with $j = 1/2$, but that is the end of the surprises – we see that it is not possible to have physical systems with weird angular momenta such as $j = 1/3$, $j = 1/4$, etc.

Using these results, we can work out the action of any component of $\mathbf{J}$ on these multiplets. Because $J_1 - iJ_2$ is a lowering operator, we must have

$$(J_1 - iJ_2)\psi_j^m = \alpha_{jm}\psi_j^{m-1} ,$$

where the α_{jm} are various constants that depend on how the wave functions are normalized. If we assume that ψ_j^m and ψ_j^{m-1} both have unit norm, then, using Eq. (5.4.21),

$$|\alpha_{jm}|^2 = \int \psi_j^{m*}(J_1 + iJ_2)(J_1 - iJ_2)\psi_j^m = \int \psi_j^{m*}(\mathbf{J}^2 - J_3^2 + \hbar J_3)\psi_j^m$$
$$= \hbar^2[j(j+1) - m(m-1)] .$$

We can adjust the phases of these states so that all α_{jm} are real and positive and so that $\alpha_{jm} = \hbar\sqrt{j(j+1) - m(m-1)}$; hence

$$(J_1 - iJ_2)\psi_j^m = \hbar\sqrt{j(j+1) - m(m-1)}\,\psi_j^{m-1} . \tag{5.4.27}$$

The same analysis shows that, with this choice of phases,

$$(J_1 + iJ_2)\psi_j^m = \hbar\sqrt{j(j+1) - m(m+1)}\,\psi_j^{m+1} . \tag{5.4.28}$$

So now we know how J_1 and J_2 as well as J_3 act on angular momentum multiplets.

A particle of species n with eigenvalue $\hbar^2 s_n(s_n+1)$ for $\mathbf{S}_n^2$ is said to have spin s_n. Electrons, muons, neutrinos, and quarks have spin 1/2; W and Z particles have spin 1; and Higgs particles have spin 0. The concept of spin is not limited to so-called elementary particles. Protons and neutrons are each composites of three quarks, and some of their intrinsic angular momentum comes from the orbital motion of these quarks, but the energies in an atomic nucleus are not high enough for us to probe the internal structure of the proton and neutron, and so we refer to their total angular momentum as spin 1/2. Likewise, the energies in an atom are not high enough for us to probe the internal structure of their nuclei, and so we refer to the intrinsic angular momentum of these nuclei as a spin. The deuteron has spin 1; the ^{3}He and ^{4}He nuclei have spins 1/2 and 0, respectively; and so on.

The wave function for a multi-particle state depends on the 3-components σ_n of the individual spin vectors $\mathbf{S}_n/\hbar$, so the wave function must be labeled by these spin 3-components as well as by coordinates and will be written

$$\psi(\mathbf{x}_1, \sigma_1;\ \mathbf{x}_2, \sigma_2;\ \ldots) .$$

The values of σ_n run over the $2s_n + 1$ values from $-s_n$ to $+s_n$. In place of Eq. (5.3.8), the scalar product of two wave functions for systems of particles with spin includes a sum over all these σ_n:

$$\int \psi_a^* \psi_b \equiv \int d^3x_1 \sum_{\sigma_1} \int d^3x_2 \sum_{\sigma_2} \cdots \psi_a^*(\mathbf{x}_1, \sigma_1, \mathbf{x}_2, \sigma_2, \ldots) \psi_b(\mathbf{x}_1, \sigma_1, \mathbf{x}_2, \sigma_1, \ldots) . \tag{5.4.29}$$

The spin operator $\mathbf{S}$ does not act on the coordinate arguments, but produces linear combinations of wave functions with various values of the σ_n. (Instead of the 3-component of angular momentum, we can label states with the helicity, the component of angular momentum in the direction of motion in units of $\hbar$. Photon states have only helicity ± 1, corresponding to the two states of circular polarization.)

Adding Angular Momenta

Physical systems typically involve angular momenta of various sorts. Even in hydrogen there are the orbital and spin angular momenta of the electron, and also a proton spin very weakly coupled to the electron. In more complicated atoms there is more than one orbital and spin electron angular momentum, as well as a nuclear spin. But rotational invariance only ensures that the states of definite energy can also be chosen to have definite values for $\mathbf{J}^2$ and J_3 (where $\mathbf{J}$ is the *total* angular momentum), which are required by rotational invariance to commute with the Hamiltonian. This is one reason why it is important to know what total angular momenta arise when we combine different angular momenta in the same system.

Suppose a system involves two different angular momenta $\mathbf{J}_a$ and $\mathbf{J}_b$. These may be the spin and/or orbital angular momenta of a single particle, or sums of various spins and/or angular momenta of a number of particles. Suppose the wave function is an eigenfunction of $\mathbf{J}_a^2$ and $\mathbf{J}_b^2$ with eigenvalues $\hbar^2 j_a(j_a + 1)$ and $\hbar^2 j_b(j_b + 1)$, respectively. We can define wave functions $\psi_{j_a, j_b}^{m_a, m_b}$ as eigenfunctions also of J_{a3} and J_{b3} with eigenvalues $\hbar m_a$ and $\hbar m_b$, respectively. We then have the following problems: what linear combinations of these wave functions are eigenfunctions of $\mathbf{J}^2$ and $\mathbf{J}_3$ (where $\mathbf{J} \equiv \mathbf{J}_a + \mathbf{J}_b$) and what are the corresponding eigenvalues $\hbar^2 j(j + 1)$ and $\hbar m$ of $\mathbf{J}^2$ and J_3?

The "stretched" wave function $\psi_{j_a, j_b}^{j_a, j_b}$ with the maximum possible eigenvalues for J_{a3} and J_{b3} is an eigenfunction of $\mathbf{J}_3$ with $m = j_a + j_b$, so it must also be an eigenfunction of $\mathbf{J}^2$ with $j = j_a + j_b$. It could not have a lower value of j with this value of m, and there are no wave functions with larger values of j since there are none with larger values of m.

Next, consider the wave functions $\psi_{j_a,j_b}^{j_a-1,j_b}$ and $\psi_{j_a,j_b}^{j_a,j_b-1}$. Both are eigenfunctions of $\mathbf{J}_3$ with $m = j_a + j_b - 1$. One linear combination of these must be the member of the $j = j_a + j_b$ multiplet with $m = j_a + j_b - 1$; the other then has to be a member of some other multiplet, which by the same reasoning as before must have $j = j_a + j_b - 1$.

We can continue in this way, with one multiplet for each $j = j_a + j_b$, $j = j_a + j_b - 1$, $j = j_a + j_b - 2$, and so on. After ν steps, with $m = j_a + j_b - \nu$, there are $\nu + 1$ choices of m_a running up from $j_a - \nu$ to j_a, and $m_b = m - m_a$ running down from j_b to $j_b - \nu$, with one new multiplet having $j = j_a + j_b - \nu$ for each increase in ν. But this ends with $\nu = 2j_b$ (taking $j_a \geq j_b$), for which m_b runs from j_b to $-j_b$. At the next step, with $\nu = 2j_b + 1$, we would only get a new multiplet if m_b could run from j_b down to $-j_b - 1$, which is impossible since we can only have $|m_b| \leq j_b$. So when $j_a \geq j_b$, the lowest value of j that is found in the addition of angular momenta j_a and j_b is $j = j_a + j_b - 2j_b = j_a - j_b$. Of course, in the same way, if $j_b \geq j_a$, the lowest value of j is $j_b - j_a$. So in this way, we construct one multiplet for each j in the range

$$j = j_a + j_b, \quad j = j_a + j_b - 1, \quad j = j_a + j_b - 2, \quad \ldots, \quad j = |j_a - j_b| \,. \tag{5.4.30}$$

This is the general rule for adding angular momenta.

The linear combination of wave functions $\psi_{j_a,j_b}^{m_a,m_b}$ with a definite value for j and m is conventionally written as

$$\psi_{j_a,\,j_b,\,j}^{m} = \sum_{m_a=-j_a}^{+j_a} \sum_{m_b=-j_b}^{+j_b} C_{j_a,\,j_b}(j,\,m\,;\,m_a,\,m_b)\psi_{j_a,\,j_b}^{m_a,\,m_b} \,, \tag{5.4.31}$$

where the $C_{j_a,\,j_b}(j,\ m\,;\,m_a,\ m_b)$ are real numerical coefficients known as Clebsch–Gordan coefficients. Because $J_3 = J_{a3} + J_{b3}$, the Clebsch–Gordan coefficients are non-zero only for $m = m_a + m_b$.

In the appendix to this section we shall work this out in a simple case, the combination in hydrogen of the electron's integer orbital angular momentum ℓ and its spin angular momentum with $s = 1/2$. We then provide a table of Clebsch–Gordan coefficients for various low values of j_a and j_b.

Fine Structure and Space Inversion

Let us apply what we have learned to alkali metals and hydrogen. In both cases the observed spectrum arises from transitions involving a single "valence" electron moving in an essentially spherically symmetric potential – the potential of the nucleus for hydrogen or the potential of the nucleus and tightly bound inner electrons for alkali metals. The total angular momentum $\mathbf{J}$ is then the sum of the valence electron's orbital angular momentum $\mathbf{L}$ and that electron's spin $\mathbf{S}$. Since $\mathbf{J}$ commutes with the Hamiltonian, we can take the wave functions of definite energy also to have a definite value $\hbar^2 j(j+1)$ of $\mathbf{J}^2$ and a value $\hbar m$ of J_3, with m running from $-j$ to $+j$. Now, as we have just seen, these wave functions would in general be linear combinations of wave functions with

$j = \ell + 1/2$ and $j = \ell - 1/2$ (where ℓ is defined by $\mathbf{L}^2 = \hbar^2 \ell(\ell+1)$), so a wave function of definite energy and j would in general be a linear combination of wave functions with both $\ell = j - 1/2$ and $\ell = j + 1/2$.

But in fact these states can be chosen to have definite values of ℓ, because there is another conserved quantity. We can define a space reflection operator Π by the condition that, for any wave function ψ,

$$[\Pi\psi](\mathbf{x}_1, \sigma_1;\ \mathbf{x}_2, \sigma_2; \dots) = \psi(-\mathbf{x}_1, \sigma_1;\ -\mathbf{x}_2, \sigma_2; \dots)\ . \tag{5.4.32}$$

This is *not* a rotation. By a rotation of 180° around the z-axis we can change the signs of x and y but there is no rotation that changes the signs of all three components of a 3-vector. It is easy to see that the operator defined in this way has the properties

$$\Pi^2 = \mathbf{1}\ , \quad \Pi^\dagger = \Pi \tag{5.4.33}$$

and (now considering just a single particle)

$$\Pi \mathbf{X} \Pi = -\mathbf{X}\ , \qquad \Pi \mathbf{P} \Pi = -\mathbf{P}\ . \tag{5.4.34}$$

So, we also have

$$\Pi \mathbf{L} \Pi = +\mathbf{L}\ . \tag{5.4.35}$$

The defining condition (5.4.6) for the total angular momentum operator $\mathbf{J}$,

$$[\mathbf{e} \cdot \mathbf{J}, \mathbf{V}] = -i\hbar\ \mathbf{e} \times \mathbf{V}\ ,$$

is also satisfied by $\Pi \mathbf{J} \Pi$ as long as $\Pi \mathbf{V} \Pi = \pm \mathbf{V}$, so

$$\Pi \mathbf{J} \Pi = +\mathbf{J} \tag{5.4.36}$$

and then also

$$\Pi \mathbf{S} \Pi = +\mathbf{S}\ . \tag{5.4.37}$$

The operator Π commutes with the Hamiltonian:

$$\Pi H = H \Pi \tag{5.4.38}$$

at least for Hamiltonians of the form encountered in atomic and molecular physics, even if we include spin–orbit coupling terms, proportional to $\mathbf{S} \cdot \mathbf{L}$. It follows that we can choose the states of definite energy so that their wave functions are also eigenfunctions of Π:

$$\Pi \psi = \pi \psi\ .$$

Because $\Pi^2 = \mathbf{1}$ the eigenvalue π, known as the *parity* of the state, can only be $+1$ or -1. Indeed, given a wave function ψ for which $H\psi = E\psi$ that is not an eigenfunction of Π, we can always write it as a superposition $\psi = \psi_+ + \psi_-$

where $\psi_\pm \equiv (1 \pm \Pi)\psi/2$. Since Π commutes with H these satisfy $H\psi_\pm = E\psi_\pm$, and since $\Pi^2 = \mathbf{1}$ they satisfy $\Pi\psi_\pm = \pm\psi_\pm$.

As we saw in Section 5.2 for general spherically symmetric potentials, a one-particle state with $\mathbf{L}^2 = \hbar^2\ell(\ell+1)$ has wave function $\psi(\mathbf{x})$ proportional to a homogeneous polynomial of order ℓ in the coordinate $\mathbf{x}$, on which the operator Π gives a factor $(-1)^\ell$, so the states with definite energy can be taken to have ℓ either even or odd. For $j - 1/2$ even or odd we have $j + 1/2$ respectively odd or even, and hence the states with definite energy and j can be taken to have a definite ℓ, either $j - 1/2$ or $j + 1/2$. These states are therefore labeled

$$1s_{1/2},\quad 2s_{1/2},\quad 2p_{1/2},\quad 2p_{3/2},\quad 3s_{1/2},\quad 3p_{1/2},\quad 3p_{3/2},\quad 3d_{3/2},\quad 3d_{5/2},\quad \ldots$$

where again the letters s, p, d, etc, stand for $\ell = 0, 1, 2$, etc.; the integer in front of the letter is the principal quantum number, defined so that the number of nodes of the wave function is $n - \ell - 1$ (and therefore $\ell \leq n - 1$); and now the subscript gives the value of j. The energy depends on j as well as on n and (except for hydrogen) on ℓ, with the j dependence arising both from relativistic corrections and the magnetic coupling of the electron's spin with the orbital motion, but this dependence is rather weak and just gives rise to the *fine structure* of the energy levels. The difference in the energies of the $3p_{1/2}$ and $3p_{3/2}$ states of sodium splits the wavelengths of the $3p_{1/2} \to 3s_{1/2}$ and $3p_{3/2} \to 3s_{1/2}$ transitions by just 1.02 parts per thousand, while the energies of the $2p_{1/2}$ and $2p_{3/2}$ states of hydrogen differ by only 4.44 parts per million.

The hydrogen fine structure was first calculated in 1928 by Dirac in a relativistic version of wave mechanics.[10] The relativistic and spin effects that he calculated left the $2p_{1/2}$ state with the same energy as the $2s_{1/2}$ state. Physicists including Hans Kramers (1894–1952) and Victor Weisskopf (1908–2002) realized in the 1930s that quantum electrodynamic effects such as the emission and reabsorption of photons by the orbiting electron would split the energies of the $2p_{1/2}$ and $2s_{1/2}$ states, but the calculation proved difficult. This splitting was first measured after the war by Willis Lamb (1903–2008) and R. C. Retherford,[11] and is known as the *Lamb shift*. It is very small, 4.3515×10^{-6} eV, about a tenth of the small fine-structure splitting between the $2p_{1/2}$ and $2p_{3/2}$ states. The successful calculation of the Lamb shift in 1949 by Norman Kroll (1922–2004) and Lamb[12] and by J. B. French and Weisskopf[13] marked the beginning of the modern understanding of quantum electrodynamics.

Any particle at rest, whether elementary or not, will have what is called an intrinsic parity π_n that depends only on the type n of the particle. If

[10] P. A. M. Dirac, Proc. Roy. Soc. A **117**, 619 (1928).

[11] W. E. Lamb, Jr. and R. C. Retherford, Phys. Rev. **72**, 241 (1947).

[12] N. M. Kroll and W. E. Lamb, Phys. Rev. **75**, 388 (1949).

[13] J. B. French and V. F. Weisskopf, Phys. Rev. **75**, 1240 (1949).

the particle is in a state with orbital angular momentum ℓ, the parity of its state is $(-1)^\ell \pi_n$. In our discussion above we have implicitly taken the electron to have positive intrinsic parity. This is a matter of definition; if the electron had negative intrinsic parity we could redefine the parity operator as $\Pi' = \exp(i\pi Q/e)\Pi$, where Q is the operator for total electric charge. The one-electron state is an eigenstate of Q with eigenvalue $-e$, so it is an eigenstate of $\exp(i\pi Q/e)$ with eigenvalue -1; if it were an eigenstate of Π with eigenvalue -1 it would be an eigenstate of Π' with eigenvalue $+1$. Since Q as well as Π commutes with the Hamiltonian, so does Π' and it can be called the operator of space inversion just as well as Π. In the same way, because of the conservation of another quantity known as baryon number (described in Section 6.2) we can define the parity of the proton as $+1$. But the intrinsic parities of most particles have to be determined experimentally.

Hyperfine Structure

We must not forget the atomic nucleus, for if it has spin this produces a magnetic field felt by orbiting electrons. This effect is most important for the s-wave electrons that are not prevented from getting close to the nucleus by the centrifugal barrier that is present for $\ell \neq 0$. In hydrogen the spin 1/2 of the nucleus combines with the spin 1/2 of the electron in its $\ell = 0$ ground state to split the energy of the ground state into components with total spin $s = 0$ and $s = 1$, separated in energy by 5.9×10^{-6} eV. The transition between these states produces the famous 21-cm absorption and emission spectral lines, discussed in Section 3.5.

Appendix: Clebsch–Gordan Coefficients

First, as an example of some intrinsic importance, let us work out how to form hydrogen wave functions with definite total angular momentum from wave functions with definite 3-components of spin and orbital angular momentum. Consider the "stretched" hydrogen wave function in which L_3 and S_3 are both as large as possible, having eigenvalues $+\hbar\ell$ and $+\hbar/2$, respectively. In general we shall label hydrogen wave functions with orbital angular momentum ℓ and spin $1/2$ and definite values $\hbar m$ and $\hbar\sigma$ for L_3 and S_3 as $\psi^{m,\,\sigma}_{\ell,\,1/2}$, so this stretched wave function is denoted $\psi^{\ell,\,+1/2}_{\ell,\,1/2}$. For this wave function, $J_3 = \hbar(\ell + 1/2)$ is also as large as possible. This is therefore a wave function with $j = \ell + 1/2$ (where as usual j is defined so that the eigenvalue of $\mathbf{J}^2$ is $\hbar^2 j(j+1)$). This wave function could not have a larger j because then there would be states with

$J_3 > \hbar(\ell + 1/2)$, and it could not have a smaller j because then J_3 could not be as large as $\hbar(\ell + 1/2)$. In general we shall label hydrogen wave functions with orbital angular momentum ℓ and spin 1/2 and definite values $\hbar^2 j(j+1)$ and $\hbar M$ for $\mathbf{J}^2$ and J_3 as $\psi^M_{\ell,\,1/2,\,j}$. So we have

$$\psi^{\ell,\,+1/2}_{\ell,\,1/2} = \psi^{\ell+1/2}_{\ell,\,1/2,\,\ell+1/2}\,. \tag{5.4.39}$$

So far, this is pretty trivial, apparently not worth the elaborate notation. But now consider the wave functions with $J_3 = \hbar(\ell - 1/2)$. There are two of these, one of them, $\psi^{\ell-1,\,+1/2}_{\ell,\,1/2}$, with $L_3 = \hbar(\ell - 1)$ and $S_3 = +\hbar/2$ and the other, $\psi^{\ell,\,-1/2}_{\ell,\,1/2}$, with $L_3 = \hbar\ell$ and $S_3 = -\hbar/2$. One linear combination of these two can be obtained by letting the lowering operator $J_1 - iJ_2$ act on the stretched wave function. This is part of the same angular momentum multiplet as the stretched wave function $\psi^{\ell+1/2}_{\ell,\,1/2,\,\ell+1/2}$, with the same eigenvalue for $\mathbf{J}^2$, so is labeled $\psi^{\ell-1/2}_{\ell,\,1/2,\,\ell+1/2}$. According to Eq. (5.4.27), if properly normalized this wave function is given by

$$\begin{aligned}\sqrt{2\ell+1}\,\psi^{\ell-1/2}_{\ell,\,1/2,\,\ell+1/2} &= (J_1 - iJ_2)\psi^{\ell+1/2}_{\ell,\,1/2,\,\ell+1/2} \\ &= (L_1 - iL_2 + S_1 - iS_2)\psi^{\ell,\,+1/2}_{\ell,\,1/2}\,.\end{aligned}$$

Orbital and spin angular momenta obey the same commutation relations as total angular momentum, so their lowering operators act the same way as given in Eq. (5.4.27) for $J_1 - iJ_2$:

$$(L_1 - iL_2)\psi^{\ell,\,+1/2}_{\ell,\,1/2} = \sqrt{2\ell}\,\psi^{\ell-1,\,+1/2}_{\ell,\,1/2}\,, \quad (S_1 - iS_2)\psi^{\ell,\,+1/2}_{\ell,\,1/2} = \psi^{\ell,\,-1/2}_{\ell,\,1/2}\,,$$

and therefore

$$\sqrt{2\ell+1}\,\psi^{\ell-1/2}_{\ell,\,1/2,\,\ell+1/2} = \sqrt{2\ell}\,\psi^{\ell-1,\,+1/2}_{\ell,\,1/2} + \psi^{\ell,\,-1/2}_{\ell,\,1/2}\,. \tag{5.4.40}$$

Since there are two independent wave functions with $J_3 = \hbar(\ell - 1/2)$, there must be another linear combination that is part of an angular momentum multiplet with no higher value of J_3 than $\hbar(\ell - 1/2)$, so this multiplet has $j = \ell - 1$, and in our notation this linear combination is $\psi^{\ell-1/2}_{\ell,\,1/2,\,\ell-1/2}$. Since it has a different value of j, this linear combination can be calculated by requiring it to be normalized and orthogonal to the one we found by acting with $J_1 - iJ_2$ on the stretched wave function with $L_3 = \hbar\ell$ and $S_3 = +\hbar/2$. That is (with a conventional choice of overall phase):

$$\sqrt{2\ell+1}\,\psi^{\ell-1/2}_{\ell,\,1/2,\,\ell-1/2} = -\psi^{\ell-1,\,+1/2}_{\ell,\,1/2} + \sqrt{2\ell}\,\psi^{\ell,\,-1/2}_{\ell,\,1/2}\,. \tag{5.4.41}$$

By continued operation of the lowering operator on the wave functions (5.4.40) and (5.4.41), we fill out two complete multiplets, one with $j = \ell + 1/2$ and one with $j = \ell - 1/2$. These results can be summarized as values for the Clebsch–Gordan coefficients in Eq. (5.4.31):

$$C_{\ell,\,1/2}(\ell+1/2,\ \ell+1/2,\ \ell,\ +1/2) = 1\ ,$$

$$C_{\ell,\,1/2}(\ell+1/2,\ \ell-1/2,\ \ell-1,\ +1/2) = \sqrt{\frac{2\ell}{2\ell+1}}\ ,$$

$$C_{\ell,\,1/2}(\ell+1/2,\ \ell-1/2,\ \ell,\ -1/2) = \sqrt{\frac{1}{2\ell+1}}\ ,$$

$$C_{\ell,\,1/2}(\ell-1/2,\ \ell-1/2,\ \ell-1,\ +1/2) = -\sqrt{\frac{1}{2\ell+1}}\ ,$$

$$C_{\ell,\,1/2}(\ell-1/2,\ \ell-1/2,\ \ell,\ -1/2) = \sqrt{\frac{2\ell}{2\ell+1}}\ .$$

All the Clebsch–Gordan coefficients can be calculated in this way, but life is too short. The best way to find Clebsch–Gordan coefficients is to look them up in a table. At the end of this section there is a table of these coefficients for small angular momenta.

There is a symmetry property of the Clebsch–Gordan coefficients in the case of adding equal angular momenta that will be important for us when we come to diatomic molecules in the next section and to nuclear forces in Section 6.2. For $j_a = j_b$,

$$C_{j_a, j_a}(jM;\ m_a\, m_b) = (-1)^{j-2j_a} C_{j_a j_a}(jM;\ m_b\, m_a). \qquad (5.4.42)$$

This is trivial for the stretched configuration, where $m_a = m_b = j_a$ and $j = 2j_a$. It is then also valid for all the Clebsch–Gordan coefficients with the same value of j, because the corresponding states are obtained by acting on the stretched configuration state with the symmetric lowering operator $J_1 - iJ_2 = J_{a1} + J_{b1} - iJ_{a2} - iJ_{b2}$. The state with $j = 2j_a - 1$ and $m_a + m_b = 2j_a - 1$ is a superposition of terms with $m_a = j_a - 1$, $m_b = j_a$ and $m_a = j_a$, $m_b = j_a - 1$, and, since it is orthogonal to the state with $j = 2j_a$ and $m_a + m_b = 2j_a - 1$, it must be antisymmetric in m_a and m_b. All the other states with $j = 2j_a - 1$ are obtained by acting on this state with the symmetric lowering operator $J_1 - iJ_2$, and so are also antisymmetric in m_a and m_b, in agreement with Eq. (5.4.42). Continuing, from the states with $m_a + m_b = j_a - 2$ we can form one antisymmetric combination, which is needed in the multiplet with $j = j_a - 1$, and two symmetric combinations, which can then only be in the multiplets with $j = j_a + j_b$ and $j = j_a + j_b - 2$. And so on.

Table 5.1 The non-vanishing Clebsch–Gordan coefficients for the addition of angular momenta j_a and j_b with 3-components m_a and m_b to give angular momentum j with 3-component M, for several low values of j_a and j_b.

j_a	j_b	j	M	m_a	m_b	$C_{j_a,j_b}(j\,M\,;\,m_a\,m_b)$
$\frac{1}{2}$	$\frac{1}{2}$	1	$+1$	$+\frac{1}{2}$	$+\frac{1}{2}$	1
$\frac{1}{2}$	$\frac{1}{2}$	1	0	$\pm\frac{1}{2}$	$\mp\frac{1}{2}$	$1/\sqrt{2}$
$\frac{1}{2}$	$\frac{1}{2}$	1	-1	$-\frac{1}{2}$	$-\frac{1}{2}$	1
$\frac{1}{2}$	$\frac{1}{2}$	0	0	$\pm\frac{1}{2}$	$\mp\frac{1}{2}$	$\pm 1\sqrt{2}$
1	$\frac{1}{2}$	$\frac{3}{2}$	$\pm\frac{3}{2}$	± 1	$\pm\frac{1}{2}$	1
1	$\frac{1}{2}$	$\frac{3}{2}$	$\pm\frac{1}{2}$	± 1	$\mp\frac{1}{2}$	$\sqrt{1/3}$
1	$\frac{1}{2}$	$\frac{3}{2}$	$\pm\frac{1}{2}$	0	$\pm\frac{1}{2}$	$\sqrt{2/3}$
1	$\frac{1}{2}$	$\frac{1}{2}$	$\pm\frac{1}{2}$	± 1	$\mp\frac{1}{2}$	$\pm\sqrt{2/3}$
1	$\frac{1}{2}$	$\frac{1}{2}$	$\pm\frac{1}{2}$	0	$\pm\frac{1}{2}$	$\mp\sqrt{1/3}$
1	1	2	± 2	± 1	± 1	1
1	1	2	± 1	± 1	0	$1/\sqrt{2}$
1	1	2	± 1	0	± 1	$1/\sqrt{2}$
1	1	2	0	± 1	∓ 1	$1/\sqrt{6}$
1	1	2	0	0	0	$\sqrt{2/3}$
1	1	1	± 1	± 1	0	$\pm 1/\sqrt{2}$
1	1	1	± 1	0	± 1	$\mp 1/\sqrt{2}$
1	1	0	0	± 1	∓ 1	$1/\sqrt{3}$
1	1	0	0	0	0	$-1/\sqrt{3}$

5.5 Bosons and Fermions

Identical Particles

Aside from their momenta and helicities, every photon in the universe is identical to every other photon. The reason is that all photons are quanta of the same field, the electromagnetic field. In the same way, aside from their momenta (or positions) and spin components, according to the modern understandings outlined in Chapter 7, every electron in the universe is identical to every other electron because they are all quanta of a single field, known as the electron field. The same is true of every other species of elementary particle – quarks, neutrinos, and so on – each is the quantum of a particular field. Indeed, our best current definition of an elementary particle is that it is the quantum of one of the fields of which the world is composed. But the same indistinguishability is true of composite systems in any one specific state. Two protons are indistinguishable because they are each composed of three quarks of the same two different types in the same bound state, and two hydrogen atoms in the same atomic state are indistinguishable because they are each composed of an electron and a proton.

In writing a wave function for identical particles as $\psi(\mathbf{x}_1, \sigma_1;\ \mathbf{x}_2, \sigma_2;\ \ldots)$, it is incorrect to say that for this wave function the first particle has position $\mathbf{x}_1$ and spin 3-component σ_1 while the second particle has position $\mathbf{x}_2$ and spin 3-component σ_2, and so on. Instead we should say that there is a particle with position $\mathbf{x}_1$ and spin 3-component σ_1 and another particle with position $\mathbf{x}_2$ and spin 3-component σ_2, and so on. Thus for identical particles, $\psi(\mathbf{x}_1, \sigma_1;\ \mathbf{x}_2, \sigma_2;\ \ldots)$ and $\psi(\mathbf{x}_2, \sigma_2;\ \mathbf{x}_1, \sigma_1;\ \ldots)$ represent the same state. Two wave functions that represent the same state can only differ by a constant factor, so

$$\psi(\mathbf{x}_2, \sigma_2;\ \mathbf{x}_1, \sigma_1;\ \ldots) = \lambda\psi(\mathbf{x}_1, \sigma_1;\ \mathbf{x}_2, \sigma_2;\ \ldots)$$

for some constant λ. Integrals don't depend on how the variables of integration are labeled, so it follows that $\int |\psi|^2 = |\lambda|^2 \int |\psi|^2$, and therefore λ can only be a phase factor, with $|\lambda| = 1$. Further, the constant λ cannot depend on position or spin 3-components without violating various symmetry principles, such as Galilean or Einsteinian relativity, rotational invariance, and translation invariance. so we can therefore repeat the same relation with identical particles 1 and 2 interchanged on both sides, but with the same λ, and write

$$\psi(\mathbf{x}_1, \sigma_1;\ \mathbf{x}_2, \sigma_2;\ \ldots) = \lambda\psi(\mathbf{x}_2, \sigma_2;\ \mathbf{x}_1, \sigma_1;\ \ldots) = \lambda^2\psi(\mathbf{x}_1, \sigma_1;\ \mathbf{x}_2, \sigma_2;\ \ldots)$$

and therefore $\lambda^2 = 1$. We have only two possibilities, $\lambda = \pm 1$. Our usual assumptions regarding locality would not allow the choice of signs to depend on whatever other particles are described by the wave function if these particles were very far away from particles 1 and 2, and the continuity of wave functions would not allow this sign to jump between $+1$ and -1 as these other particles come close. We conclude that the value of λ encountered when we exchange a pair of indistinguishable particles can depend only on the species of these particles.

Particles for which $\lambda = 1$, so that the wave function is symmetric in the labels of these particles, are known as *bosons*. They are named after Satyendra

Nath Bose (1894–1974), who first described multi-photon states, imposing this symmetry condition.[14] Einstein had Bose's paper translated into German and published, and then applied these ideas to material particles.[15]

Particles for which $\lambda = -1$, so that the wave function is antisymmetric in the labels of these particles, are known as *fermions*, named after Enrico Fermi (1901–1954). Fermi[16] and Dirac[17] at about the same time described multi-electron states, imposing this antisymmetry condition.

It is another consequence of the relativistic quantum theory of fields that elementary particles (the quanta of fields) are bosons or fermions according to whether their spin is an integer or half an odd integer.[18] The reason for this is outlined in Section 7.4, but a complete proof is beyond the scope of this book. It is easy, though, to see that if this correlation with spin is valid for some set of elementary particles then it is valid for any composites of these particles. If we interchange two identical composite particles then we are interchanging all their constituents, so the interchange gives a minus sign multiplying the wave function if each of the composites contains an odd number of fermions and a plus sign otherwise, no matter how many bosons it contains. But, according to the rules for adding angular momenta described in the previous section, a composite has a half odd integer spin if it contains an odd number of half odd integer spin particles, and integer spin otherwise, no matter how many integer spin particles it contains. So a composite with half odd integer spin contains an odd number of fermions, and is therefore a fermion, while if it has integer spin it contains an even number of fermions (perhaps zero) and is therefore a boson. No other correlation of boson/fermion character with spin would have this consistency.

So electrons, quarks, protons, and neutrons, which have spin 1/2, are fermions. The spin of massless particles like photons requires special consideration, but as noted in Section 7.5 the components of their angular momentum in the direction of travel can only be $\pm\hbar$, corresponding to left and right circular polarization, and they are bosons. Indeed, as already mentioned, Bose's original introduction of symmetric states had to do with photons. Hydrogen and helium atoms are bosons, while ^{6}Li atoms (with three protons, three neutrons, and three electrons) are fermions.

Statistics

The distinction between bosons and fermions has a profound impact on the properties of gases in thermal equilibrium. As we did for photons in Section 3.2,

[14] S. N. Bose, Z. Phys. **26**, 178 (1924).
[15] A. Einstein, Sitz. Preuss. Akad. Wiss. **1**, 3 (1926).
[16] E. Fermi, Rend. Lincei **3**, 145 (1926).
[17] P. A. M. Dirac, Proc. Roy. Soc. A **112**, 661 (1926).
[18] This was first stated as a general rule by M. Fierz, Helv. Phys. Acta **12**, 3 (1939) and W. Pauli, Phys. Rev. **58**, 716 (1940).

to calculate the densities of particles with various momenta we can imagine a gas of any identical particles in a cube of volume L^3. Since the particles in a gas are essentially free particles, their momenta are quantized like photon momenta, with $\mathbf{p} = 2\pi \mathbf{n}\hbar/L$, where $\mathbf{n}$ is a 3-vector with integer components. As we saw in our discussion of momentum space in Section 5.3, the number of these allowed momentum values in a momentum-space volume d^3p is

$$dN = d^3n = (L/2\pi\hbar)^3 \times d^3p \,. \tag{5.5.1}$$

The number of particles per volume with momentum in a volume d^3p of momentum space around one of these allowed momentum values is then

$$d\mathcal{N}(\mathbf{p}) = g\, dN\, \bar{N}_{\mathbf{p}}/L^3 = g(2\pi\hbar)^{-3} d^3p\, \bar{N}_{\mathbf{p}} \,, \tag{5.5.2}$$

where $\bar{N}_{\mathbf{p}}$ is the mean number of particles in these states with momentum $\mathbf{p}$ and any given spin 3-component or helicity. (In Eq. (5.5.2), we include a factor g equal to the number of spin or helicity states for each allowed momentum value. For massive particles of spin s, we have $g = 2s + 1$ states, which are characterized by different values of S_3, while for photons $g = 2$.) The mean number $\bar{N}_{\mathbf{p}}$ in a gas with temperature T and chemical potential μ is given by the grand canonical ensemble discussed in Section 2.4:

$$\bar{N}_{\mathbf{p}} = \frac{\sum_N N \exp(-N[E(\mathbf{p}) - \mu]/kT)}{\sum_N \exp(-N[E(\mathbf{p}) - \mu]/kT)} \tag{5.5.3}$$

the sums running over the allowed numbers of particles with momentum $\mathbf{p}$. It is in these sums that there appears a distinction between bosons and fermions.

For bosons N runs over all integers from zero to infinity, and we have

$$\bar{N}_{\mathbf{p}} = \frac{1}{\exp([E(\mathbf{p}) - \mu]/kT) - 1} \,. \tag{5.5.4}$$

This is known as the case of *Bose–Einstein statistics*. The chemical potential μ can be non-zero only if the total number of particles is conserved, so for photons $\mu = 0$ and the result of using Eq. (5.5.4) in Eq. (5.5.2) (with the number of polarization states $g = 2$) is equivalent to the Planck distribution (3.1.14). For material particles such as atoms whose number is conserved under ordinary conditions we can have $\mu > 0$, and then at very low temperature $\bar{N}_{\mathbf{p}}$ is very sharply peaked at momenta for which the energy $E(\mathbf{p})$ is close to μ. It is even possible to have a macroscopic number of particles with energy μ, a phenomenon known as Bose–Einstein condensation, first seen by Eric Cornell and Carl Wiemann and their collaborators in a gas of rubidium atoms in 1995.[19] (There is also a sort of Bose–Einstein condensation in liquid helium, but it is not a good approximation to treat liquid helium as a gas.)

[19] M. H. Anderson *et al.*, Science **269**, 198 (1995).

For fermions it is not possible to have more than one particle with a given momentum $\mathbf{p}$ (and a given spin 3-component), because the wave function for two such particles would be proportional to

$$\exp(i\mathbf{p}\cdot\mathbf{x}_1/\hbar)\exp(i\mathbf{p}\cdot\mathbf{x}_2/\hbar),$$

which is symmetric rather than antisymmetric in the two particles. Hence the sums in Eq. (5.5.3) run only over the values $N = 0$ and $N = 1$:

$$\bar{N}_{\mathbf{p}} = \frac{1}{\exp([E(\mathbf{p}) - \mu]/kT) + 1} . \tag{5.5.5}$$

This is known as the case of *Fermi–Dirac statistics*. For very low temperatures this takes the form

$$\bar{N}_{\mathbf{p}} \to \begin{cases} 1 & E(\mathbf{p}) < \mu \\ 0 & E(\mathbf{p}) > \mu . \end{cases} \tag{5.5.6}$$

This is used to derive a relation between the number densities and energy densities in white dwarf stars, whose high density requires electrons to have energies much larger than chemical binding energies, though they are essentially at zero temperature. For white dwarfs of relatively low mass, μ is much less than $m_e c^2$ but much larger than chemical binding energies, so the number density of electrons is given by

$$n_e = \int_{p<p_F} d\mathcal{N}(\mathbf{p}) = \frac{2}{(2\pi\hbar)^3}\int_0^{p_F} 4\pi p^2\,dp = \frac{8\pi p_F^3}{3(2\pi\hbar)^3}$$

where p_F is the *Fermi momentum* defined by $E(p_F) = \mu$. (In practice, we use a known or assumed value of n to calculate p_F.) The corresponding kinetic energy density is

$$\mathcal{E} = \int_{|\mathbf{p}|<p_F} E(\mathbf{p})d\mathcal{N}(\mathbf{p}) = \frac{2}{(2\pi\hbar)^3}\int_0^{p_F} 4\pi p^2\,dp \times \frac{p^2}{2m_e} = \frac{8\pi p_F^5}{10m_e(2\pi\hbar)^3}$$
$$= (8\pi)^{-2/3}(2\pi\hbar)^2(3n_e)^{5/3}/10m_e .$$

As shown in Eq. (1.1.3), the pressure of any non-relativistic monatomic gas is $p = 2\mathcal{E}/3$, so this gives an equation of state for low-mass white dwarfs:

$$p = K\rho^{5/3} , K = (2/3)(8\pi)^{-2/3}(2\pi\hbar)^2(3Z/Am_1)^{5/3}/10m_e ,$$

where $\rho = n_e A m_1/Z$ is the mass density.

The Hartree Approximation

In multi-electron atoms it is often a good approximation to treat each electron as moving in a spherically symmetric (but not Coulomb!) effective potential arising from the atomic nucleus and from all the other electrons. This is known as the *Hartree approximation*, introduced by Douglas Hartree (1897–1958) in 1928.[20] Each electron occupies some one-particle state of definite energy in this effective potential with corresponding wave functions $\psi_1(\mathbf{x}, \sigma)$, $\psi_2(\mathbf{x}, \sigma)$, etc. If electrons were distinguishable, the wave function for an atomic state with N electrons – electron 1 in state 1, electron 2 in state 2, etc. – would be the product

$$\psi_1(\mathbf{x}_1, \sigma_1)\psi_2(\mathbf{x}_2, \sigma_2) \cdots \psi_N(\mathbf{x}_N . \sigma_N) .$$

But, because electrons are indistinguishable fermions, this must be antisymmetrized. The true wave function (up to a normalization constant) is

$$\psi = \sum_P \delta_P \psi_1(\mathbf{x}_{P1}, \sigma_{P1})\psi_2(\mathbf{x}_{P2}, \sigma_{P2}) \cdots \psi_N(\mathbf{x}_{PN}, \sigma_{PN}) , \tag{5.5.7}$$

the sum running over all permutations P of $1, 2, \ldots, N$ into $P1, P2, \ldots, PN$, with $\delta_P = +1$ or $\delta_P = -1$ for P an even or odd permutation, respectively. For instance, for a two-electron state there are two permutations P, the identity $1 \to 1,\ 2 \to 2$ with $\delta_P = +1$, and the interchange $1 \leftrightarrow 2$, with $\delta_P = -1$, so

$$\psi = \psi_1(\mathbf{x}_1, \sigma_1)\psi_2(\mathbf{x}_2, \sigma_2) - \psi_1(\mathbf{x}_2, \sigma_2)\psi_2(\mathbf{x}_1, \sigma_1) .$$

In general, the wave function (5.5.7) can be written as a determinant, known as a *Slater determinant*:[21]

$$\psi = \begin{vmatrix} \psi_1(\mathbf{x}_1, \sigma_1) & \psi_1(\mathbf{x}_2, \sigma_2) & \cdots & \psi_1(\mathbf{x}_N, \sigma_N) \\ \psi_2(\mathbf{x}_1, \sigma_1) & \psi_2(\mathbf{x}_2, \sigma_2) & \cdots & \psi_2(\mathbf{x}_N, \sigma_N) \\ \psi_3(\mathbf{x}_1, \sigma_1) & \psi_3(\mathbf{x}_2, \sigma_2) & \cdots & \psi_3(\mathbf{x}_N, \sigma_N) \\ \cdots & \cdots & \cdots & \cdots \end{vmatrix} \tag{5.5.8}$$

The Pauli Exclusion Principle

None of the one-particle states occupied by electrons in the Hartree approximation can be the same, for if they were then two rows of the Slater determinant would be identical, and the wave function would vanish. This principle was first stated by Pauli,[22] on the basis of efforts to understand the periodic table of the elements, before it became understood that multi-electron wave functions have to be antisymmetric. The number of values of L_3 for a given ℓ is $2\ell + 1$, so Pauli at first thought that not more than $2\ell + 1$ electrons can have the same n and ℓ, but as we shall see, to get the chemistry right it is necessary to assume that the maximum number of electrons with a given n and ℓ is $2(2\ell + 1)$. For this reason Pauli introduced a new quantum number that takes just two values, which as discussed in the previous section were identified by Goudsmit and Uhlenbeck as the two values $S_3 = \pm\hbar/2$ of the 3-component of electron spin.

---

20 D. H. Hartree, Proc. Camb. Phil. Soc. **24**, 111 (1928).
21 J. C. Slater, Phys. Rev. **34**, 1293 (1929).
22 W. Pauli, Z. Physik **31**, 763 (1925).

Pauli reasoned that as we increase the number Z of electrons in atoms each added electron must occupy the one-particle state of next largest energy. This is why electrons in atoms do not all fall into the two $1s$ states of lowest energy, so that atoms with $Z > 2$ do not all behave chemically just like helium.

The Periodic Table

The Pauli exclusion principle provides an explanation for the periodic table of elements, first described in purely chemical terms by Dmitri Ivanovich Mendeleev (1834–1907) in 1869, long before atomic structure was understood. Of course, Mendeleev knew nothing about electrons but he knew the values of atomic weights and could list the elements in order of increasing atomic weight. As we saw in Section 5.2, in the twentieth century it became clear that the atomic number, defined as the place of an atom in this list, is the same (with a few exceptions) as the charge Z of the atomic nucleus in units of e and is hence equal to the number of electrons in the atom, on which the chemical properties of elements chiefly depend.

Detailed calculations show that the one-electron states are filled (with sporadic exceptions) in the order

$$\begin{aligned} &1s, \\ &2s,\ 2p, \\ &3s,\ 3p, \\ &4s,\ 3d,\ 4p, \\ &5s,\ 4d,\ 5p, \\ &6s,\ 4f,\ 5d,\ 6p, \\ &7s,\ 5f,\ 7p,\ \ldots \end{aligned} \tag{5.5.9}$$

(We are here ignoring the small fine-structure splitting in the energies of these states, and so are leaving out subscripts giving the values of j.) For a given ℓ, increasing n increases the number of nodes of the wave function, so that the wave function oscillates more with r, which increases the kinetic energy. This is the main reason why electron energies increase going down the list. But, for a given n, the increase in centrifugal force with increasing ℓ decreases the wave function at small r where the charge interior to r is largest, which decreases the effective absolute value of the negative potential energy, increasing the state's total energy. Hence, although the one-electron states listed above on the same line have approximately equal energy, the energies increase somewhat from left to right. In the case of $3d$, $4d$, $4f$, $5d$, and $5f$ states and many states with $n \geq 6$, the dependence of the energy on ℓ turns out to overcome its dependence on n.

Taking spin into account, the total number of states for the energy levels listed on each line of Eq. (5.5.9) are 2, $2+6=8$, $2+6=8$, $2+10+6=18$, $2+10+6=18$, $2+14+10+6=32$, and so on. These are substantially the same periodicities that had been discovered chemically by Mendeleev. For instance, electrons that fill up any one of the lines of the table are said to form a closed shell. It is energetically unfavorable for atoms whose electrons just fill closed shells to gain or lose electrons, so these atoms are chemically inert. They are the *noble gases*: there is helium with $Z=2$, neon with $Z=2+8=10$, argon with $Z=10+8=18$, krypton with $Z=18+18=36$, xenon with $Z=36+18=54$, and radon with $Z=54+32=86$. Elements with one electron outside closed shells find it easy to lose that electron, which can move freely through the crystal lattice carrying currents of electricity or of heat. These are the *alkali metals*: lithium with $Z=2+1=3$, sodium with $Z=10+1=11$, potassium with $Z=18+1=19$, and so on. Elements with one electron missing from the highest energy closed shell react strongly in chemical reactions in which they can gain an electron. These are the *halogens*: there is fluorine with $Z=10-1=9$, chlorine with $Z=18-1=17$, bromine with $Z=36-1=35$, and so on.

More generally, if an atom has a few electrons outside closed shells, it has what chemists call a positive valence, equal to that number of extra electrons; if it has a few electrons less than needed to fill closed shells, then it has negative valence, equal to that number of missing electrons. Thus alkali metals have valence $+1$; the so-called alkali earths beryllium, magnesium, calcium, etc. have valence $+2$; the halogens have valence -1; oxygen, sulfur, etc. have valence -2; and so on. The molecules of many simple chemical compounds (not all!) are held together by electrostatic attraction between ions of elements with positive and negative valence that have traded electrons. Since electrons are neither created nor destroyed in chemistry, in such molecules if electrically neutral the total valence must be zero. These include such compounds as salts composed of metal and halogen atoms, like sodium chloride, oxides like calcium oxide, etc. Hydrogen can act as if it has valence $+1$, as in water or ammonia, or valence -1, as in metal hydrides.

Diatomic Molecules

The rotational energy spectrum of molecules like H_2, N_2, O_2, etc. that are composed of two identical atoms is profoundly affected by the bosonic or fermionic nature of the atomic nuclei. The energy required to excite rotational states of molecules is less than the energy required to excite vibrational states by factors of order $(m_e/Am_1)^{1/2}$, and less than the energy required to excite electronic states by even smaller factors, of order m_e/Am_1, so the lowest energy states of molecules are rotational states in which the separations of atomic nuclei and the state of atomic electrons can be regarded as fixed. In such states the wave function of a molecule consisting of two identical atoms is proportional to

$$c_s(\sigma_1, \sigma_2) Y_\ell^m(\hat{n}) \pm c_s(\sigma_2, \sigma_1) Y_\ell^m(-\hat{n}) , \qquad (5.5.10)$$

where $\hat{n}$ is a unit vector in the direction from nucleus 1 to nucleus 2; Y_ℓ^m is the usual spherical harmonic, of the sort discussed in Section 5.2; $c_s(\sigma_1, \sigma_2)$ is a spin wave function that depends on the total spin s of the two nuclei and their individual spins $s_1 = s_2$, as well as on the spin 3-components σ_1 and σ_2, about which more later; and the sign is $+1$ or -1 if the nuclei are bosons or fermions, respectively. The energy of the rotational states with a given ℓ is given in quantum mechanics by replacing $\mathbf{L}^2$ in the classical formula $E = \mathbf{L}^2/2I$ with $\hbar^2 \ell(\ell + 1)$, so that

$$E_\ell = \frac{\hbar^2 \ell(\ell + 1)}{2I} \qquad (5.5.11)$$

with almost no dependence on total spin. Here I is the moment of inertia of the molecule around a line perpendicular to $\hat{n}$ through the center of mass of the molecule. Now, $Y_\ell^m(-\hat{n}) = (-1)^\ell Y_\ell^m(\hat{n})$. Also, the spin wave functions have the important symmetry property

$$c_s(\sigma_2, \sigma_1) = \pm(-1)^s c_s(\sigma_1, \sigma_2) , \qquad (5.5.12)$$

where the sign $\pm$ is $(-1)^{2s_1}$; that is, $+1$ for adding two equal integer spins, and -1 for adding two equal half odd integer spins. (In terms of the Clebsch–Gordan coefficients described in the previous section,

$$c_s(\sigma_1, \sigma_2) = C_{s_1, s_1}(s\ \sigma; \sigma_1 \sigma_2) ,$$

where $\sigma = \sigma_1 + \sigma_2$. Equation (5.4.42) with $j_a = s_1$, $J = s$, $m_a = \sigma_1$, $m_b = \sigma_2$, $M = \sigma$ gives

$$C_{s_1 s_1}(s\ \sigma;\ \sigma_2\, \sigma_1) = (-1)^{s - 2s_1} C_{s_1 s_1}(s\ \sigma;\ \sigma_1\, \sigma_2) ,$$

which is the same as Eq. (5.5.12).) Because of the spin–statistics connection, the $\pm$ sign in Eq. (5.5.12) is the same as in Eq. (5.5.10). We see then that these $\pm$ signs cancel, and the only states in which the wave function does not vanish are therefore those in which

$$(-)^\ell = (-1)^s . \qquad (5.5.13)$$

Either s and ℓ are even, in which case the molecule is distinguished by the prefix *para*, or both are odd, and the prefix is *ortho*. For instance, in H_2 we have parahydrogen, with $s = 0$ and ℓ even, and orthohydrogen, with $s = 1$ and ℓ odd. The degeneracy of the states is then $(2\ell + 1)$ for parahydrogen and $3(2\ell + 1)$ for orthohydrogen.

The forces acting on spins are so weak that radiative transitions do not change s and therefore can only change ℓ by an even number. The dominant transitions are those in which ℓ changes by two units, giving a radiated energy

$$E_{\ell+2} - E_\ell = \frac{\hbar^2}{2I}[(\ell+2)(\ell+3) - \ell(\ell+1)] = \frac{\hbar^2}{2I}[4\ell+6] . \tag{5.5.14}$$

For para molecules the energies (5.5.14) are $3\hbar^2/I$, $7\hbar^2/I$, $11\hbar^2/I$, etc., while for ortho molecules they are $5\hbar^2/I, 9\hbar^2/I$, $13\hbar^2/I$, etc. Observing this pattern of energies, with proportions $3 : 7 : 11 : \cdots$ or $5 : 9 : 13 : \cdots$, it is possible to judge which transitions are in para and which in ortho molecules, even if one does not know the moment of inertia I.

The energy $\hbar^2/2I$ is typically much less than kT, so the abundance of diatomic molecules in a state with given s and ℓ is simply proportional to the degeneracy $(2s+1)(2\ell+1)$. (For instance, for hydrogen $\hbar^2/2I = k \times 45$ K.) The observed transitions are typically between states with $\ell \gg 1$, and the intensity of the radiation emitted is mostly a matter of the number of spin states for ℓ and s even or odd, as follows. If the spin s_1 of each nucleus is an integer, so that they are bosons, then the allowed even values of s are $2s_1,\ 2s_1 - 2,\ \ldots, 0$, and the allowed odd values of s are $2s_1 - 1,\ 2s_1 - 3,\ \ldots, 1$. Hence in this case the total number of spin states for para and ortho molecules is

$$\#_{\text{para}} = \sum_{n=0}^{s_1} 2(2n) + 1 = (s_1+1)(2s_1+1) ,$$

$$\#_{\text{ortho}} = \sum_{n=1}^{s_1} 2(2n-1) + 1 = s_1(2s_1+1) ,$$

and the ratio of the intensities of para and ortho transitions is

$$\frac{\text{para}}{\text{ortho}} = \frac{s_1+1}{s_1} \quad \text{(bosons)} . \tag{5.5.15}$$

On the other hand, if s_1 is half an odd integer, so that the nuclei are fermions, then the allowed even values of s are $2s_1 - 1, 2s_1 - 3, \ldots, 0$ and the allowed odd values of s are $2s_1, 2s_1 - 2, \ldots, 1$. Hence for s_1 a half odd integer the total number of spin states for ortho and para molecules is the same as the number of spin states for para and ortho molecules in the case where s_1 is an integer and the ratio of the intensities of para and ortho molecules is the reciprocal of the ratio (5.5.15):

$$\frac{\text{para}}{\text{ortho}} = \frac{s_1}{s_1+1} \quad \text{(fermions)} . \tag{5.5.16}$$

(For example, for hydrogen $s_1 = 1/2$, so the abundance of parahydrogen is about one-third that of orthohydrogen, and the total intensity of radiation emitted or absorbed in transitions in parahydrogen is about one-third the ratio for orthohydrogen.) Evidently one can tell whether nuclei are bosons or fermions just by observing whether radiation from the para or ortho transitions is stronger. In the next chapter we will see that observations of the diatomic nitrogen molecule presented a puzzle regarding the nature of the nitrogen nucleus that was only resolved with the discovery of the neutron.

Clouds of interstellar diatomic molecules can cool to quite low temperatures by collisional excitation of rotational energy levels, after which the excitation energy is emitted as radiation that leaves the cloud. This is an important feature in the formation of stars by gravitational condensation of interstellar matter, which requires low temperatures to mitigate pressure forces that can prevent condensation. But for cooling, it is necessary that radiation should often be emitted before the molecule gives its excitation energy back to the cloud in another collision.

This is an obstacle to cooling by diatomic molecules with identical atoms. As discussed in Section 7.5, the fastest radiative transitions in atoms and molecules generally are *electric dipole* transitions, in which there is a non-zero value for $\int \psi^*_{\rm final}\mathbf{P}\psi_{\rm initial}$ (where **P** is the momentum of the radiating particle). Since **P** is a three-vector that changes sign under reflection of coordinates, this integral vanishes unless certain selection rules are obeyed: when spin effects are neglected, the initial and final states must have opposite signs for the parity $(-)^\ell$ and must have values of ℓ that differ by no more than one unit. Neither selection rule is satisfied by the transitions in diatomic molecules with identical atoms, in which ℓ changes by two units. These are what in Section 7.5 are called *electric quadrupole* transitions, which are much slower than electric dipole transitions. Thus, although H_2 is by far the most common molecule in interstellar space, it contributes little to the cooling of molecular clouds.

On the other hand, in diatomic molecules with distinguishable atoms radiative transitions can occur rapidly as electric dipole transitions in which ℓ changes by one unit, and these molecules when excited by collisions often lose energy by radiation rather than in further collisions. Of the more abundant molecules of this sort, the most effective at cooling interstellar clouds is CO. This molecule has a large moment of inertia, with $\hbar^2/Ik \simeq 5.5$ K, so it can cool clouds to very low temperatures. The hydroxyl molecule OH is more abundant but has a smaller moment of inertia and hence larger excitation energies, so it cannot cool clouds to temperatures as low as can CO.

5.6 Scattering

Much of atomic, nuclear, and elementary particle physics is based on data gained from the scattering of particles in collisions with other particles. In the main body of this section we will consider scattering processes only in the case that is simplest kinematically: the scattering of a particle by a much heavier particle, such as the scattering of alpha particles by nuclei of various metals in the 1911 experiment that led Rutherford to the discovery of the atomic nucleus. In this case we can approximate the effect of the heavy target particle by taking it to be at rest at the origin of coordinates, and representing its interaction with the scattered particle as a fixed external potential $V(\mathbf{x})$ that depends only on the coordinate of the scattered particle. Not only is this a good approximation for some scattering processes of historical importance – as we shall see, it was the study of scattering using this approximation that led to the probabilistic interpretation of quantum mechanics. An appendix to this section considers the calculation of more general scattering and decay processes with any number of particles of any type in the initial and final states.

Scattering Wave Function

We again use the time-independent Schrödinger equation

$$-\frac{\hbar^2}{2m}\nabla^2\psi(\mathbf{x}) + V(\mathbf{x})\psi(\mathbf{x}) = E\psi(\mathbf{x}) \,, \tag{5.6.1}$$

where $V(\mathbf{x}) \to 0$ for $|\mathbf{x}| \to \infty$. But now instead of treating bound states with $E < 0$, we here consider a particle with $E > 0$ that comes into the range of the potential from an infinite distance and then recedes to infinity. We define a wave number $k > 0$ by

$$E = \frac{\hbar^2 k^2}{2m} \,, \tag{5.6.2}$$

and we rewrite the Schrödinger equation as

$$(\nabla^2 + k^2)\psi(\mathbf{x}) = \frac{2m}{\hbar^2} V(\mathbf{x})\psi(\mathbf{x}) \,. \tag{5.6.3}$$

When $\mathbf{x}$ is far outside the range of the potential there is an asymptotic solution of Eq. (5.6.3) that approaches a plane wave $\exp(ikx_3)/(2\pi\hbar)^{3/2}$ (conveniently normalized), which represents a particle coming in from infinity along the 3-axis. We seek a solution of Eq. (5.6.3) with this asymptotic form:

$$\psi(\mathbf{x}) \to \frac{e^{ikx_3}}{(2\pi\hbar)^{3/2}} \tag{5.6.4}$$

for $|\mathbf{x}| \to \infty$. To find such a solution, we replace Eq. (5.6.3) with an integral equation that incorporates the boundary condition (5.6.4):

$$\psi(\mathbf{x}) = \frac{e^{ikx_3}}{(2\pi\hbar)^{3/2}} + \frac{2m}{\hbar^2}\int d^3x'\, G_k(\mathbf{x}-\mathbf{x}')V(\mathbf{x}')\psi(\mathbf{x}') \,, \tag{5.6.5}$$

where $G_k(\mathbf{x}-\mathbf{x}')$ is a Green's function (named after the nineteenth century mathematician George Green (1793–1841)) satisfying the conditions

$$(\nabla^2 + k^2)G_k(\mathbf{x}-\mathbf{x}') = \delta^3(\mathbf{x}-\mathbf{x}')\,, \tag{5.6.6}$$

$$G_k(\mathbf{x}-\mathbf{x}') \to 0 \ \text{ for } \ |\mathbf{x}-\mathbf{x}'| \to \infty\,. \tag{5.6.7}$$

Here $\delta^3(\mathbf{x}-\mathbf{x}')$ is the Dirac delta function, defined by the condition

$$\int d^3x'\; \delta^3(\mathbf{x}-\mathbf{x}')\, f(\mathbf{x}') = f(\mathbf{x}) \tag{5.6.8}$$

for any sufficiently smooth function $f(\mathbf{x})$.

Representations of the Delta Function

Of course, there is no function for which Eq. (5.6.8) is literally satisfied but, by taking $\delta^3(\mathbf{x}-\mathbf{x}')$ to be very large when $\mathbf{x}$ is very close to $\mathbf{x}'$ and very small otherwise, we can come arbitrarily near to satisfying Eq. (5.6.8). For example, we can take

$$\delta^3(\mathbf{x}-\mathbf{x}') = \frac{1}{(\pi d)^3}\exp\big(-(\mathbf{x}-\mathbf{x}')^2/d^2\big)\,,$$

where d is some very small length. It is more convenient here to use another well-known representation of the delta function:

$$\delta^3(\mathbf{x}-\mathbf{x}') = \frac{1}{(2\pi)^3}\int d^3q\; \exp\big(\mathbf{q}\cdot(\mathbf{x}-\mathbf{x}')\big)\,. \tag{5.6.9}$$

With this representation, Eq. (5.6.8) is the fundamental theorem of Fourier analysis: if

$$g(\mathbf{q}) = \int d^3x' e^{-i\mathbf{q}\cdot\mathbf{x}'} f(\mathbf{x}')$$

then

$$f(\mathbf{x}) = \frac{1}{(2\pi)^3}\int d^3q\; e^{i\mathbf{q}\cdot\mathbf{x}}\, g(\mathbf{q}) = \frac{1}{(2\pi)^3}\int d^3q\; e^{i\mathbf{q}\cdot\mathbf{x}}\int d^3x' e^{-i\mathbf{q}\cdot\mathbf{x}'} f(\mathbf{x}')\,,$$

which with an interchange of the order of integration (discarding mathematical rigor) is the same as Eq. (5.6.8). If the wave function $\varphi_{\mathbf{p}}(\mathbf{x})$ for a free particle of momentum $\mathbf{p}$ is defined so that

$$\varphi_{\mathbf{p}}(\mathbf{x}) = (2\pi\hbar)^{-3/2}\exp(i\mathbf{p}\cdot\mathbf{x}/\hbar)$$

then Eq. (5.6.9) gives these wave functions a simple delta-function normalization

$$\int d^3x\; \varphi^*_{\mathbf{p}'}(\mathbf{x})\varphi_{\mathbf{p}}(\mathbf{x}) = \delta^3(\mathbf{p}'-\mathbf{p})\,,$$

which is why we inserted a denominator $(2\pi\hbar)^{3/2}$ in Eq. (5.6.4). As we shall see, we can derive valid results by manipulating $\delta^3(\mathbf{x}-\mathbf{x}')$ as if it were a well-defined function.

Calculation of the Green's Function

Using Eq. (5.6.9), we can easily write a solution of the differential equation (5.6.6):

$$G_k(\mathbf{x}-\mathbf{x}') = \frac{1}{(2\pi)^3}\int d^3q\,\frac{\exp\left(i\mathbf{q}\cdot(\mathbf{x}-\mathbf{x}')\right)}{k^2-\mathbf{q}^2+i\epsilon} \tag{5.6.10}$$

where ϵ is a positive infinitesimal that makes the integral well-defined despite the singularity at $|\mathbf{q}|=k$. (The reason for taking ϵ positive will be made clear below.)

The integral over the directions of $\mathbf{q}$ in Eq. (5.6.10) gives

$$\begin{aligned} G_k(\mathbf{x}-\mathbf{x}') &= \frac{4\pi}{(2\pi)^3}\int_0^\infty q^2\,dq\,\frac{\sin q|\mathbf{x}-\mathbf{x}'|}{q|\mathbf{x}-\mathbf{x}'|}\,\frac{1}{k^2-q^2+i\epsilon} \\ &= \frac{4\pi}{(2\pi)^3}\frac{1}{2i|\mathbf{x}-\mathbf{x}'|}\int_{-\infty}^\infty q\,dq\,\frac{\exp\left(iq|\mathbf{x}-\mathbf{x}'|\right)}{k^2-q^2+i\epsilon}\ . \end{aligned} \tag{5.6.11}$$

We can evaluate the integral over q by closing the contour with a very large semicircle in the upper half of the complex plane, on which the integrand is exponentially small. Since ϵ is infinitesimal, we can write

$$\frac{1}{k^2-q^2+i\epsilon} = \frac{1}{k+i\epsilon/2k-q}\,\frac{1}{k+i\epsilon/2k+q}$$

and evaluate the integral as $2i\pi$ times the residue of the pole inside the contour, at $q=k+i\epsilon/2k$, and then take $\epsilon\to 0$:

$$G_k(\mathbf{x}-\mathbf{x}') = -\frac{1}{4\pi}\frac{1}{|\mathbf{x}-\mathbf{x}'|}\exp\left(ik|\mathbf{x}-\mathbf{x}'|\right)\ , \tag{5.6.12}$$

so that $G_k(\mathbf{x}-\mathbf{x}')$ satisfies the boundary condition (5.6.7).

The Scattering Amplitude

Using Eq. (5.6.12) in Eq. (5.6.5) gives

$$\psi(\mathbf{x}) = \frac{e^{ikx_3}}{(2\pi\hbar)^3} - \frac{1}{2\pi}\frac{m}{\hbar^2}\int d^3x'\,\frac{1}{|\mathbf{x}-\mathbf{x}'|}\exp\left(ik|\mathbf{x}-\mathbf{x}'|\right)V(\mathbf{x}')\psi(\mathbf{x}')\ . \tag{5.6.13}$$

In the limit when $|\mathbf{x}|$ is much larger than the values of $\mathbf{x}'$ at which $V(\mathbf{x}')$ is appreciable, we can use the approximation

$$|\mathbf{x} - \mathbf{x}'| \to r\sqrt{1 - 2\mathbf{x}\cdot\mathbf{x}'/r^2} \to r - \hat{x}\cdot\mathbf{x}'$$

where $r = |\mathbf{x}|$ and $\hat{x} = \mathbf{x}/r$. This gives

$$\psi(\mathbf{x}) \to \frac{1}{(2\pi\hbar)^{3/2}}\left[e^{ikx_3} + \frac{e^{ikr}}{r}f(\hat{x})\right] \tag{5.6.14}$$

where $f(\hat{x})$ is the *scattering amplitude*

$$f(\hat{x}) = -\frac{(2\pi\hbar)^{3/2}}{2\pi}\frac{m}{\hbar^2}\int d^3x'\,\exp(-ik\hat{x}\cdot\mathbf{x}')V(\mathbf{x}')\psi(\mathbf{x}')\,. \tag{5.6.15}$$

Probabilistic Interpretation

At a distance r from the scattering center that is not only large compared with the range of the potential but also much greater than the wavelength $2\pi/k$, the second term in Eq. (5.6.14) at any given direction $\hat{x}$ behaves like a plane wave moving outward with wave vector $k\hat{x}$. This is a familiar behavior for all sorts of waves. A plane ocean wave encountering an obstacle in the water will break up and spread out in all directions, just as in Eq. (5.6.14). But a particle like an alpha particle in Rutherford's laboratory encountering a target like a gold nucleus does not break up. It hangs together, and is scattered in some definite direction, though not a direction that can be predicted in advance. This showed that $\psi(\mathbf{x})$ or $|\psi(\mathbf{x})|^2$ cannot represent how much of the scattered particle is at $\mathbf{x}$. It was this remark about scattering that led Max Born (1882–1970) in 1926 to propose[23] that if ψ is suitably normalized then $|\psi(\mathbf{x})|^2$ is the probability density at $\mathbf{x}$ – that is, $|\psi(\mathbf{x})|^2 d^3x$ is the probability that the particle is in a small volume d^3x around $\mathbf{x}$.

For a proper treatment of what happens in scattering it is necessary to consider a wave function that at early times is a packet of free-particle waves, as in Eq. (5.1.3), and use the time-dependent Schrödinger equation to follow the subsequent scattering. This is the approach followed in the appendix to this section. But, with a moderate amount of hand-waving, we can derive the most important results more simply, just using Eq. (5.6.14).

Suppose that at some early time before the scattering the incoming particle is in a thin disk of area A and thickness L at right angles to the path of the particle. In order for $|\psi|^2$ to serve as a probability density, we have to arrange that e^{ikx_3} comes with a factor $1/\sqrt{AL}$ instead of $1/(2\pi\hbar)^{3/2}$, so that the integral of $|\psi|^2$ over the disk at early times is unity. The scattering wave function (5.6.14) will then also be multiplied by $(2\pi\hbar)^{3/2}/\sqrt{AL}$. At a late time t after the collisions a scattered particle will be in a thin disk of the same thickness L at a distance

[23] M. Born, Z. Phys. **38**, 803 (1926).

$r = vt$ from the scattering center (where $v = \sqrt{2E/m}$). The probability density at position $r\hat{x}$ will be

$$|\psi(r\hat{x})|^2 = \frac{|f(\hat{x})|^2/r^2}{AL}$$

and the probability dP that the particle will be in a small solid angle $d\Omega$ around $\hat{x}$ is this probability density times the volume of a disk of thickness L and area $r^2\, d\Omega$:

$$dP = \frac{|f(\hat{x})|^2/r^2}{AL} \times Lr^2 d\Omega = |f(\hat{x})|^2\, d\Omega/A\,. \tag{5.6.16}$$

This is the same probability as if the particle by chance had to hit a tiny target area $d\sigma = |f(\hat{x})|^2 d\Omega$ somewhere within the larger area A in order to be scattered into a solid angle $d\Omega$ around $\hat{x}$. The ratio of the target area $d\sigma$ to the solid angle $d\Omega$ is then

$$\frac{d\sigma}{d\Omega} = |f(\hat{x})|^2 = \frac{2\pi m^2}{\hbar}\left|\int d^3x' \exp(-ik\hat{x}\cdot\mathbf{x}')V(\mathbf{x}')\psi(\mathbf{x}')\right|^2 \tag{5.6.17}$$

and is known as the *differential cross section*. Much of modern theoretical and experimental physics consists of the calculation and measurement of differential cross sections.

Now we can see why it was necessary to take ϵ positive-definite in Eq. (5.6.10) for the Green's function. With ϵ negative-definite the integral (5.6.11) over q would still be well-defined and we could still evaluate it by closing the contour of integration with a large semicircle in the upper half complex plane of q, on which the factor $\exp(iq|\mathbf{x}-\mathbf{x}'|)$ is exponentially small. Only now, with ϵ negative, the pole in the integrand in the upper half of the complex plane would be at $q = -k - i\epsilon/2k$, and in the asymptotic form of the wave function the factor $\exp(ikr)$ would be replaced with $\exp(-ikr)$. Instead of a wave going out in all directions to large distances, as in Eq. (5.6.14), this would represent a wave coming *in*to the potential along all directions from a great distance, which is not what happens in any scattering process.

The Born Approximation

Equation (5.6.5) is of course not in itself a solution of the differential equation (5.6.3), because ψ appears on the right-hand side of the equation, as well as on the left. But it does suggest a solution that is a good approximation if $2m|V|/\hbar^2$ is everywhere much less than k^2. In this case we can approximate ψ on the right-hand side of Eq. (5.6.5) with the term of zeroth order in V, that is, with $e^{ikx_3}/(2\pi\hbar)^{3/2}$:

$$\psi(\mathbf{x}) \simeq \frac{1}{(2\pi\hbar)^{3/2}}\left[e^{ikx_3} + \frac{2m}{\hbar^2}\int d^3x' G_k(\mathbf{x}-\mathbf{x}')V(\mathbf{x}')e^{ikx_3'}\right]. \tag{5.6.18}$$

This is known as the Born approximation. Repeating our earlier calculation of the scattering amplitude, or just jumping back to Eq. (5.6.15) and replacing $\psi(\mathbf{x}')$ with $e^{ikx'_3}/(2\pi\hbar)^{3/2}$, gives the corresponding approximation for the scattering amplitude:

$$f(\hat{x}) \simeq -\frac{m}{2\pi\hbar^2}\int d^3x' \exp(-ik\hat{x}\cdot\mathbf{x}')V(\mathbf{x}')\,e^{ikx'_3}\ . \tag{5.6.19}$$

This formula becomes particularly simple in the frequently encountered case in which the potential is spherically symmetric. We can write Eq. (5.6.19) in this case as

$$f(\hat{x}) \simeq -\frac{m}{2\pi\hbar^2}\int d^3x' \exp(i\mathbf{K}\cdot\mathbf{x}')V(|\mathbf{x}'|)\ ,$$

where

$$\mathbf{K} = k(\hat{z}-\hat{x})\ .$$

Here $\hat{z}$ is a unit vector in the 3-direction, the direction of the original particle velocity. The integral over the direction of $\mathbf{x}'$ is then easy:

$$f(\hat{x}) \simeq -\frac{m}{2\pi\hbar^2}\int_0^\infty 4\pi r'^2\,\frac{\sin Kr'}{Kr'}\,V(r') = -\frac{2m}{K\hbar^2}\int_0^\infty V(r')\,\sin(Kr')\,r'\,dr'\ , \tag{5.6.20}$$

where

$$K \equiv |\mathbf{K}| = k\sqrt{2-2\cos\theta} = k\sqrt{2-2[1-2\sin^2(\theta/2)]} = 2k\,\sin(\theta/2)\ , \tag{5.6.21}$$

and θ is the angle between the incident direction $\hat{z}$ and the scattered direction $\hat{x}$. It is a special feature of the Born approximation for spherically symmetric potentials that the scattering amplitude depends on k and θ only in the combination K.

Coulomb Scattering

For an important example of the Born approximation, consider a shielded Coulomb potential

$$V(r) = \frac{Z_1Z_2e^2}{r}\exp(-\kappa r)\ . \tag{5.6.22}$$

This is a rough approximation to the Coulomb energy of a scattered particle of charge Z_2e in the electric field of an atom whose nucleus has charge Z_1e. The full electrostatic potential of the nucleus is felt by the scattered particle when the particle is closer to the nucleus than the electronic orbits, taken to have typical radii of order $1/\kappa$, but the potential vanishes when the scattered particle is far enough from the atom for the orbiting electrons to completely shield the charge of the nucleus. (This potential is also known as a Yukawa potential, because, as we will see in Section 7.3, in 1935 Hideki Yukawa(1907–1981) showed that the exchange of a meson of mass $\hbar\kappa/c$ between two nuclear particles would produce such a potential, though of course with some other constant factor in place of $Z_1 Z_2 e^2$.) Using this in Eq. (5.6.20) gives a scattering amplitude

$$f(\hat{x}) \simeq -\frac{2mZ_1Z_2e^2}{\hbar^2}\frac{1}{K^2+\kappa^2}\ , \tag{5.6.23}$$

with K given by Eq. (5.6.21). We can find the scattering amplitude for a pure Coulomb potential by just taking $\kappa = 0$ in Eq. (5.6.23). This result is only valid to first order in $Z_1 Z_2 e^2$, but a calculation of higher-order corrections shows that for $\kappa = 0$ these higher-order corrections change the scattering amplitude only by a phase factor, which has no effect on the differential cross section (5.6.17), so in this case

$$\frac{d\sigma}{d\Omega} = \frac{4m^2 Z_1^2 Z_2^2 e^4}{\hbar^4 K^4}\ , \tag{5.6.24}$$

which holds even beyond the Born approximation. This is the same as the formula calculated classically in 1911 by Rutherford, following the hyperbolic trajectory of the alpha particle to find the area $d\sigma$ that it must hit to reach a given direction within a solid angle $d\Omega$. Rutherford's calculation would not have given the correct scattering probability for a general potential, except at very short wavelength. It was just good luck that for Coulomb scattering the classical calculation gives the right answer for general wavelengths.

Appendix: General Transition Rates

So far we have considered only the scattering of a single non-relativistic particle by a fixed scattering center. Nature presents us with a much wider variety of processes, in which any number of particles coming together from large separations in an initial state interact, producing some number of particles (not necessarily the same number) that then go out to large separations in a final state. These processes range from the decay of a single particle to the collision of any number of relativistic or non-relativistic particles, producing any other particles. This appendix describes a very general formalism for the calculation of the rates of all such processes.

We consider a Hamiltonian of the general form

$$H = H_0 + V \tag{5.6.25}$$

in which the two terms are distinguished by the condition that the eigenfunctions of H_0 represent states of free particles, such as those that are present long before or long after a collision, while V is an interaction that becomes negligible when these particles are very far apart. For instance, for non-relativistic processes H_0 is the operator representing the total kinetic energy. The eigenfunctions φ_α of H_0 satisfy

$$H_0\varphi_\alpha = E_\alpha\varphi_\alpha \,. \tag{5.6.26}$$

Here α labels the species, three-momenta, and spin z-components (or helicities) of all the particles in the state α represented by φ_α, and E_α is the sum of the kinetic plus mass energies of these particles. These wave functions can be normalized so that

$$\int \varphi_\beta^*\varphi_\alpha = \delta(\beta-\alpha) \,, \tag{5.6.27}$$

with the understanding that $\delta(\beta-\alpha)$ vanishes unless the numbers of particles in the states α and β and the species and spin components of the corresponding particles in these states are all equal, and where they are equal it is given by a product of Dirac delta functions for the three-momentum of each particle. (In Eq. (5.6.27) we continue to use the abbreviation, that in $\int \varphi_\beta^*\varphi_\alpha$ we integrate over all coordinates and sum over all spin 3-components on which both wave functions depend.) To be explicit, for wave functions representing free-particle states containing respectively N and N' particles, we have

$$\begin{aligned}\int \varphi^*_{n'_1,\sigma'_1,\mathbf{p}'_1;\ldots;n'_{N'},\sigma'_{N'},\mathbf{p}'_{N'}}\,\varphi_{n_1,\sigma_1,\mathbf{p}_1;\ldots;n_N,\sigma_N,\mathbf{p}_N} \\ = \delta_{N'N}\delta_{\sigma'_1,\sigma_1}\cdots\delta_{\sigma'_N,\sigma_N}\times\delta_{n'_1n_1}\cdots\delta_{n'_Nn_N} \\ \times\,\delta^3(\mathbf{p}'_1-\mathbf{p}_1)\cdots\delta^3(\mathbf{p}'_N-\mathbf{p}_N)\,,\end{aligned}$$

with the ns labeling species and the σs labeling spin z-components or helicities. (For identical bosons or fermions it is necessary to respectively symmetrize or antisymmetrize the products on the right-hand side.) We seek to calculate the probability that the interaction V will cause a state that looks at very early times like the free-particle state α to look at very late times like some other free-particle state β.

To pursue this calculation, we consider an eigenfunction ψ_α of the full Hamiltonian (5.6.25) with energy E_α:

$$H\psi_\alpha = E_\alpha\psi_\alpha \,. \tag{5.6.28}$$

We can incorporate this condition along with our initial condition in what is known as the *Lippmann–Schwinger equation*[24]:

$$\psi_\alpha = \varphi_\alpha + (E_\alpha - H_0 + i\epsilon)^{-1}V\psi_\alpha \,, \tag{5.6.29}$$

[24] B. Lippmann and J. Schwinger, Phys. Rev. **79**, 469 (1950).

with ϵ a positive-definite infinitesimal quantity that makes $(E_\alpha - H_0 + i\epsilon)^{-1}$ well-defined even though E_α is within the spectrum of eigenvalues of H_0. (The general reason for taking ϵ positive will be revealed shortly.) Multiplying Eq. (5.6.29) with the operator $E_\alpha - H_0$ and using Eq. (5.6.26), we see that $(E_\alpha - H_0)\psi_\alpha = V\psi_\alpha$, so any ψ_α that satisfies Eq. (5.6.29) also satisfies Eq. (5.6.28).

To check the initial condition, we need to consider the time dependence of a packet of the wave functions ψ_α. If we expand $V\psi_\alpha$ as an integral over free-particle wave functions φ_β and use Eqs. (5.6.26) and (5.6.27), Eq. (5.6.29) becomes

$$\psi_\alpha = \varphi_\alpha + \int d\beta \, \frac{\int \varphi_\beta^* \, V\psi_\alpha}{E_\alpha - E_\beta + i\epsilon} \, \varphi_\beta \,, \tag{5.6.30}$$

where the integral over β includes an integration over all three-momenta in the state represented by φ_β and a sum over all species and spin labels on the particles in this state. The time dependence of a packet of these wave functions is given in the Schrödinger picture by

$$\begin{aligned} \psi^{(g)}(t) &= e^{-iHt/\hbar} \int g(\alpha)\psi_\alpha \, d\alpha = \int g(\alpha) e^{-iE_\alpha t/\hbar} \psi_\alpha \, d\alpha \\ &= \int g(\alpha) e^{-iE_\alpha t/\hbar} \varphi_\alpha \, d\alpha + \int d\beta \int g(\alpha) \, d\alpha \; e^{-iE_\alpha t/\hbar} \, \frac{\int \varphi_\beta^* \, V\psi_\alpha}{E_\alpha - E_\beta + i\epsilon} \, \varphi_\beta \end{aligned} \tag{5.6.31}$$

where $g(\alpha)$ is some smooth function of the momenta that may also depend on the spin and species labels. It will be convenient to separate an integral over energy from the second integral over α, writing Eq. (5.6.31) as

$$\psi^{(g)}(t) = \int g(\alpha) e^{-iE_\alpha t/\hbar} \varphi_\alpha \, d\alpha + \int d\beta \; \varphi_\beta \int_{-\infty}^{+\infty} dE \; e^{-iEt/\hbar} \, \frac{G_\beta(E)}{E - E_\beta + i\epsilon} \,, \tag{5.6.32}$$

where

$$G_\beta(E) = \int d\alpha \; g(\alpha) \, \delta(E_\alpha - E) \int \varphi_\beta^* \, V\psi_\alpha \,. \tag{5.6.33}$$

Now let us take $t \to -\infty$. For $t < 0$ we can close the contour of integration over E in Eq. (5.6.32) with a very large semicircle in the upper half of the complex plane, on which the factor $e^{-iEt/\hbar}$ makes the integrand negligible. The integral over E is then given by a sum of the residues of any singularities of the integrand in the upper half of the complex plane. There may well be such singularities, but for $t \to -\infty$ their residues are exponentially suppressed by the same factor $e^{-iEt/\hbar}$. A singularity infinitesimally above the real axis would not be suppressed in this way, but the energy at which the denominator

$E - E_\beta + i\epsilon$ vanishes is just *below* the real axis, and so does not contribute to this contour integral. (This reveals why we took ϵ to be positive.) Hence the integral over E vanishes for $t \to -\infty$, so for very early times only the first term in Eq. (5.6.32) survives:

$$\psi^{(g)}(t) \to \int g(\alpha) e^{-iE_\alpha t/\hbar} \varphi_\alpha \, d\alpha \; . \tag{5.6.34}$$

This is what we mean when we say that at very early times the state represented by ψ_α looks like the free-particle state represented by φ_α, as was to be shown.

What does this state look like at very late times? For $t > 0$ we can only close the contour of integration over E with a very large contour in the *lower* half of the complex plane, on which the factor $e^{-iEt/\hbar}$ is now negligible. The residues of any singularities of $G_\beta(E)$ at a finite distance below the real axis are exponentially suppressed for $t \to +\infty$ by the same factor. But now the singularity at $E = E_\beta - i\epsilon$ does contribute to the integral. The contour of integration goes clockwise around this singularity, so this integral equals $-2\pi i G_\beta(E_\beta - i\epsilon) \exp([-iE_\beta - \epsilon]t/\hbar)$. As long as we take $\epsilon \to 0$ before we take $t \to +\infty$, we can drop the ϵ here, so the integral over E in Eq, (5.6.32) equals $-2\pi i G_\beta(E_\beta) \exp(-iE_\beta t/\hbar)$, and Eq. (5.6.32) then gives

$$\psi^{(g)}(t) \to \int g(\alpha) e^{-iE_\alpha t/\hbar} \varphi_\alpha \, d\alpha - 2\pi i \int d\beta \; G_\beta(E_\beta) \varphi_\beta \; \exp(-iE_\beta t/\hbar)$$

for $t \to +\infty$. Using Eq. (5.6.33), this is

$$\psi^{(g)}(t) \to \int g(\alpha) \, d\alpha \int d\beta \; S_{\beta\alpha} \exp(-iE_\beta t/\hbar) \varphi_\beta \; , \tag{5.6.35}$$

where

$$S_{\beta\alpha} = \delta(\beta - \alpha) - 2\pi i \delta(E_\beta - E_\alpha) \int \varphi_\beta^* \, V \psi_\alpha \; . \tag{5.6.36}$$

So, in the same sense as in the case $t \to -\infty$, Eq. (5.6.35) shows that the state represented by ψ_α looks at $t \to +\infty$ as a superposition $\int d\beta \; S_{\beta\alpha} \varphi_\beta$. The coefficient (5.6.36) is known as the *S-matrix* and is the central object of study in modern scattering theory.

But experiments do not measure probability amplitudes. They measure *probabilities*, or the rates at which probabilities change. However, we cannot just set the probability for the transition $\alpha \to \beta$ equal to $|S_{\beta\alpha}|^2$. Even if we consider a process for which $\alpha \neq \beta$, so that we can drop the term $\delta(\beta - \alpha)$ in $S_{\beta\alpha}$, the S-matrix element will still be proportional to the energy-conservation delta function $\delta(E_\beta - E_\alpha)$, whose square is not well-defined. Also, in the most common case, where no external fields affect the transition $\alpha \to \beta$, momentum is conserved, so

$$\int \varphi^*_\beta \, V \psi_\alpha = \delta^3(\mathbf{P}_\beta - \mathbf{P}_\alpha) M_{\beta\alpha} \; ; \tag{5.6.37}$$

where $\mathbf{P}$ here denotes the total momentum of the state and $M_{\beta\alpha}$ is some amplitude that is not singular when $\mathbf{P}_\beta = \mathbf{P}_\alpha$. So we have to worry about the square of $\delta^3(\mathbf{P}_\beta - \mathbf{P}_\alpha)$ as well as the square of $\delta(E_\beta - E_\alpha)$.

For a completely convincing way of dealing with these problems, we would need to take superpositions of states with a range of energies *and* momenta, and follow the evolution of these wave packets from very early to very late times. We will adopt a much simpler approach that gives the right answers with a minimum of trouble.

First, to deal with the inevitable energy-conservation delta function, we adopt the fiction that the interaction V acts only for a long but finite time interval of duration T. This should not introduce significant errors if this interval extends back in time to long before the particles in state α become close to one another, and extends forward in time to long after the particles in state β have been close to one another.[25] In this case, the one-dimensional version of the representation (5.6.9) of the delta function becomes instead

$$\delta_T(E_\beta - E_\alpha) = \frac{1}{2\pi\hbar} \int_T dt \; \exp(-it(E_\beta - E_\alpha)/\hbar) \, , \tag{5.6.38}$$

the integral extending over the time interval of duration T. The square of the delta function is then

$$\left[\delta_T(E_\beta - E_\alpha)\right]^2 = \delta_T(0)\delta_T(E_\beta - E_\alpha) = \left(\frac{T}{2\pi\hbar}\right) \delta_T(E_\beta - E_\alpha) \, .$$

As long as we do not attempt to measure energies to an uncertainty less than the tiny amount $\hbar/T$, we can drop the subscript T on the final delta function, and write this as

$$\left[\delta_T(E_\beta - E_\alpha)\right]^2 = \left(\frac{T}{2\pi\hbar}\right) \delta(E_\beta - E_\alpha) \, . \tag{5.6.39}$$

Likewise, in the absence of external fields momentum is conserved; to deal with the momentum-conservation delta function we imagine that the system is enclosed in a box of large but finite volume V. The representation (5.6.9) of the momentum-conservation delta function in Eq. (5.6.37) (now with momentum and position taking the place of position and wave vector) is then replaced with

[25] For a decay process with a single-particle initial state we must take the duration T of the time interval sufficiently large that the interval extends back in time close enough to the time when the particle was produced, so that it had not yet had time to decay, and far enough forward in time that if the particle has decayed by then its decay products will have had time to separate far enough that they are no longer interacting.

$$\delta_V^3(\mathbf{P}_\beta - \mathbf{P}_\alpha) = \frac{1}{(2\pi\hbar)^3}\int_V d^3x \, \exp\left(i\mathbf{x}\cdot(\mathbf{P}_\beta - \mathbf{P}_\alpha)/\hbar\right) , \tag{5.6.40}$$

the integral running over the interior of the box. The square of this delta function is

$$\left[\delta_V^3(\mathbf{P}_\beta - \mathbf{P}_\alpha)\right]^2 = \delta_V^3(0)\delta_V^3(\mathbf{P}_\beta - \mathbf{P}_\alpha) = \frac{V}{(2\pi\hbar)^3}\delta^3(\mathbf{P}_\beta - \mathbf{P}_\alpha) , \tag{5.6.41}$$

in which we drop the subscript V in the final expression because the uncertainty in measurements of momenta is generally larger than the tiny amount $\hbar V^{-1/3}$. Hence, putting together Eqs. (5.6.36), (5.6.37), (5.6.39), and (5.6.41), the probability of a transition $\alpha \to \beta$ with $\alpha \neq \beta$ occurring in a time T in the volume V is

$$\begin{aligned} P(\alpha \to \beta) &= \left|S^{\text{box}}_{\beta\alpha}\right|^2 \\ &= \left(\frac{T}{2\pi\hbar}\right)\left(\frac{V}{(2\pi\hbar)^3}\right)\delta(E_\beta - E_\alpha)\delta^3(\mathbf{P}_\beta - \mathbf{P}_\alpha)\left|2\pi M^{\text{box}}_{\beta\alpha}\right|^2 . \end{aligned} \tag{5.6.42}$$

A superscript "box" has been attached to the matrix elements $S_{\beta\alpha}$ and $M_{\beta\alpha}$ because putting the system in a box changes the way that we must normalize the wave functions φ_α and φ_β. Without a box, the wave function for a particle of momentum $\mathbf{p}$ far from any interaction is taken as $\varphi_\mathbf{p}(\mathbf{x}) = \exp(i\mathbf{p}\cdot\mathbf{x}/\hbar)/(2\pi\hbar)^{3/2}$, so that $\int d^3x\varphi^*_{\mathbf{p}'}(\mathbf{x})\varphi_\mathbf{p}(\mathbf{x}) = \delta^3(\mathbf{p}-\mathbf{p}')$, but in a box of volume V we must instead take $\varphi_\mathbf{p}(\mathbf{x}) = \exp(i\mathbf{p}\cdot\mathbf{x})/\sqrt{V}$, so that the integral of $|\varphi_\mathbf{p}(\mathbf{x})|^2$ over the volume of the box is unity. Thus the matrix element for the transition $\alpha \to \beta$ in a box is related to the usual matrix element by

$$M^{\text{box}}_{\beta\alpha} = \left(\frac{(2\pi\hbar)^3}{V}\right)^{(N_\alpha+N_\beta)/2} M_{\beta\alpha} , \tag{5.6.43}$$

where N_α and N_β are the numbers of particles in the initial and final states. There is a further complication, that in a large box the final states are very close together. According to Eq. (5.5.1), the number of allowed momentum values for a single particle in a range d^3p of momenta is $(V/(2\pi\hbar)^3)d^3p$, so the number of momentum states in the range of final states is

$$d\mathcal{N}(\beta) = (V/(2\pi\hbar)^3)^{N_\beta}d\beta \tag{5.6.44}$$

where $d\beta$ denotes a product of momentum-space volume elements d^3p for each particle in the final state. Using Eqs. (5.6.43) and (5.6.44) in Eq. (5.6.42), the differential rate for transitions from an initial state α into a range $d\beta$ of final states is

$$d\Gamma(\alpha \to \beta) = \frac{|S^{\text{box}}_{\beta\alpha}|^2 d\mathcal{N}(\beta)}{T}$$
$$= \left(\frac{2\pi}{\hbar}\right)\left(\frac{V}{(2\pi\hbar)^3}\right)^{1-N_\alpha} \left|M_{\beta\alpha}\right|^2 \delta(E_\beta - E_\alpha)\delta^3(\mathbf{P}_\beta - \mathbf{P}_\alpha)\, d\beta\ . \tag{5.6.45}$$

This is the master formula for calculating the rates for all sorts of transitions between free-particle states.

The factor $\left(V/(2\pi\hbar)^3\right)^{1-N_\alpha}$ in Eq. (5.6.45) may look peculiar, but it is in fact just what is needed to account for what is measured. For a decay process with $N_\alpha = 1$ this factor is of course absent, corresponding to the obvious fact that the decay rate of a particle does not depend on the size of the box in which it is contained. For a two-particle initial state α, the differential rate of the scattering $\alpha \to \beta$ into an arbitrary final state β is proportional to the flux, the product of the relative velocity u_α and the number density $1/V$ of either particle as seen from the other, and is therefore written as the flux times a differential cross section $d\sigma(\alpha \to \beta)$. (For a pair of non-relativistic particles $u_\alpha = |\mathbf{p}_1/m_1 - \mathbf{p}_2/m_2|$, while if one of the particles is a photon then $u_\alpha = c$.) Hence Eq. (5.6.45) gives

$$d\sigma(\alpha \to \beta) \equiv \frac{d\Gamma(\alpha \to \beta)}{u_\alpha/V} = \frac{(2\pi)^4\hbar^2}{u_\alpha}\left|M_{\beta\alpha}\right|^2 \delta(E_\beta - E_\alpha)\delta^3(\mathbf{P}_\beta - \mathbf{P}_\alpha)\, d\beta\ . \tag{5.6.46}$$

To clarify the meaning of the closing factor $\delta(E_\beta - E_\alpha)\delta^3(\mathbf{P}_\beta - \mathbf{P}_\alpha)\, d\beta$ in Eqs. (5.6.45) and (5.6.46), consider a process $\alpha \to \beta$ in the center-of-mass system, with $\mathbf{P}_\alpha = 0$, where β is a state of two particles with momenta $\mathbf{p}'_A$ and $\mathbf{p}'_B$ and masses m'_A and m'_B. The closing factor in Eq. (5.6.46) is here

$$\delta(E_\beta - E_\alpha)\delta^3(\mathbf{P}_\beta - \mathbf{P}_\alpha)\, d\beta = \delta(E'_A + E'_B - E_\alpha)\delta^3(\mathbf{p}'_A + \mathbf{p}'_B)\, d^3p'_A\, d^3p'_B\ .$$

When we integrate over the final momenta the momentum-conservation delta function directs us to set $\mathbf{p}'_A = -\mathbf{p}'_B \equiv \mathbf{p}$, so

$$\delta(E_\beta - E_\alpha)\delta^3(\mathbf{P}_\beta - \mathbf{P}_\alpha)\, d\beta$$
$$\to p^2\, dp\, d\Omega\, \delta\big((p^2c^2 + m'^2_A c^4)^{1/2} + (p^2c^2 + m'^2_B c^4)^{1/2} - E_\alpha\big)\ ,$$

where $\mathbf{p}$ is in the solid angle $d\Omega$. There is a general rule that since for an arbitrary increasing function $f(p)$ which takes a value f_0 at a single point p_0 we have $1 = \int \delta(f(p) - f_0)\, df(p)$, it follows that

$$\delta(f(p) - f_0) = \delta(p - p_0)/f'(p_0)\ . \tag{5.6.47}$$

In our case, this means that when we integrate over p, we are directed to set $p = p_\beta$, where

$$(p_\beta^2 c^2 + m_A'^2 c^4)^{1/2} + (p_\beta^2 c^2 + m_B'^2 c^4)^{1/2} = E_\alpha \ , \tag{5.6.48}$$

and Eq. (5.6.46) becomes

$$d\sigma(\alpha \to \beta) = \frac{(2\pi)^4 \hbar^2 p_\beta^2}{u_\alpha u_\beta} \left|M_{\beta\alpha}\right|^2 d\Omega \ , \tag{5.6.49}$$

in which it is understood that, in the center-of-mass system, $M_{\beta\alpha}$ is to be evaluated by placing $\mathbf{p}'_A = -\mathbf{p}'_B$ in the infinitesimal solid angle $d\Omega$, with $|\mathbf{p}'_A| = |\mathbf{p}'_B| = p_\beta$, and

$$u_\beta \equiv \frac{p_\beta c^2}{(p_\beta^2 c^2 + m_A'^2 c^4)^{1/2}} + \frac{p_\beta c^2}{(p_\beta^2 c^2 + m_B'^2 c^4)^{1/2}} \ . \tag{5.6.50}$$

Of course in the center-of-mass system the initial relative velocity u_α in Eq. (5.6.47) is given by similar formulas but with β replaced with α and the final masses m'_A and m'_B replaced with initial masses m_A and m_B:

$$u_\alpha \equiv \frac{p_\alpha c^2}{(p_\alpha^2 c^2 + m_A^2 c^4)^{1/2}} + \frac{p_\alpha c^2}{(p_\alpha^2 c^2 + m_B^2 c^4)^{1/2}} \ , \tag{5.6.51}$$

where

$$(p_\alpha^2 c^2 + m_A^2 c^4)^{1/2} + (p_\alpha^2 c^2 + m_B^2 c^4)^{1/2} = E_\alpha \ . \tag{5.6.52}$$

We can now see how our earlier results for scattering by a fixed potential emerge from this general formalism. Consider an elastic non-relativistic scattering process, in which $m'_A = m_A \equiv m$ and $m'_B = m_B \gg m$. In this case $p_\alpha = p_\beta$, $u_\alpha = u_\beta = p_\alpha/m$, and $E_\alpha - m_A c^2 - m_B c^2 = p_\alpha^2/2m$. Equation (5.6.49) then gives the differential cross section

$$\frac{d\sigma(\alpha \to \beta)}{d\Omega} = (2\pi)^4 \hbar^2 m^2 \left|M_{\beta\alpha}\right|^2 \ . \tag{5.6.53}$$

To calculate the matrix element $M_{\beta\alpha}$, we note that the final free-particle wave function is

$$\varphi_\beta(\mathbf{x}_A, \mathbf{x}_B) = \frac{e^{i\mathbf{p}'_A \cdot \mathbf{x}_A/\hbar}}{(2\pi\hbar)^{3/2}} \frac{e^{i\mathbf{p}'_B \cdot \mathbf{x}_B/\hbar}}{(2\pi\hbar)^{3/2}}$$

and in the center-of-mass system the initial interacting wave function takes the form

$$\psi_\alpha(\mathbf{x}_A, \mathbf{x}_B) = \psi(\mathbf{x}_A - \mathbf{x}_B) \times \frac{1}{(2\pi\hbar)^{3/2}} \ ,$$

where ψ is the wave function discussed in the main body of this section (which already includes a normalization factor $(2\pi\hbar)^{-3/2}$), and the second factor takes care of the normalization of the heavy particle wave function. Then, setting $\mathbf{x}_A = \mathbf{x} + \mathbf{x}_B$ and integrating over $\mathbf{x}_B$,

$$\int \varphi_\beta^* V \psi_\alpha \equiv \int \varphi_\beta^*(\mathbf{x}_A, \mathbf{x}_B) V(\mathbf{x}_A - \mathbf{x}_B) \psi_\alpha(\mathbf{x}_A, \mathbf{x}_B)\, d^3x_A\, d^3x_B$$
$$= \delta^3(\mathbf{p}'_A + \mathbf{p}'_B) \int d^3x \frac{e^{-i\mathbf{p}'_A\cdot\mathbf{x}/\hbar}}{(2\pi\hbar)^{3/2}} V(\mathbf{x})\psi(\mathbf{x})$$

so

$$M_{\beta\alpha} = \int d^3x \frac{e^{-i\mathbf{p}'_A\cdot\mathbf{x}/\hbar}}{(2\pi\hbar)^{3/2}} V(\mathbf{x})\psi(\mathbf{x})\,. \tag{5.6.54}$$

Using Eq. (5.6.54) in Eq. (5.6.53) gives the same differential cross section (5.6.17) as found earlier.

It is frequently observed that the cross section for some reaction is a function of energy with a sharp peak. This is a sign of a *resonance*, the formation of a slowly decaying intermediate state in the scattering process. Suppose the integral $\int \psi_\beta^* V \psi_\alpha$ in Eq. (5.6.36) for the S-matrix has a term with an energy dependence proportional to $(E_\alpha - E_R + i\Gamma\hbar/2)^{-1}$, with E_R and Γ real and $\Gamma > 0$. This yields a term in the function $G_\beta(E)$ defined by Eq. (5.6.33) with energy dependence proportional to $(E - E_R + i\hbar\Gamma/2)^{-1}$, which has a pole in the lower half of the complex E plane. Although, as noted in the derivation of Eq. (5.6.35), the contribution of any singularity in $G_\beta(E)$ at an energy E at a finite distance below the real axis vanishes for $t \to +\infty$, if the singularity is close to the real axis then this contribution lasts a long time. So if Γ is relatively small then the integral over E in Eq. (5.6.32) contains a term that decays slowly, with a time dependence proportional to $\exp(-iE_Rt/\hbar)\exp(-\Gamma t/2)$, giving a term in $|\psi^{(g)}(t)|^2$ that decays as $\exp(-\Gamma t)$, indicating the presence of an intermediate state whose probability decays at a rate Γ. The singular term in $\int \psi_\beta^* V \psi_\alpha$ gives a term in the cross section with energy dependence

$$\sigma \propto \left|\frac{1}{E - E_R + i\hbar\Gamma/2}\right|^2 = \frac{1}{(E - E_R)^2 + \hbar^2\Gamma^2/4}\,. \tag{5.6.55}$$

So, this is the general rule for resonances: the decay rate Γ of the intermediate state is the full width in energy of the resonant peak in the cross section at half maximum, divided by $\hbar$.

5.7 Canonical Formalism

Until now we have followed de Broglie in representing the momentum of a particle as $-i\hbar$ times the gradient with respect to the particle's position, so that the wave function representing a state with definite momentum $\mathbf{p}$ is $\propto \exp(i\mathbf{p}\cdot\mathbf{x}/\hbar)$. From this, we obtained the commutation relation among the operators $\mathbf{X}$ and $\mathbf{P}$ that represent position and momentum, for instance, for a single particle,

$$[X_i, P_j] = i\hbar\delta_{ij}\ , \quad [X_i, X_j] = [P_i.P_j] = 0\ , \tag{5.7.1}$$

where in the Heisenberg picture $\mathbf{P} = m\dot{\mathbf{X}}$. This has been adequate in dealing with charged particles moving in an electrostatic potential but not in more complicated contexts, such as the case of charged particles moving in general classical electromagnetic fields, discussed in the next section, much less for a quantum theory of fields. Also, in using commutation relations like Eq. (5.7.1), we must wonder (or at least we *should* wonder) why these relations are valid.

Hamiltonian Formalism

There is a more general approach, known as the *canonical formalism*, according to which the continuous degrees of freedom (excluding spin) of any system are represented by a set of canonical variables Q_a (such as all the components of the positions of all the particles in a system) and an equal number of "canonical conjugates" P_a. Like any operators, in the Heisenberg picture these operators satisfy the equations of motion (5.3.34):

$$i\hbar\frac{d}{dt}Q_a(t) = [Q_a(t), H]\ , \quad i\hbar\frac{d}{dt}P_a(t) = [P_a(t), H]\ , \tag{5.7.2}$$

where $H = H\big(Q(t), P(t)\big)$ is the Hamiltonian of the system. On the basis of previous experience with classical phenomena, we commonly need to require that these equations of motion take the same form as the Hamiltonian equations of motion in classical mechanics:

$$\frac{d}{dt}Q_a(t) = \frac{\partial}{\partial P_a(t)}H\big(Q(t), P(t)\big)\ , \tag{5.7.3}$$

$$\frac{d}{dt}P_a(t) = -\frac{\partial}{\partial Q_a(t)}H\big(Q(t), P(t)\big)\ . \tag{5.7.4}$$

For instance, for a particle of mass m in a potential $V(\mathbf{X})$, the variables Q_a are the components of the position vector $\mathbf{X}$, the Hamiltonian is

$$H(\mathbf{X}, \mathbf{P}) = \frac{\mathbf{P}^2}{2m} + V(\mathbf{X})\ ,$$

and the equations of motion (5.7.3) and (5.7.4) are

$$\frac{d}{dt}\mathbf{X} = \frac{\mathbf{P}}{m}\ , \quad \frac{d}{dt}\mathbf{P} = -\nabla V(\mathbf{X})$$

as in Newtonian mechanics. In order to guarantee that the equations of motion (5.7.3) and (5.7.4) follow from the equations (5.7.2) of the Heisenberg picture, we impose the canonical commutation relations

$$[Q_a(t), P_b(t)] = i\hbar\delta_{ab} , \quad [Q_a(t), Q_b(t)] = [P_a(t), P_b(t)] = 0 . \qquad (5.7.5)$$

To see that this works, recall that as remarked in Section 5.3 commutation is algebraically like differentiation. It follows from the commutation relations (5.7.5) that for *any* function $F(Q, P)$ of the Qs and Ps,

$$\big[Q_a(t), F\big(Q(T), P(t)\big)\big] = i\hbar\frac{\partial}{\partial P_a(t)} F\big(Q(T), P(t)\big) , \qquad (5.7.6)$$

$$\big[P_a(t), F\big(Q(T), P(t)\big)\big] = -i\hbar\frac{\partial}{\partial Q_a(t)} F\big(Q(T), P(t)\big) . \qquad (5.7.7)$$

So, by taking $F = H$ it follows trivially from the Heisenberg picture equations (5.7.2) and the commutation relations (5.7.5) that the Qs and Ps satisfy the Hamiltonian equations of motion (5.7.3) and (5.7.4). This is why we impose these commutation relations.

Of course, since operators in the Heisenberg and Schrödinger pictures are related by Eq. (5.3.35), the commutation relations for the Schrödinger-picture operators Q_a and P_a are the same as for the Heisenberg-picture operators $Q_a(t)$ and $P_a(t)$.

It is in order to satisfy the canonical commutation relations (5.7.5) that in wave mechanics we represent the momentum vector by the operator $-i\hbar\nabla$. What for de Broglie and Schrödinger was just a guess is a necessary consequence of the canonical formalism. But there are cases where the canonical conjugates P_a are not simply masses times velocities but take a different form, as dictated by the Hamiltonian equation (5.7.3). In such cases, it is the quantities P_a and not masses times velocities that must be represented as gradients.

For instance, consider a particle that experiences a momentum-dependent interaction, with Hamiltonian

$$H = \frac{\mathbf{P}^2}{2m} + \frac{1}{2}\mathbf{P}\cdot\mathbf{V}(\mathbf{X}) + \frac{1}{2}\mathbf{V}(\mathbf{X})\cdot\mathbf{P} , \qquad (5.7.8)$$

where $\mathbf{V}$ is some vector function of position. (Since P_i does not commute with X_i, we need to average over orderings of $\mathbf{P}$ and $\mathbf{V}(\mathbf{X})$ in order for the Hamiltonian to be self-adjoint.) Here Eq. (5.7.3) tells us that the momentum is *not* just the mass times the velocity, but instead

$$\mathbf{P}(t) = m\left[\frac{d}{dt}\mathbf{X}(t) - \mathbf{V}(\mathbf{X}(t))\right] . \qquad (5.7.9)$$

Nevertheless, it is $\mathbf{P}$ and not $m\, d\mathbf{X}/dt$ that must be represented in wave mechanics by $-i\hbar\nabla$, in order to satisfy the first commutation relation (5.7.5). In particular, the time-dependent Schrödinger equation here reads

$$i\hbar\frac{\partial}{\partial t}\psi(\mathbf{x},t) = -\frac{\hbar^2}{2m}\nabla^2\psi(\mathbf{x},t) - \frac{i\hbar}{2}\boldsymbol{\nabla}\cdot[\mathbf{V}(\mathbf{x})\psi(\mathbf{x},t)] - \frac{i\hbar}{2}\mathbf{V}(\mathbf{x})\cdot\boldsymbol{\nabla}\psi(\mathbf{x},t)\;. \tag{5.7.10}$$

Lagrangian Formalism

There is another version of the canonical formalism, in quantum mechanics as well as classical mechanics, based on a Lagrangian $L(Q,\dot{Q})$ taken as a function of canonical variables $Q_a(t)$ and their time derivatives $\dot{Q}_a(t)$ rather than a Hamiltonian function of canonical variables and their canonical conjugates. The fundamental assumption of the Lagrangian formalism is that a quantity known as the *action*

$$I \equiv \int_{-\infty}^{+\infty} L(Q(t),\dot{Q}(t))\,dt \tag{5.7.11}$$

is unaffected by infinitesimal shifts in the functions $Q_a(t)$ that vanish at $t \to \pm\infty$. To use this assumption, note that when $Q_a(t)$ is changed to $Q_a(t) + \delta Q_a(t)$ with $\delta Q_a(t)$ infinitesimal, the change in the action is

$$\delta I = \sum_a \int_{-\infty}^{+\infty}\left[\frac{\partial L(Q(t),\dot{Q}(t))}{\partial Q_a(t)}\delta Q_a(t) + \frac{\partial L(Q(t),\dot{Q}(t))}{\partial \dot{Q}_a(t)}\frac{d}{dt}\delta Q_a(t)\right]dt\;.$$

In the case where $\delta Q_a(t)$ vanishes at $t \to \pm\infty$, integrating the second term in the integrand by parts gives

$$\delta I = \sum_a \int_{-\infty}^{+\infty}\left[\frac{\partial L(Q(t),\dot{Q}(t))}{\partial Q_a(t)} - \frac{d}{dt}\frac{\partial L(Q(t),\dot{Q}(t))}{\partial \dot{Q}_a(t)}\right]\delta Q_a(t)\,dt\;,$$

and since this is assumed to vanish for arbitrary variations $\delta Q_a(t)$ that vanish at $t \to \pm\infty$, we must have

$$\frac{d}{dt}\left(\frac{\partial L(Q(t),\dot{Q}(t))}{\partial \dot{Q}_a(t)}\right) = \frac{\partial L(Q(t),\dot{Q}(t))}{\partial Q_a(t)}\;. \tag{5.7.12}$$

These are the equations of motion in the Lagrangian formalism.

From this, we can go over to the classical Hamiltonian formalism, defining

$$P_a(t) = \frac{\partial L(Q(t),\dot{Q}(t))}{\partial \dot{Q}_a(t)} \tag{5.7.13}$$

with Hamiltonian

$$H(Q,P) = \sum_a \dot{Q}_a P_a - L(Q,\dot{Q})\;. \tag{5.7.14}$$

(Taken literally, this may not put the Qs and Ps in the right order for H to be self-adjoint, in which case we must average over their ordering to make H self-adjoint as we did in Eq. (5.7.8).) In Eq. (5.7.14) we should regard $\dot{Q}$ as a function of the Qs and Ps, given by solving Eq. (5.7.13) for $\dot{Q}$. We can then check that the Qs and Ps satisfy the Hamiltonian equations of motion

$$\frac{\partial H(Q,P)}{\partial Q_a} = \sum_b \frac{\partial \dot{Q}_b}{\partial Q_a} P_b - \frac{\partial L(Q,\dot{Q})}{\partial Q_a} - \sum_b \frac{\partial L(Q,\dot{Q})}{\partial \dot{Q}_b} \frac{\partial \dot{Q}_b}{\partial Q_a}$$
$$= -\frac{\partial L(Q,\dot{Q})}{\partial Q_a} = -\dot{P}_a$$

and

$$\frac{\partial H(Q,P)}{\partial P_a} = \sum_b \frac{\partial \dot{Q}_b}{\partial P_a} P_b + \dot{Q}_a - \sum_b \frac{\partial L(Q,\dot{Q})}{\partial \dot{Q}_b} \frac{\partial \dot{Q}_b}{\partial P_a} = \dot{Q}_a \, ,$$

as was to be shown.

Noether's Theorem

The chief reason for using the Lagrangian formalism to construct a Hamiltonian is that there is a deep relation between conservation laws and symmetries of the Lagrangian, first stated in classical physics[26] by Amalie Emmy Noether (1882–1935). Let us consider a symmetry of the Lagrangian under an infinitesimal transformation that for simplicity takes the Qs into functions of Qs:

$$Q_a \to Q_a + \epsilon f_a(Q) \, , \quad \dot{Q}_a \to \dot{Q}_a + \epsilon \sum_b \frac{\partial f_a(Q)}{\partial Q_b} \dot{Q}_b \, , \tag{5.7.15}$$

where the $f_a(Q)$ are some functions only of the Qs that are dictated, up to a constant factor, by the nature of the symmetry principle, and ϵ is an infinitesimal parameter. (Time-independent rotations and translations of coordinates are of this general form.) The invariance of L under this transformation tells us that

$$0 = \sum_a \frac{\partial L}{\partial Q_a} f_a(Q) + \sum_a \frac{\partial L}{\partial \dot{Q}_a} \frac{d}{dt} f_a(Q) \, .$$

Using Eqs. (5.7.12) and (5.7.13), we see that this is a conservation law:

$$\frac{dF(Q,P)}{dt} = 0 \quad \text{where} \quad F(Q,P) \equiv \sum_a P_a f_a(Q) \, . \tag{5.7.16}$$

[26] E. Noether, Nachr. König Gesell. Wiss. zu Göttingenm Math.-Phys. Klasse 235 (1918).

Not only is F conserved – in quantum mechanics it generates the symmetry with which we began, in the sense that

$$[F, Q_a] = -i\hbar f_a(Q) \tag{5.7.17}$$

or, equivalently, for infinitesimal ϵ,

$$\exp[i\epsilon F/\hbar] Q_a \exp[-i\epsilon F/\hbar] = Q_a + \epsilon f_a(Q) \,, \tag{5.7.18}$$

which is just the transformation (5.7.15).

For instance, if we take the canonical variables Q as the ith components X_{ni} of the coordinate vectors $\mathbf{X}_n$ of particles distinguished by a label n, and if as usual the Lagrangian for a multi-particle system depends only on velocities and differences of coordinate vectors, then L is invariant under the transformation $X_{ni} \to X_{ni} + \epsilon_i$, with the same infinitesimal vector $\boldsymbol{\epsilon}$ for each particle label n, and Eq. (5.7.16) gives a conserved quantity,

$$\mathbf{P} = \sum_n P_n \,.$$

This of course is the total momentum, and generates the translation symmetry, in the sense that

$$[\boldsymbol{\epsilon} \cdot \mathbf{P}, X_{ni}] = -i\hbar\epsilon_i \,.$$

A similar analysis uses the assumed rotational invariance of the Lagrangian to give the usual formula for the total angular momentum of any system that does not involve spin. But note that invariance under the Galilean transformation $\mathbf{X} \to \mathbf{X} + \mathbf{u}t$ does *not* lead to a conservation law because, unlike translation or rotation, this transformation involves the time.

5.8 Charged Particles in Electromagnetic Fields

We now turn to the quantum theory of a charged particle moving in classical electric and magnetic fields. This theory will provide us in this section with a good example of the use of the canonical formalism, and as we will see in the following section this theory played an important part in understanding the effect of external magnetic fields on atomic spectra.

Scalar and Vector Potentials

It is frequently convenient in classical electrodynamics to write the electric and magnetic fields as linear combinations of derivatives of a vector potential $\mathbf{A}(\mathbf{x}, t)$ and a scalar potential $\phi(\mathbf{x}, t)$:

$$\mathbf{E} = -\frac{1}{c}\dot{\mathbf{A}} - \nabla\phi \,, \quad \mathbf{B} = \nabla \times \mathbf{A} \,. \tag{5.8.1}$$

This ensures that the fields satisfy the homogeneous Maxwell equations

$$\nabla \times \mathbf{E} + \dot{\mathbf{B}}/c = 0\,, \quad \nabla \cdot \mathbf{B} = 0\,, \tag{5.8.2}$$

and leads to simplifications in the other Maxwell equations.

What in classical physics is merely a convenience, in quantum mechanics is a necessity. It is not possible to write a simple local Hamiltonian for a charged particle in general electric and magnetic fields using just the fields **E** and **B**. But we can write such Hamiltonians in terms of **A** and ϕ. For a single non-relativistic particle of mass m and charge e, the Hamiltonian is

$$H(\mathbf{X}, \mathbf{P}) = \frac{1}{2m}\left[\mathbf{P} - \frac{e}{c}\mathbf{A}(\mathbf{X}, t)\right]^2 - e\phi(\mathbf{X}, t)\,. \tag{5.8.3}$$

Whether or not we derive this Hamiltonian from a Lagrangian, its real justification is that it leads to the correct equations of motion. The Hamiltonian equations of motion (5.7.3) and (5.7.4) here take the form

$$\dot{X}_i(t) = \frac{\partial H}{\partial P_i(t)} = \frac{1}{m}\left[P_i(t) - \frac{e}{c}A_i(\mathbf{X}, t)\right],$$

$$\dot{P}_i(t) = -\frac{\partial H}{\partial X_i(t)} = \frac{e}{mc}\left[P_j(t) - \frac{e}{c}A_j(\mathbf{X}, t)\right]\frac{\partial A_j(\mathbf{X}, t)}{\partial X_i} - e\frac{\partial \phi(\mathbf{X}, t)}{\partial X_i}\,,$$

where the indices i, j, etc. run over the values 1, 2, 3, and repeated indices are summed. Eliminating the momentum from these two equations (and dropping arguments), we have an equation of motion for the position:

$$m\ddot{X}_i = \frac{e}{c}\dot{X}_j\frac{\partial A_j}{\partial X_i} - e\frac{\partial \phi}{\partial X_i} - \frac{e}{c}\left[\frac{\partial A_i}{\partial t} + \dot{X}_j\frac{\partial A_i}{\partial X_j}\right]$$

$$= \frac{e}{c}\dot{X}_j\left(\frac{\partial A_j}{\partial X_i} - \frac{\partial A_i}{\partial X_j}\right) - e\frac{\partial \phi}{\partial X_i} - \frac{e}{c}\frac{\partial A_i}{\partial t}\,.$$

To put this in a more familiar form, note that

$$\dot{X}_j\left(\frac{\partial A_j}{\partial X_i} - \frac{\partial A_i}{\partial X_j}\right) = \left[\dot{\mathbf{X}} \times (\nabla \times \mathbf{A})\right]_i\,.$$

(For instance, for $i = 3$ the left-hand side is

$$\dot{X}_1\left(\frac{\partial A_1}{\partial X_3} - \frac{\partial A_3}{\partial X_1}\right) + \dot{X}_2\left(\frac{\partial A_2}{\partial X_3} - \frac{\partial A_3}{\partial X_2}\right) = \dot{X}_1(\nabla \times \mathbf{A})_2 - \dot{X}_2(\nabla \times \mathbf{A})_1$$

$$= \left[\dot{\mathbf{X}} \times (\nabla \times \mathbf{A})\right]_3$$

and likewise for $i = 1$ and $i = 2$.) Using the formulas (5.8.1) for **E** and **B**, the equation of motion takes the form

$$m\ddot{\mathbf{X}} = e\mathbf{E} + \frac{e}{c}[\dot{\mathbf{X}} \times \mathbf{B}]\,, \tag{5.8.4}$$

which we recognize as the equation of motion (4.6.23) dictated by Lorentz invariance, to first order in $|\dot{\mathbf{X}}|/c$.

Gauge Transformations

There is more than one set of potentials $\mathbf{A}$ and ϕ that give the same fields $\mathbf{E}$ and $\mathbf{B}$. Given a set of potentials $\mathbf{A}$ and ϕ that yield a set of fields $\mathbf{E}$ and $\mathbf{B}$, we can always find other potentials

$$\mathbf{A}^{\#} = \mathbf{A} + \nabla\xi \ , \quad \phi^{\#} = \phi - \frac{1}{c}\frac{\partial \xi}{\partial t} \ , \tag{5.8.5}$$

which give the same fields for an arbitrary function $\xi(\mathbf{x}, t)$. A given choice of potentials is called a choice of gauge, and Eq. (5.8.5) is known as a *gauge transformation*. Even though the equation of motion (5.8.4) derived from the Hamiltonian (5.8.3) involves only the fields $\mathbf{E}$ and $\mathbf{B}$, the Hamiltonian depends on $\mathbf{A}$ and ϕ and is not gauge invariant. So it is important to observe that no physical implications of this Hamiltonian depend on the choice of gauge.

Let us check this for the simple case of a time-independent gauge transformation function $\xi(\mathbf{X})$, which has no effect on ϕ. The gauge-transformed Hamiltonian is

$$H^{\#} = \frac{1}{2m}\left[\mathbf{P} - \frac{e}{c}\mathbf{A} - \frac{e}{c}\nabla\xi\right]^2 - e\phi \ . \tag{5.8.6}$$

Define an operator

$$U(\mathbf{X}) \equiv \exp\left(-\frac{ie}{\hbar c}\xi(\mathbf{X})\right) \ .$$

According to Eq. (5.7.7),

$$[\mathbf{P}, U(\mathbf{X})] = -\frac{e}{c}\nabla\xi(\mathbf{X})U(\mathbf{X})$$

and therefore

$$U^{-1}(\mathbf{X})\mathbf{P}U(\mathbf{X}) = \mathbf{P} - \frac{e}{c}\nabla\xi(\mathbf{X}) \ .$$

It follows that

$$H^{\#}(\mathbf{X}, \mathbf{P}) = U^{-1}(\mathbf{X})H(\mathbf{X}, \mathbf{P})U(\mathbf{X}) \ . \tag{5.8.7}$$

So if $\psi(\mathbf{x})$ satisfies the time-independent Schrödinger equation $H\psi = E\psi$ for energy E, then the gauge-transformed Schrödinger equation $H^{\#}\psi^{\#} = E\psi^{\#}$ is satisfied for the same energy, with gauge-transformed wave function

$$\psi^{\#}(\mathbf{x}) = \exp\left(-\frac{ie}{\hbar c}\xi(\mathbf{x})\right)\psi(\mathbf{x}) \ . \tag{5.8.8}$$

Not only the energy but also the probability density $|\psi|^2$ is unchanged by this transformation.

Magnetic Interactions

Now let us take the simplest example of magnetic interactions, a one-electron atom in a uniform time-independent magnetic field **B**. We can take the vector potential here as

$$\mathbf{A} = -\frac{1}{2}\mathbf{X} \times \mathbf{B} ,$$

for which $\nabla \times \mathbf{A} = \mathbf{B}$. Of course this is not unique, but as we have seen this makes no difference.

The factor $1/c$ multiplying the vector potential in Eq. (5.8.3) makes the magnetic term in the Hamiltonian generally very small. To first order in this term, it shifts the Hamiltonian (5.8.3) by

$$\Delta H = \frac{e}{m_e c}\mathbf{A}(\mathbf{X}) \cdot \mathbf{P} = -\frac{e}{2m_e c}[\mathbf{X} \times \mathbf{B}] \cdot \mathbf{P} = \frac{e}{2m_e c}\mathbf{B} \cdot \mathbf{L} , \quad (5.8.9)$$

where $\mathbf{L} = \mathbf{X} \times \mathbf{P}$ is the orbital angular momentum operator. (Here e has been changed to $-e$, because in the usual notation this is the charge of the electron. Also, we have not had to worry about the order of the operators $\mathbf{A}(\mathbf{X})$ and $\mathbf{P}$, because in this choice of gauge, $\nabla \cdot \mathbf{A} = 0$.)

Spin Coupling

What about spin? The form of the interaction (5.8.9) suggests that there should also be a similar term in the magnetic interaction Hamiltonian with the spin operator **S** in place of **L**, and not necessarily with the same coefficient. The magnetic interaction is therefore taken to be in the form

$$\Delta H = \frac{e}{2m_e c}\mathbf{B} \cdot [\mathbf{L} + g_e \mathbf{S}] , \quad (5.8.10)$$

where g_e is a dimensionless coefficient known as the *gyromagnetic ratio* of the electron. It was first calculated in 1928 on the basis of a relativistic theory of the electron by Dirac,[27] who found the value $g_e = 2$. The development of quantum electrodynamics after World War II led to a calculation[28] of a radiative correction due to the emission and reabsorption of a photon by the electron while it is interacting with the magnetic field. This gave $g_e = 2 \times 1.00162$, in good agreement with experiment.

The effect of the interaction (5.8.10) on atomic energy levels in a magnetic field is described in the next section.

[27] P. A. M. Dirac, Proc. Roy. Soc. A **117**, 610 (1928).
[28] J. Schwinger, Phys. Rev. **73**, 416 (1948).

5.9 Perturbation Theory

There are few problems in quantum mechanics that can be solved exactly. Fortunately it is often possible to find useful approximate solutions by a technique known as perturbation theory. Sometimes it happens that the results obtained in this way are more revealing than would be provided by a more complicated exact solution, even where one is available.

The basis of perturbation theory is the assumption that the Hamiltonian can be divided into two parts:

$$H = H_0 + H' \,, \tag{5.9.1}$$

where H_0 is simple enough to allow exact solutions of the Schrödinger equation, and H' is in some sense small. We have already used a Hamiltonian of this type to derive the Born approximation for scattering amplitudes in Section 5.6. In this section we shall concentrate on deriving approximations for energy levels and the corresponding wave functions, assuming that H' is small enough to allow the eigenfunctions and eigenvalues of H to be usefully expressed as power series in H'. That is, in the Schrödinger equation $H\psi = E\psi$ we write

$$\psi = \psi_0 + \psi_1 + \psi_2 + \cdots \,, \qquad E = E_0 + E_1 + E_2 + \cdots \,, \tag{5.9.2}$$

where ψ_N and E_N are of Nth order in H'. The Schrödinger equation then takes the form

$$\begin{aligned}(H_0 + H')&(\psi_0 + \psi_1 + \psi_2 + \cdots) \\ &= (E_0 + E_1 + E_2 + \cdots)(\psi_0 + \psi_1 + \psi_2 + \cdots) \,.\end{aligned} \tag{5.9.3}$$

In the Nth order of perturbation theory we keep all terms in Eq. (5.9.3) up to Nth order in H'. To zeroth order in H', this is the unperturbed Schrödinger equation

$$H_0\psi_0 = E_0\psi_0 \,, \tag{5.9.4}$$

whose solutions we assume are known.

First-Order Perturbation Theory

Keeping only terms in Eq. (5.9.3) of first order in H' and taking ψ_0 to satisfy the zeroth-order equation Eq. (5.9.4), the Schrödinger equation becomes

$$H_0\psi_1 + H'\psi_0 = E_0\psi_1 + E_1\psi_0 \,. \tag{5.9.5}$$

To find the first-order term E_1 in the energy, multiply Eq. (5.9.5) with ψ_0^* and integrate and sum over all coordinates and spin 3-components. Because H_0 is a Hermitian operator, we have

$$\int \psi_0^* H_0 \psi_1 = \int (H_0 \psi_0)^* \psi_1 = E_0 \int \psi_0^* \psi_1 ,$$

so the terms in this integral involving ψ_1 cancel, and we have

$$E_1 \int \psi_0^* \psi_0 = \int \psi_0^* H' \psi_0 ,$$

or, if ψ_0 is normalized,

$$E_1 = \int \psi_0^* H' \psi_0 . \tag{5.9.6}$$

Very nice, but this does not necessarily work in the case where E_0 is a degenerate energy eigenvalue, with several independent eigenfunctions $\psi^{(n)}$:

$$H_0 \psi^{(n)} = E_0 \psi^{(n)} . \tag{5.9.7}$$

It is convenient to choose these eigenfunctions to be orthonormal:

$$\int \psi^{(n)^*} \psi^{(m)} = \delta_{nm} . \tag{5.9.8}$$

Multiply Eq. (5.9.5) with any of the $\psi^{(n)*}$, integrate, and sum over all coordinates and spin 3-components, and again use the fact that H_0 is Hermitian, so that $\int \psi^{(n)*} H_0 \psi_1 = E_0 \int \psi^{(n)*} \psi_1$. The terms in this integral involving ψ_1 again cancel, and we have

$$\int \psi^{(n)*} H' \psi_0 = E_1 \int \psi^{(n)*} \psi_0 . \tag{5.9.9}$$

The difficulty is that with more than one independent solution $\psi^{(n)}$ of Eq. (5.9.7), whatever we choose for our unperturbed wave function ψ_0, we can always choose some linear combination $\sum_n c_n \psi^{(n)}$ of these eigenfunctions to be orthogonal to ψ_0, in the sense that $\int \left[\sum_n c_n \psi^{(n)} \right]^* \psi_0 = 0$, so that the same linear combination of Eq. (5.9.9) gives a condition on H':

$$\int \left[\sum_n c_n \psi^{(n)} \right]^* H' \psi_0 = 0 , \tag{5.9.10}$$

which in general need not be the case.

To avoid this contradiction, we must make an appropriate choice of the zeroth-order eigenfunction ψ_0. What we need is to choose ψ_0 so that any linear combination of the degenerate wave functions $\psi^{(n)}$ that is orthogonal to ψ_0 will also be orthogonal to $H' \psi_0$. Because H' is a Hermitian operator, the integrals $\mathcal{H}_{nm} \equiv \int \psi^{(n)*} H' \psi^{(m)}$ form a Hermitian matrix, in the sense that $\mathcal{H}_{mn}^* = \mathcal{H}_{nm}$. According to a general theorem of matrix algebra, it is always possible to replace the $\psi^{(n)}$ with linear combinations for which the orthonormality condition (5.9.8) is still satisfied, and now $\mathcal{H}_{nm}$ is diagonal:

$$\int \psi^{(n)*} H' \psi^{(m)} = \begin{cases} \mathcal{E}_n & n = m \\ 0 & n \neq m \,, \end{cases} \tag{5.9.11}$$

for some real $\mathcal{E}_n$. We must take the zeroth-order solution to be one of these redefined eigenfunctions, say $\psi^{(m)}$, so that if we multiply Eq. (5.9.5) with the complex conjugate of any linear combination $\sum_{n \neq m} c_n \psi^{(n)}$ of the other degenerate eigenfunctions that is orthogonal to $\psi^{(m)}$, Eq. (5.9.11) implies that Eq. (5.9.10) is also necessarily satisfied, and there is no contradiction. (We will see an example of this procedure in our treatment below of the Zeeman effect.) With the zeroth-order wave function $\psi_0 = \psi^{(m)}$, Eq. (5.9.6) gives $E_1 = \mathcal{E}_m$.

We can get a further insight into the necessity of a suitable choice of the zeroth-order wave function by considering a problem of some importance in its own right, the calculation of the first-order contribution to the wave function. Let us introduce a complete orthonormal set of solutions φ_a of the zeroth-order Schrödinger equation

$$H_0 \varphi_a = E_a \varphi_a \,, \qquad \int \varphi_a^* \varphi_b = \delta_{ab} \,. \tag{5.9.12}$$

Multiply Eq. (5.9.5) by φ_a^* and integrate and sum over all coordinates and spins. Since H_0 is Hermitian the first term gives $\int \varphi_a^* H_0 \psi_1 = E_a \int \varphi_a^* \psi_1$, and so

$$(E_0 - E_a) \int \varphi_a^* \psi_1 = \int \varphi_a^* H' \psi_0 - E_1 \int \varphi_a^* \psi_0 \,. \tag{5.9.13}$$

For $E_a = E_0$, Eq. (5.9.13) makes no sense unless $\int \varphi_a^* H' \psi_0$ vanishes for every such wave function orthogonal to ψ_0, which is accomplished by taking ψ_0 to be one of the wave functions $\psi^{(m)}$ for which Eq. (5.9.11) is satisfied. On the other hand, for $E_a \neq E_0$, φ_a is orthogonal to ψ_0 so Eq. (5.9.13) gives a formula that is valid for any φ_a for which $E_a \neq E_0$:

$$\int \varphi_a^* \psi_1 = \frac{\int \varphi_a^* H' \psi_0}{E_0 - E_a} \quad \text{for} \quad E_a \neq E_0 \,. \tag{5.9.14}$$

In the case where the eigenvalue E_0 of H_0 is not degenerate, ψ_0 and the functions φ_a with $E_a \neq E_0$ form a complete set, so we can expand ψ_1 as

$$\psi_1 = \alpha \psi_0 + \sum_{a: E_a \neq E_0} \varphi_a \int \varphi_a^* \psi_1 = \alpha \psi_0 + \sum_{a: E_a \neq E_0} \varphi_a \frac{\int \varphi_a^* H' \psi_0}{E_0 - E_a} \,,$$

with the complex number α the only component of ψ_1 that is still unknown. We can always take α to be real, because any change in the imaginary part of

α needed to make α real has no effect on $\psi_0 + \psi_1$ if it is compensated by a first-order change in the phase of ψ_0, which we are free to choose as we like. With α real, to first order the norm of $\psi_0 + \psi_1$ is

$$\int |\psi_0 + \psi_1|^2 = (1 + 2\alpha) \int |\psi_0|^2 \ .$$

So, if we normalize ψ_0 and require the wave function to remain normalized in first order, then we must have $\alpha = 0$. The first-order shift in the wave function is then finally

$$\psi_1 = \sum_{a:E_a \neq E_0} \varphi_a \frac{\int \varphi_a^* H' \psi_0}{E_0 - E_a} \ . \tag{5.9.15}$$

Note that if the parameters of the theory are changed so that one of the E_a approaches E_0, then the corresponding component of the wave function becomes very large, invalidating perturbation theory, unless in this limit $\int \varphi_a^* H' \psi_0$ becomes very small. So even approximate degeneracy can be a problem.

In the case of degeneracy Eq. (5.9.13) tells us nothing about the components of ψ_1 along the φ_a with $E_a = E_0$, and the normalization condition on $\psi_0 + \psi_1$ does not determine these components either. For this, it is necessary to invoke the condition that the changes of the wave function in higher orders of perturbation theory are small. We will not pursue this aspect here.

Zeeman Effect

For an example of the use of perturbation theory, let us return to the Zeeman effect, mentioned at the end of the previous section. Here H_0 is the Hamiltonian of an alkali metal atom, considering the outermost electron to move in an effective potential arising from the charges of the nucleus and all other electrons, with no external fields. To calculate the effect of a weak external magnetic field $\mathbf{B}$, we consider a first-order perturbation given by Eq. (5.8.10):

$$H' = \frac{e}{2m_e c} \mathbf{B} \cdot [\mathbf{L} + g_e \mathbf{S}] \ , \tag{5.9.16}$$

where $g_e \simeq 2$ is the gyromagnetic ratio of the electron. The eigenfunctions of H_0 may be labeled $\psi_{nj\ell M}$. Here

$$\mathbf{J}^2 \psi_{nj\ell M} = \hbar^2 j(j+1) \psi_{nj\ell M} \ , \mathbf{L}^2 \psi_{nj\ell M} = \hbar^2 \ell(\ell+1) \psi_{nj\ell M} \ ,$$
$$J_z \psi_{nj\ell M} = \hbar M \psi_{nj\ell M} \ , \tag{5.9.17}$$

where M runs by unit steps from $-j$ to $+j$, and $n - \ell - 1$ is the number of nodes of the wave function. The states with a given n, j, and ℓ but varying M

all have the same energy, so the eigenstates of H_0 are all degenerate, except for those with $j = 0$. For a magnetic field in an arbitrary direction the operator H' in general includes terms proportional to L_z and S_z, which do commute with J_z, but also L_x, L_y, S_x, and S_y, which do not commute with J_z, so there will be non-vanishing components of $\int \psi^*_{nj\ell M'} H' \psi_{nj\ell M}$ with $M' \neq M$, and first-order perturbation theory will not work if we take the zeroth-order wave function to be one of the $\psi_{nj\ell M}$.[29]

The cure is obvious. Take the zeroth-order wave function to be an eigenstate of the component of $\mathbf{J}$ in the direction of $\mathbf{B}$. Or, to save writing, just continue to use the $\psi_{nj\ell M}$ as zeroth-order wave functions but from the beginning choose the coordinate system so that the z-axis is in the direction of $\mathbf{B}$. In this case, the first-order shift in the energy is given by Eq. (5.9.6) as

$$E_1(n\ell jM) = \frac{eB}{2m_e c} \int \psi^*_{nj\ell M}(L_z + g_e S_z)\psi_{nj\ell M} \,. \tag{5.9.18}$$

It is easiest to evaluate E_1 for s-wave states with $\ell = 0$, for which $j = 1/2$ and $M = \pm 1/2$. In this case Eq. (5.9.18) gives immediately

$$E_1(n \;\; 0 \;\; 1/2 \;\; \pm 1/2) = \pm \frac{e g_e B\hbar}{4m_e c} \,. \tag{5.9.19}$$

To deal with the general case with $\ell \neq 0$, we use a general property of angular momentum multiplets. Let φ_{jM} be any multiplet of $2j + 1$ wave functions, with $\mathbf{J}^2\varphi_{jM} = \hbar^2 j(j+1)\varphi_{jM}$ and $J_z\varphi_{jM} = \hbar M\varphi_{jM}$, formed as described in Section 5.4 by letting lowering operators $J_x - iJ_y$ act on a state with $M = j$. For any vector operator $\mathbf{V}$, the integrals $\int \varphi^*_{jM'} V_i \varphi_{jM}$ can all be calculated from any one of them by using the commutation relations of the raising and lowering operators $J_x \pm iJ_y$ with the V_i and the effect of these operators on the multiplet φ_{jM}, none of which depends on the choice of the operator $\mathbf{V}$ or the wave functions φ_{jM}, so in general the integrals $\int \varphi_{jM'} V_i \varphi_{jM}$ can depend only on the specific choice of the operator $\mathbf{V}$ or the wave functions φ_{jM} through an overall factor. In particular, we have

$$\int \varphi^*_{jM'} V_i \varphi_{jM} = \alpha_V \int \varphi^*_{jM'} J_i \varphi_{jM} \,, \tag{5.9.20}$$

[29] If it were not for the fine structure produced by spin–orbit coupling there would be an additional degeneracy: the energies for states with the same n and ℓ but different j would be equal. The discussion here of the Zeeman effect assumes that the magnetic field is sufficiently weak that the energy shift it produces is small compared with the fine-structure splitting, in which case states with the same n and ℓ but different j are not effectively degenerate. But we are ignoring the even smaller hyperfine energy shifts due to the interaction of the electron with the magnetic field of the nucleus.

In hydrogen there is a further degeneracy of states with the same n and j but different ℓ, such as the $2s_{1/2}$ and $2p_{1/2}$ states, which are separated only by the very small Lamb shift described in Section 5.4. The treatment here applies to hydrogen only when the energy shift due to the interaction of the electron with the external magnetic field is less than the Lamb shift but greater than the hyperfine splitting.

where the factor α_V will in general depend on the nature of the operator **V** and the wave functions φ_{jM}, but *not* on the vector index i nor on the angular momentum z-components M and M'. This is an example of a general quantum-mechanical result known as the *Wigner–Eckart theorem*.[30]

In its application to the Zeeman effect, Eq. (5.9.20) gives

$$\int \psi^*_{nj\ell M'} L_i \psi_{nj\ell M} = \alpha_L(nj\ell) \int \psi^*_{nj\ell M'} J_i \psi_{nj\ell M} \; ,$$
$$\int \psi^*_{nj\ell M'} S_i \psi_{nj\ell M} = \alpha_S(nj\ell) \int \psi^*_{nj\ell M'} J_i \psi_{nj\ell M} \; . \tag{5.9.21}$$

To calculate the coefficients α_L and α_S, we use a trick. The wave functions $J_k \psi_{nj\ell M}$ are linear combinations of the wave functions $\psi_{nj\ell M''}$ in the same multiplet, so we can apply Eq. (5.9.21) also to these functions:

$$\int \psi^*_{nj\ell M'} L_i J_k \psi_{nj\ell M} = \alpha_L(nj\ell) \int \psi^*_{nj\ell M'} J_i J_k \psi_{nj\ell M} \; ,$$
$$\int \psi^*_{nj\ell M'} S_i J_k \psi_{nj\ell M} = \alpha_S(nj\ell) \int \psi^*_{nj\ell M'} J_i J_k \psi_{nj\ell M} \; . \tag{5.9.22}$$

Taking the wave functions $\psi_{nj\ell M}$ to be orthonormal, we have

$$\int \psi^*_{nj\ell M'} \mathbf{J}^2 \psi_{nj\ell M} = \hbar^2 j(j+1)\hbar^2 \delta_{M'M} \; .$$

Hence, setting $i = k$ and summing over i in Eq. (5.9.22), we have

$$\hbar^2 j(j+1)\alpha_L(nj\ell) = \int \psi^*_{nj\ell M'} \mathbf{L}\cdot\mathbf{J}\psi_{nj\ell M} \; ,$$
$$\hbar^2 j(j+1)\alpha_S(nj\ell) = \int \psi^*_{nj\ell M'} \mathbf{S}\cdot\mathbf{J}\psi_{nj\ell M} \; .$$

Note that

$$\mathbf{L}\cdot\mathbf{J} = \frac{1}{2}\big[-(\mathbf{J}-\mathbf{L})^2 + \mathbf{J}^2 + \mathbf{L}^2\big] = \frac{1}{2}\big[-\mathbf{S}^2 + \mathbf{J}^2 + \mathbf{L}^2\big]$$

and likewise

$$\mathbf{S}\cdot\mathbf{J} = \frac{1}{2}\big[-\mathbf{L}^2 + \mathbf{J}^2 + \mathbf{S}^2\big] \; ,$$

so

$$\alpha_L(nj\ell) = \frac{-3/4 + j(j+1) + \ell(\ell+1)}{2j(j+1)} \; ,$$
$$\alpha_S(nj\ell) = \frac{-\ell(\ell+1) + j(j+1) + 3/4}{2j(j+1)} \; . \tag{5.9.23}$$

[30] For a statement of this theorem and a detailed proof, see Section 4.1 of Weinberg, *Lectures on Quantum Mechanics*, listed in the bibliography.

Using Eqs. (5.9.23) and (5.9.21) in Eq. (5.9.18) then gives the first-order Zeeman energy shift:

$$E_1(n\ell jM)$$
$$= \frac{eB\hbar M}{4m_e cj(j+1)}\big[-3/4 + j(j+1) + \ell(\ell+1) + g_e[-\ell(\ell+1) + j(j+1) + 3/4]\big]\,. \tag{5.9.24}$$

Second-Order Perturbation Theory

In some cases the interesting effects of a perturbation H' arise only in second or even higher order. The terms in the Schrödinger equation (5.9.3) of second order in H' give

$$H_0\psi_2 + H'\psi_1 = E_0\psi_2 + E_1\psi_1 + E_2\psi_0\,. \tag{5.9.25}$$

To find E_2, multiply with ψ_0^* and integrate and sum over all coordinates and spins. Again using the fact that H_0 is Hermitian, we have $\int \psi_0^* H_0 \psi_2 = E_0 \int \psi_0^* \psi_2$, so the terms involving ψ_2 cancel. Also, as we have seen, the normalization condition for ψ requires that $\int \psi_0^* \psi_1 = 0$, so the term proportional to E_1 vanishes. This leaves

$$E_2 = \int \psi_0^* H' \psi_1\,. \tag{5.9.26}$$

In the case where the eigenfunction of H_0 with energy E_0 is not degenerate, we can use Eq. (5.9.15), so that E_2 is given by a sum over all the other eigenfunctions of H_0:

$$E_2 = \sum_{a:E_a \neq E_0} \frac{\left|\int \varphi_a^* H' \psi_0\right|^2}{E_0 - E_a}\,. \tag{5.9.27}$$

When field theorists say that the Lamb shift is due to the emission and reabsorption of a photon by the electron in hydrogen they mean that this is a second-order effect, in which the wave functions φ_a in Eq. (5.9.27) represent states containing an electron and a photon. Since these states form a continuum, the sum over states involves an integral over the photon momentum, which introduces infinities into the calculation. This calculation was completed only in 1949, when it was recognized that the same second-order processes require a redefinition of the mass and charge of the electron and of the photon and electron fields, which leads to a cancellation of infinities.[31]

[31] N. M. Kroll and W. E. Lamb, Phys. Rev. **75**, 388 (1949); J. B. French and V. F. Weisskopf, Phys. Rev. **75**, 1240 (1949).

5.10 Beyond Wave Mechanics

Our discussion of quantum mechanics in this chapter has so far been based on wave mechanics, in which physical states are represented by functions of particle positions and spins. This is too parochial a formalism. Why position, among all observable physical quantities? Indeed, we have already seen in Section 5.3 that a physical state can just as well be represented by a wave function depending on momenta (such as (5.3.20) for a one-particle system) as by a wave function depending on position.

The study of other physical systems forces us much farther away from wave mechanics than merely substituting momenta for position as the argument of wave functions. The state of a field, such as the electromagnetic field, cannot be described in terms of the positions or the momenta of any fixed number of particles. It is partly as a preparation for our account of quantum field theory in Chapter 7 that we need to consider a formulation of quantum mechanics, due chiefly to Dirac,[32] that is general enough to apply to any physical system.

In this general formulation, physical states are represented by *state vectors* in an infinite-dimensional space, known as Hilbert space. Like ordinary vectors in three dimensions, a linear combination $a_1\Psi_1 + a_2\Psi_2$ of two state vectors Ψ_1 and Ψ_2 is also a state vector, only here the numerical coefficients a_1 and a_2 can be complex. Addition here has the same properties as the addition of complex numbers, including associativity and commutativity and the existence of a zero for which $0 + \Psi = \Psi + 0 = \Psi$. Also, as in Euclidean space, for any two state vectors Ψ and Φ there is a scalar product denoted (Φ, Ψ), here a complex number, with the properties

$$(\Psi, \Phi) = (\Phi, \Psi)^* \,, \tag{5.10.1}$$

$$(\Phi, a_1\Psi_1 + a_2\Psi_2) = a_1(\Phi, \Psi_1) + a_2(\Phi, \Psi_2) \,, \tag{5.10.2}$$

$$(\Psi, \Psi) \geq 0 \tag{5.10.3}$$

and $(\Psi, \Psi) = 0$ if and only if $\Psi = 0$. As we shall see, wave functions are the components of these state vectors in one basis or another, and the integrals (5.3.8) of products of these wave functions, abbreviated as $\int \psi^*\varphi$, are the scalar products (Ψ, Φ) of the state vectors of which they are the components.

Observable quantities are represented in this formulation by linear operators that act on state vectors rather than on wave functions. Here an operator A being "linear" means that for any state vectors Ψ_1 and Ψ_2 and complex numbers a_1 and a_2, we have

$$A(a_1\Psi_1 + a_2\Psi_2) = a_1 A\Psi_1 + a_2 A\Psi_2 \,. \tag{5.10.4}$$

[32] This approach is the basis of Dirac's 1930 treatise, *The Principles of Quantum Mechanics*, listed in the bibliography.

The adjoint of an operator A is defined as an operator $A^\dagger$ for which

$$(\Phi, A^\dagger \Psi) = (A\Phi, \Psi) . \tag{5.10.5}$$

Real observables are represented by operators that are self-adjoint, in the sense that $A^\dagger = A$.

The first interpretive postulate of quantum mechanics is that a state represented by a non-zero state vector Ψ has a definite value α for an observable represented by an operator A if and only if Ψ is an eigenvector of A with eigenvalue α – that is,

$$A\Psi = \alpha\Psi . \tag{5.10.6}$$

If Ψ_1 and Ψ_2 are non-zero eigenvectors of a self-adjoint operator A with eigenvalues α_1 and α_2 then

$$\alpha_1(\Psi_2, \Psi_1) = (\Psi_2, A\Psi_1) = (A\Psi_2, \Psi_1) = \alpha_2^*(\Psi_2, \Psi_1) . \tag{5.10.7}$$

Taking $\Psi_1 = \Psi_2$ and then of course $\alpha_1 = \alpha_2$, we see that eigenvalues of self-adjoint operators are real, while taking $\alpha_1 \neq \alpha_2$ and then of course $\Psi_1 \neq \Psi_2$, we see that eigenvectors of a self-adjoint operator with different eigenvalues are orthogonal, in the sense that $(\Psi_2, \Psi_1) = 0$.

The second interpretive postulate of quantum mechanics is that in a state represented by a state vector Ψ, the observable quantity represented by an operator A has the expectation value

$$\langle A \rangle_\Psi = \frac{(\Psi, A\Psi)}{(\Psi, \Psi)} . \tag{5.10.8}$$

Obviously it follows that if Ψ is normalized so that $(\Psi, \Psi) = 1$, then the expectation value is $(\Psi, A\Psi)$.

Suppose an observable is represented by an operator A with discrete eigenvalues α_n and eigenvectors Φ_n,

$$A\Phi_n = \alpha_n \Phi_n \tag{5.10.9}$$

and we normalize these eigenvectors so that

$$(\Phi_n, \Phi_m) = \delta_{nm} . \tag{5.10.10}$$

(If there is only one eigenvector for each eigenvalue it follows from Eq. (5.10.7) and the reality of eigenvalues that the different eigenvectors are orthogonal, and we can always multiply them by numerical factors so that they satisfy Eq. (5.10.10). Even in the case of degeneracy, with several eigenvectors for the same eigenvalue, we can always define linear combinations of these eigenvectors to satisfy Eq. (5.10.10).) If we expand an arbitrary state vector Ψ in a series of these eigenvectors $\Psi = \sum_n c_n \Phi_n$, by taking the scalar product with any of the Φ_m and using Eq. (5.10.10) we find that $c_n = (\Phi_n, \Psi)$, so that

$$\Psi = \sum_n (\Phi_n, \Psi)\, \Phi_n\, . \tag{5.10.11}$$

Inserting this into Eq. (5.10.8) gives the expectation value of the observable represented by A:

$$\langle A \rangle_\Psi = \frac{\sum_n \alpha_n\, |(\Phi_n, \Psi)|^2}{\sum_n |(\Phi_n, \Psi)|^2}\, . \tag{5.10.12}$$

Since a corresponding result applies for any function of this observable, it follows from Eq. (5.10.12) that the probability of finding a value α_m when we measure the observable represented by A is

$$P_m(\Psi) = \frac{|(\Phi_m, \Psi)|^2}{\sum_n |(\Phi_n, \Psi)|^2}\, . \tag{5.10.13}$$

Note in particular that the sum of these probabilities is one.

As in Section 5.3, we can pass over to the case of an operator A with a continuum of eigenvalues by supposing that it has a very large number of very close discrete eigenvalues. If there are $\mathcal{N}(\alpha)\, d\alpha$ eigenvalues between α and $\alpha + d\alpha$ then, in the limit of close packing, we can evaluate sums over n by replacing then with integrals over α:

$$\sum_n \cdots \to \int d\alpha\, \mathcal{N}(\alpha) \cdots\, . \tag{5.10.14}$$

Making this replacement, and defining renormalized eigenvectors

$$\Upsilon_\alpha \equiv \sqrt{\mathcal{N}(\alpha)}\Phi_n \quad \text{for} \quad \alpha = \alpha_n\, , \tag{5.10.15}$$

Eqs. (5.10.11) and (5.10.14) become

$$\Psi = \int d\alpha\, \Upsilon_\alpha (\Upsilon_\alpha, \Psi) \tag{5.10.16}$$

and Eq. (5.10.12) gives

$$\langle A \rangle_\Psi = \frac{\int \alpha\, |(\Upsilon_\alpha, \Psi)|^2\, d\alpha}{\int |(\Upsilon_\alpha, \Psi)|^2\, d\alpha}\, . \tag{5.10.17}$$

We conclude that the probability that a measurement of the observable represented by A will give a value in the range α to $\alpha + d\alpha$ is $\mathcal{P}(\alpha)\, d\alpha$, where $\mathcal{P}(\alpha)$ is the probability density:

$$\mathcal{P}(\alpha) = \frac{|(\Upsilon_\alpha, \Psi)|^2}{\int |(\Upsilon_\alpha, \Psi)|^2\, d\alpha} \tag{5.10.18}$$

with the normalization of the state vectors Υ_α fixed by the condition (5.10.15). In particular, if in Eq. (5.10.16) we take $\Psi = \Upsilon_{\alpha'}$, we find

$$\Upsilon_{\alpha'} = \int d\alpha\ \Upsilon_\alpha (\Upsilon_\alpha, \Upsilon_{\alpha'}) ,$$

so with this normalization the scalar product of these eigenvectors is the Dirac delta function discussed in Section 5.6,

$$(\Upsilon_\alpha, \Upsilon_{\alpha'}) = \delta(\alpha - \alpha') . \tag{5.10.19}$$

Of course, if we also normalize the state vector Ψ so that

$$\int |(\Upsilon_\alpha, \Psi)|^2\ d\alpha = 1$$

then the probability density is

$$\mathcal{P}(\alpha) = |(\Upsilon_\alpha, \Psi)|^2 . \tag{5.10.20}$$

It should by now be clear that the wave function $\psi(x)$ (for instance, for a single particle in one dimension) is nothing but the scalar product

$$\psi(x) = (\Upsilon_x, \Psi) \tag{5.10.21}$$

where Ψ is the state vector representing the physical state and Υ_x is a state vector, normalized to satisfy Eq. (5.10.15) or equivalently Eq. (5.10.19), representing a state in which the particle is at x. We can use suitably normalized eigenvectors of operators representing any other observables to define corresponding wave functions (Υ_α, Ψ), such as the momentum-space wave function introduced in Section 5.3.

In general, eigenvalues and probabilities are to be calculated using relations among operators that represent physical observables, including commutation relations and formulas giving the operators that represent conserved quantities such as the Hamiltonian and angular momentum in terms of other operators. These relations embody the physical content of any particular quantum-mechanical theory.

6
Nuclear Physics

Atoms were at the center of physicists' interest in the 1920s. It was largely from the effort to understand atomic properties that modern quantum mechanics emerged in this decade. In this work physicists did not have to concern themselves much with the nature of the atomic nucleus. It had been known since Rutherford's interpretation in 1911 of the scattering experiments in his laboratory that almost all the mass of atoms is contained in a tiny positively charged nucleus, but all that the atomic physicist needed to know about this nucleus was its electric charge, mass, and (to account for hyperfine splitting) its spin and magnetic moment.

In the 1930s physicists' concerns expanded to include the nature of atomic nuclei. The constituents of the nucleus were identified, and a start was made in learning what held them together. And, as everyone knows, world history was changed in subsequent decades by the military application of nuclear physics.

6.1 Protons and Neutrons

Discovery of the Proton

The first known constituent of the atomic nucleus was the proton. In a series of experiments in 1919 on the passage of alpha particles from radioactive nuclei through various gases, Rutherford found that collisions of alpha particles with nitrogen atoms produced penetrating rays of particles whose range and deflection by electric and magnetic fields seemed identical to what would be expected for hydrogen nuclei.[1] The reaction is now known to be $^{14}\text{N} + {}^{4}\text{He} \rightarrow {}^{17}\text{O} + {}^{1}\text{H}$, and is shown on a seven cent postage stamp of New Zealand, the country of Rutherford's birth. Rutherford at first called these "H particles," and he speculated that they were constituents of all atomic nuclei. In the following year he gave them their modern name, *protons*.

[1] E. Rutherford, Phil. Mag. Series 6 **37**, 381 (1919); reproduced in Beyer, *Foundations of Nuclear Physics*, listed in the bibliography.

It was clear from the beginning that protons could not be the only constituents of atomic nuclei. This would have been close to a realization of a hypothesis in 1815 of the chemist William Prout (1785–1860). Observing that known atomic weights were generally close to whole number multiples of the atomic weight of hydrogen, Prout proposed that all atoms are composites of hydrogen atoms. Applying Prout's hypothesis to nuclei rather than to atoms would have done well in accounting for nuclear masses (which provide almost all of the masses of atoms). It would even work when applied to isotopes, sets of atoms that have an equal number of electrons and hence display the same chemical behavior but differ in their atomic weights. Measurements at the Cavendish Laboratory by Francis William Aston (1877–1945) had shown by 1919 that the atomic weights of various isotopes of hydrogen, carbon, oxygen, chlorine, etc. were all close to whole number multiples of the atomic weight of the lightest isotope of hydrogen. But to suppose that nuclei are made up only of protons would have entirely failed in dealing with nuclear electric charges. If nuclei were composed only of protons their atomic weights in units of the atomic weight of hydrogen would all be close to their atomic numbers, which as we saw in Section 3.4 were by 1919 already known to equal their electric charges in units of the proton charge. But light nuclei such as helium, carbon, nitrogen, oxygen, etc. typically have atomic weights close to twice their atomic numbers.

Electrons in the Nucleus?

In his celebrated Bakerian lecture to the Royal Society of London in 1920,[2] Rutherford proposed that nuclei consist of two kinds of particle: protons and electrons. He was undecided about how these particles might be grouped within nuclei, though he tentatively proposed that nuclei consist of alpha particles (known to be ^{4}He nuclei), supposed to consist of four protons and two electrons, and nuclei of the isotope ^{3}He, which Rutherford had discovered in the collisions of alpha particles with nuclei of nitrogen and oxygen, supposed to consist of three protons and an electron. In his lecture he also proposed the existence of neutral particles later called *neutrons*, with a mass similar to the proton's, and with no electric charge. But for Rutherford the neutron was not a new particle – it was a composite of a proton and one strongly bound electron.

The theory that nuclei consist of protons and electrons had some plausibility. Because electrons have so much less mass than protons, this theory implied that all atomic weights would be close to whole number multiples of the atomic weight of a single proton, the nucleus of hydrogen, as had been noticed by Prout, Also, some nuclei were known to emit electrons in beta radioactivity. But it was hard to see how this could work dynamically. In particular, if there are states of an electron and a proton that are much more deeply bound than a hydrogen

[2] E. Rutherford, Proc. Roy. Soc. A **97**, 374 (1920).

atom, then why do the electrons in ordinary atoms including hydrogen atoms not all fall into these states, emitting the released energy as radiation?

There was an even stronger argument coming from molecular physics against supposing nuclei to consist only of protons and electrons. As we saw in Section 5.5, we can tell whether the identical nuclei in a diatomic molecule are bosons or fermions from the ratio of intensities of transitions in the para and ortho states, which have orbital angular momentum ℓ respectively even and odd. At temperatures T for which the energies of these transitions are much less than kT, the total intensity of the para lines is greater than for the ortho lines by a factor $(s_1 + 1)/s_1$ if the spin s_1 of each nucleus is an integer and the nuclei are bosons, while the total intensity of the para lines is less than for the ortho lines by a factor $s_1/(s_1 + 1)$ if the spin s_1 of each nucleus is a half odd integer and the nuclei are fermions. In 1929 Walter Heitler (1904–1981) and Gerhard Herzberg (1904–1999) observed that the total intensity of the para lines in the diatomic nitrogen molecule is greater than the intensity of the ortho lines, indicating that the nucleus of the most common nitrogen isotope, ^{14}N, is a boson.[3] (In fact, we now know that it has spin 1.) But if nuclei consist of protons and electrons, then the ^{14}N nucleus would consist of 14 protons to give atomic weight 14, and seven electrons, to give atomic number $14 - 7 = 7$, adding up to $14 + 7 = 21$ fermions, and the ^{14}N nucleus would be a fermion.

Discovery of the Neutron

This puzzle began to be resolved in 1932 with the discovery of the neutron[4] by James Chadwick (1891–1974), Rutherford's second in command at the Cavendish Laboratory at Cambridge. Chadwick had learned about observations in Paris[5] that showed that collisions of energetic alpha particles with beryllium atoms produce highly penetrating electrically neutral rays, which when directed into a hydrogen-rich substance like paraffin produce protons that recoil with very high energy. Experiments at the Cavendish Laboratory showed that these neutral rays would also cause heavier nuclei to recoil, though with smaller recoil velocities, and from the ratios of the recoil velocities he was able to calculate the mass of the particles making up the neutral rays. It follows from Eq. (3.3.1) that if a particle B moving with velocity v_B strikes a particle A at rest, and A recoils in the same direction as the initial direction of motion of B, then its recoil velocity will be

$$v'_A = \frac{2m_B}{m_A + m_B} v_B \ .$$

3 W. Heitler and G. Herzberg, Naturwiss. **17**, 673 (1929).

4 J. Chadwick, Proc. Roy. Soc. A **136**, 692 (1932), reproduced in Beyer, *Foundations of Nuclear Physics*, listed in the bibliography.

5 I. Curie and F. Joliot, Compt. Rend. Acad. Sci. Paris **194**, 273 (1932).

Chadwick did not know the initial velocity v_B, but he could eliminate it by taking the ratio of recoil velocities for different target nuclei of known atomic weights, and from this ratio he could calculate the atomic weight A_n of the particle comprising the neutral ray. For instance, measurements showed that the same neutral ray from beryllium that causes hydrogen nuclei to recoil straight back with speed 3.3×10^7 m/sec would cause nitrogen nuclei to recoil straight back with speed 4.7×10^6 m/sec, so

$$\frac{3.3 \times 10^7}{4.7 \times 10^6} = \frac{A_n/(1 + A_n)}{A_n/(14 + A_n)} = \frac{(14 + A_n)}{(1 + A_n)}$$

from which it follows that $A_n \simeq 1.16$. Chadwick concluded that these neutral rays consist of particles he called neutrons, with mass close to that of hydrogen.

Chadwick assumed that this was the neutron that Rutherford had anticipated in his 1920 Bakerian lecture, and he followed Rutherford in supposing that the neutron is a proton–electron bound state. He knew about the problem that study of the diatomic nitrogen molecule indicated that the ^{14}N nucleus is a boson, which is not possible if it consists of 14 protons and seven electrons (whether or not combined into nuclei of ^{4}He or ^{3}He or proton–electron composites), but at first he decided to ignore the problem. This may have been due to a widespread reluctance at the time to contemplate any new fundamental particles besides the proton, electron, and photon, or perhaps it was just the influence of the formidable Lord Rutherford. The status of the neutron as a fermion that is every bit as elementary as the proton only became clear with studies of the forces between these particles, to be discussed in the next section. As a result of these studies, neutrons and protons became regarded as two members of a family of particles known as *nucleons*.

Nuclear Radius and Binding Energy

Like the states of electrons in atoms, the states of nucleons in all but the lightest nuclei can be described approximately by the Hartree approximation: each nucleon can be supposed to move in a potential due to all the other nucleons. Because nuclear forces have short range, each nucleon is chiefly affected by nucleons with the same one-nucleon orbital wave function. And, because nucleons are spin 1/2 fermions satisfying the Pauli exclusion principle, there are just three of these: for a proton (or neutron) state there is another proton (or neutron) state with opposite spin 3-component, and two neutron or proton states with each value for the spin 3-component. Thus, whatever the total number $\mathcal{A}$ of nucleons, as a first approximation the binding energy per nucleon and the volume per nucleon tend to be similar for all nuclei. This is known as the *saturation of nuclear forces*.

With a constant volume per nucleon, the volume of a nucleus is proportional to the number $\mathcal{A}$ of nucleons, so the nuclear radius R is proportional to $\mathcal{A}^{1/3}$.

These radii can be calculated from measurements of the effect of the nuclear electric quadrupole moment on atomic spectra; from measurements of the scattering of electrons in the Coulomb field of the nucleus; and from the measured rates of alpha decays, to be discussed in Section 6.4. A consensus of these measurements gives a nuclear radius

$$R \simeq 1.3 \times 10^{-13}\ \text{cm} \times \mathcal{A}^{1/3}\,. \tag{6.1.1}$$

The binding energy of a nucleus is the energy required to take all of its nucleons to rest at a great distance. It can easily be calculated from measurements of atomic weights: it is the sum of the atomic weights of all the nucleons in the nucleus minus the atomic weight of the nucleus, times the mass energy $m_1c^2 = 931.494$ MeV of unit atomic weight.

Liquid Drop Model

According to the idea of the saturation of nuclear force, the dominant term in the binding energy per nucleon is a constant, estimated to be about 15.8 MeV.[6] There are several corrections to this simple rule, which taken together provide the *liquid drop model* of the nucleus.

Surface Tension

With a nuclear radius proportional to $\mathcal{A}^{1/3}$ the surface area of the nucleus is proportional to $\mathcal{A}^{2/3}$, so a fraction proportional to $\mathcal{A}^{-1/3}$ of the $\mathcal{A}$ nucleons is closer to the surface than the range of the nuclear force and therefore feels less attraction to other nucleons. This decreases the nuclear binding energy per nucleon by a term proportional to $\mathcal{A}^{-1/3}$, estimated from measured atomic weights as $-18.3\,\mathcal{A}^{-1/3}$ MeV.

Coulomb Repulsion

The electrostatic repulsion of Z protons introduces a negative term in the total binding energy proportional to Z^2 and to the inverse nuclear radius, which is proportional to $\mathcal{A}^{-1/3}$. The Coulomb contribution to the binding energy per nucleon is therefore proportional to $Z^2\mathcal{A}^{-4/3}$. It is approximately $-0.71\,Z^2\,\mathcal{A}^{-4/3}$ MeV. (The energy coefficient here is smaller than for the other terms in the binding energy because electric forces are intrinsically weaker than nuclear forces. For instance, the Coulomb energy of a uniformly charged sphere with charge Ze and radius (6.1.1) is $3Z^2e^2/5R = 0.66Z^2\mathcal{A}^{-1/3}$ MeV.)

---

[6] The numerical values of coefficients of various terms in the nuclear binding energy are rounded off here from values derived from a fit to measured binding energies by A. H. Wapstra and N. B. Gove, Nuclear Data Tables **9**, 267 (1971).

Neutron–Proton Inequality

The Pauli exclusion principle leads to a decrease in the binding energy for nuclei with unequal numbers of protons and neutrons. Given a nucleus with equal numbers of protons and neutrons, if we imagine a proton changed into a neutron the new neutron would be forced by the exclusion principle to occupy a state of energy higher than any of the originally occupied neutron states. Because of the symmetry between protons and neutrons (discussed in the next section), with equal numbers of protons and neutrons the highest energy of the originally occupied neutron states equals the highest energy of the originally occupied proton states, so changing this proton into a neutron necessarily increases its energy. The same is true if we change a neutron into a proton. This decrease in the total binding energy is approximately proportional to $(N - Z)^2/\mathcal{A}$, where $N = \mathcal{A} - Z$ is the number of neutrons. It is taken as proportional to $1/\mathcal{A}$ to take account of the decrease in the spacing of nuclear energy levels with increasing $\mathcal{A}$. Observed binding energies indicate a term in the binding energy per nucleon of -23.2 MeV $\times$ $(\mathcal{A} - 2Z)^2/\mathcal{A}^2$.

Putting this together, the binding energy per nucleon goes as follows;

$$\begin{aligned}\text{binding energy}/\mathcal{A} \simeq 15.8 - 18.3\,\mathcal{A}^{-1/3} - 0.71\,Z^2\,\mathcal{A}^{-4/3} \\ - 23.2\,(\mathcal{A} - 2Z)^2\mathcal{A}^{-2}\ \text{MeV}\,. \qquad (6.1.2)\end{aligned}$$

There are also sporadic bumps in the binding energy. Nuclei with even or odd numbers both of protons and of neutrons have an additional term in the binding energy that is about $12/\sqrt{\mathcal{A}}$ MeV or $-12/\sqrt{\mathcal{A}}$ MeV, respectively. Also, the binding energy is increased for certain "magic" numbers of protons or neutrons, to be discussed in Section 6.3.

Stable Valley and Decay Modes

For a given value of $\mathcal{A}$, the most deeply bound nucleus has a value of Z given by the stationary point of the binding energy per nucleon (6.1.2):

$$Z \simeq \frac{A}{2 + 0.015A^{2/3}}\,. \qquad (6.1.3)$$

Nuclei with smaller or larger values of Z for a given $\mathcal{A}$ tend to decay into the nucleus whose Z is given approximately by Eq. (6.1.3), with the emission of an electron or its antiparticle, the process known as beta decay, to be discussed in Section 6.5. In a contour map of nuclear masses plotted against $\mathcal{A}$ and Z, the nuclei satisfying Eq. (6.1.3) form a valley of relatively high binding energy and hence low mass, known as the *stable valley*.

For $\mathcal{A} < 50$ Eq. (6.1.3) gives Z close to $\mathcal{A}/2$, as was noticed with the earliest measurements of the atomic numbers of nuclei such as ^{4}He, ^{12}C, ^{14}N, ^{16}O, etc. As we consider nuclei with increasing values of $\mathcal{A}$ the Coulomb repulsion among the protons becomes more and more important, and the nuclei with the lowest ground state energy tend to have an increasing ratio of neutrons to protons. For instance, for $\mathcal{A} = 56$ the nucleus with the lowest ground state energy is ^{56}Fe, with 26 protons and 30 neutrons. The atomic numbers of the stable valley fall increasingly below the line $Z = \mathcal{A}/2$ for larger values of $\mathcal{A}$, to a value $Z = 92$ for $\mathcal{A} = 238$.

In the stable valley, Eqs. (6.1.2) and (6.1.3) give a binding energy per nucleon that increases with increasing $\mathcal{A}$ for lighter nuclei, owing to the decreasing effect of surface tension, reaches a maximum of about 9 MeV for iron and nickel, and then, because of the Coulomb term, decreases slowly for larger $\mathcal{A}$, taking a value of about 7.5 MeV for ^{238}U. The decrease with $\mathcal{A}$ of the binding energy per nucleon for heavy nuclei makes it energetically favorable for these nuclei to decay by splitting into fragments, either by spontaneous fission into two nuclei of much lower $\mathcal{A}$, or more often by emitting an alpha particle. After emitting one or a few alpha particles a nucleus becomes excessively neutron-rich for the new, lower, value of $\mathcal{A}$, and it becomes energetically favorable for the nucleus to lower the neutron–proton ratio by one or more beta decays, moving back toward the stable valley. These alpha and beta decay processes sometimes yield nuclei in excited states, which then undergo gamma decay to the ground state, emitting an energetic photon. A succession of alpha, beta, and gamma decays continues until the nucleus transforms into a non-radioactive nucleus, such as one of the stable isotopes of lead.

For instance, in the decay chain that is most important in the history of physics, uranium 238 alpha-decays to thorium 234 with a half life of 4.47×10^9 years, and then, with much shorter half lives, thorium 234 beta-decays to protactinium 234, which beta-decays to uranium 234, which alpha-decays to thorium 230, which alpha-decays to radium 226, which alpha-decays to radon 222 (an example of alpha decay considered in detail in Section 6.4), which alpha-decays to polonium 218, which alpha-decays to lead 214, which beta-decays to bismuth 214, which beta-decays to polonium 214, which alpha-decays to lead 210, which beta-decays to bismuth 210, which beta-decays to polonium 210, which alpha-decays to the stable isotope lead 206, which makes up 24% of natural lead.

6.2 Isotopic Spin Symmetry

There is a deep symmetry between protons and neutrons, which made it evident that neutrons are fermions and just as elementary as protons. Knowledge of this symmetry emerged in the late 1930s from a study of the forces among protons and neutrons.

Nuclear Forces

The first of the nuclear forces to be studied was that between a proton and a neutron, which could be measured by observing the scattering of neutrons on the protons in a hydrogen-rich substance such as paraffin. As in all scattering processes, the scattering amplitude $f(\hat{x})$ introduced in Section 5.6 may be expanded as a sum over terms with angular dependence proportional to the spherical harmonic functions $Y_\ell^m(\hat{x})$ defined in Section 5.2. The terms with $\ell > 0$ are suppressed at low energy by a centrifugal barrier, which makes the wave function vanish for vanishing separation r as r^ℓ, so at the energies available in the 1930s the scattering was dominated by the term with $\ell = 0$, for which the scattering amplitude f is independent of direction. But it is important here to keep track of the dependence of the scattering amplitude on spin, which we ignored in Section 5.6. With the neutron taken like the proton to have spin 1/2, there are now two terms in the amplitude for neutron–proton scattering, with total spin $s = 0$ or $s = 1$. In the absence of orbital angular momentum the total spin is conserved in the scattering process, so the total scattering cross section takes the form $\sigma_0 + \sigma_1$, where σ_s is the cross section in the $\ell = 0$ proton–neutron state with total spin s. It is possible to separate the contributions of spin zero and spin one by using data on the deuteron, a proton–neutron bound state with $\ell = 0$ (and a small admixture of $\ell = 2$) and with total angular momentum $j = 1$ and hence total spin $s = 1$. There is a classic relation[7] that to a good approximation gives $\sigma_1 = 2\pi\hbar^2/\mu B$, where μ is the reduced mass of a proton and a neutron and B is the deuteron binding energy, so using scattering data and the deuteron binding energy one can separately find σ_0 and σ_1.

This is important because protons and neutrons are fermions, so the $\ell = 0$ state of two protons or two neutrons must be antisymmetric in the particles' spin 3-components. As can be seen from either Eq. (5.4.42) or Table 5.1, this requires the state to have total spin zero. It is therefore of interest to compare the value of σ_0 deduced for proton–neutron scattering for $s = 0$ with the observed total low-energy proton–proton scattering cross section.

Unfortunately there is no way to make a target out of the electrically neutral (and, as we shall see, unstable) neutron, so it was not possible to make a direct measurement of neutron–neutron scattering. There is no similar obstacle to the measurement of proton–proton scattering for, as in Rutherford's 1919 experiments, one can make a target of a hydrogen gas or a proton-rich substance like paraffin. Here the problem is that at low energy the scattering is almost entirely due to the Coulomb potential, and reveals nothing about the nuclear forces.

---

[7] For a textbook derivation, see Section 8.8 of Weinberg, *Lectures on Quantum Mechanics*, listed in the bibliography.

The measurements in Rutherford's laboratory of the scattering of alpha particles by various nuclei had indicated that the range of nuclear forces is no larger than about $R \approx 10^{-13}$ cm. In order for two protons to approach to a distance less than this, it is necessary for their kinetic energy to be greater than $e^2/R \approx 1.4$ MeV. High-energy proton beams became available in the 1930s with the invention of accelerators with potential differences produced electrostatically, which were used[8] to make accurate measurements of proton–proton scattering. It turned out that when the scattering amplitude due to Coulomb forces was subtracted, the $\ell = 0$ part of the purely nuclear proton–proton scattering amplitude was equal to the previously measured $\ell = 0$ proton–neutron scattering amplitude in the state with total spin zero.

Isotopic Spin Rotations

The equality of forces soon led two pairs of theorists[9] to propose that the laws governing nuclear forces (whatever they are) respect a symmetry among neutrons and protons. It is not just that these laws do not change if everywhere in the equations we change neutrons into protons and protons into neutrons. That would imply that the proton–proton nuclear force is the same as the neutron–neutron force but would say nothing about their relation to the proton–neutron force. Rather, according to the proposed symmetry principle, the laws governing nuclear (but not electromagnetic) forces are invariant under what is called an *isotopic spin rotation*, which acts not on momenta or ordinary spin but on the labels of the nuclear particles. The neutron and proton are supposed to form a doublet, called the *nucleon*:

$$\begin{pmatrix} p \\ n \end{pmatrix}$$

on which isotopic spin rotations act in the same way mathematically that ordinary rotations act on the two ordinary spin states of any particle with $s = 1/2$. (Specifically, isotopic spin rotations act on the nucleon doublet as a 2×2 matrix U having the property $U^\dagger = U^{-1}$, known as unitarity, and having determinant unity. But we won't need to use this information here.) Just as we saw in Section 5.4 that the effect of infinitesimal ordinary rotations on physical states is given by an angular momentum operator $\mathbf{J}$, whose components satisfy the commutation relations $[J_i, J_j] = i\hbar\epsilon_{ijk}J_k$ (where ϵ_{ijk} is the totally antisymmetric quantity with $\epsilon_{123} = +1$, and repeated indices are summed), in the same way infinitesimal isotopic spin rotations are generated by a three-component operator $\mathbf{T}$, whose components satisfy the commutation relations

---

8 M. A. Tuve, N. Heydenberg, and L. Hafstad, Phys. Rev. **50**, 850 (1936).

9 B. Cassen and E. U. Condon, *Phys. Rev.* **50**, 846 (1936); G. Breit and E. Feenberg, *Phys. Rev.* **50**, 850 (1936).

$[T_a, T_b] = i\epsilon_{abc}T_c$. (The a, b, c indices can be taken like i, j, k to run over the values 1, 2, 3, but of course for the isotopic spin these values have nothing to do with directions in ordinary space. Repeated indices are again summed, and ϵ_{abc} like ϵ_{ijk} is a totally antisymmetric quantity with $\epsilon_{123} = 1$.) The proton and neutron are taken as the states with $T_3 = +1/2$ and $T_3 = -1/2$, respectively.

Just as two particles with ordinary spin 1/2 can combine to form a compound state with total spin s equal to 0 or 1, two nucleons can combine to form a compound state with total isotopic spin 0 or 1, which transforms under isotopic spin rotations in the same way that states with ordinary total spin 0 or 1 transform under ordinary rotations. The invariance of nuclear forces under isotopic spin rotations tells us that total isotopic spin is conserved, so the cross section for the scattering of two nucleons is the sum of a cross section for isotopic spin 1 and a cross section for isotopic spin *zero*. The states with total isotopic spin 1 form a triplet, just like orbital angular momentum states with $\ell = 1$, whose components are a proton+proton state with $T_3 = +1$, a proton+neutron state with $T_3 = 0$, and a neutron+neutron state with $T_3 = -1$. The proton+proton and neutron+neutron $\ell = 0$ states must be antisymmetric in the nucleon spin 3-components, and therefore have total ordinary spin 0. Since spin and isotopic spin commute, the proton+neutron component of this triplet must then also have spin zero. Since these three s-wave nucleon–nucleon states with total ordinary spin 0 form a triplet, the scattering cross sections are the same for each.

On the other hand, an s-wave state of two nucleons with ordinary spin 1 is symmetric in the spin 3-components, so it cannot be a proton+proton or neutron+neutron state, and can therefore only be a proton+neutron $T_3 = 0$ state of a singlet with total isotopic spin zero. This is the deuteron, with total angular momentum and total ordinary spin both equal to one.

Multiplets

The implications of isotopic spin symmetry go far beyond the equality of s-wave nucleon–nucleon cross sections for total spin zero. Before we go into this, it is necessary to say something about the relation of isotopic spin quantum numbers and electric charge. For the proton–neutron doublet, it is obvious that the electric charge of a nucleon is

$$Q = e[T_3 + 1/2] \tag{6.2.1}$$

so that protons and neutrons will have charges respectively e and 0. In a nucleus with B nucleons, the charge is the sum of (6.2.1) for all the nucleons, so

$$Q = e[T_3 + B/2] \,, \tag{6.2.2}$$

where now $\mathbf{T}$ is the isotopic spin operator of the whole nucleus and B is the number of nucleons. As we have seen, B is very close to the atomic weight A of the element, but we use the symbol B instead of A because they are not precisely equal, and in order that Eq. (6.2.2) should apply for some of the particles discovered after World War II that are not composed of protons and neutrons. In this more general context, B is known as the *baryon number*. (For some unstable particles a quantity S known as *strangeness* that is conserved in strong and electromagnetic interactions must be added to B in Eq. (6.2.2).)

Of course, electromagnetism does not respect isotopic spin symmetry: protons are charged while neutrons are not. Equation (6.2.2) shows that in electromagnetic phenomena involving the charge operator the 3-component of the isotopic spin operator plays a different role from the 1- and 2- components. There is also a nucleon mass difference, $m_n - m_p = 1.293\ \text{MeV}/c^2$, which contributes a term in the total rest mass proportional to T_3. For relatively light nuclei, with atomic numbers less than about 20 to 30, Coulomb forces are less important than nuclear forces and isotopic spin symmetry is fairly well respected, but this is not true for heavy nuclei, where the Coulomb repulsion of protons in the nucleus comes close to tearing the nucleus apart. It makes no sense to talk about isotopic spin symmetry when we are dealing with uranium.

Relatively light nuclei must form isotopic spin multiplets. We characterize any multiplet by a total isotopic spin quantum number t, defined so that (just as for ordinary spin multiplets) the multiplet consists of $2t + 1$ nuclei with T_3 equal to $t, t-1, \ldots, -t$, all with the same ordinary spin (that is, total angular momentum) and with close to the same energy. Acting on the multiplet the isotopic spin operator $\mathbf{T}$ satisfies $\mathbf{T}^2 = t(t+1)$, the proton and neutron form a $t = 1/2$ doublet, and the deuteron is a $t = 0$ singlet. There are many $t = 1/2$ doublets of complex nuclei; the lightest consists of the light isotope ^{3}He of helium, whose discovery was announced by Rutherford in his 1920 Bakerian lecture, and tritium, the radioactive isotope ^{3}H of hydrogen discovered at the Cavendish Laboratory[10] in 1934. The ^{3}He nucleus consists of two protons and one neutron and has atomic weight 3.01605, while the ^{3}H nucleus is composed of one proton and two neutrons and has atomic weight 3.01603. Both nuclei have spin 1/2.

There are also triplets of nuclear states with $t = 1$, which show again that this is a symmetry under transformations that go beyond the mere interchange of protons and neutrons. A famous example includes the ground states of the nuclei of ^{12}B and ^{12}N, which have $B = 12$ and charges $5e$ and $7e$, and hence according to Eq. (6.2.2) have $T_3 = -1$ and $T_3 = +1$. The $T_3 = 0$ member of the triplet would then be ordinary carbon, ^{12}C, with nuclear charge $6e$. But it is not the ground state of ^{12}C, which has total angular momentum $j = 0$, while the ground states of ^{12}B and ^{12}N both have $j = 1$. Also, although the ^{12}B and ^{12}N ground states have nearly equal atomic weights, 12.0144 and 12.0186, respectively, the ^{12}C ground state by definition has atomic weight 12.0000. (The greater binding energy of ^{12}C is due to two effects mentioned in the previous

[10] M. Oliphant, E. Harteck, and E. Rutherford, Nature **133**, 413 (1934); Proc. Roy. Soc. A **144**, 692 (1934).

section: the numbers of protons and neutrons in ^{12}C are equal, and both numbers are even.) The small difference in atomic weights of ^{12}B and ^{12}N is due to the greater Coulomb repulsion among the seven protons of ^{12}N than among the five protons of ^{12}B, but this cannot account for the large difference from the atomic weight of the ground state of carbon. In order to provide the $T_3 = 0$ member of a triplet with ^{12}B and ^{12}N, there would have to be a spin 1 state of ^{12}C with an excitation energy well above the ground state. Since the number of protons in ^{12}C is the average of the numbers in ^{12}B and ^{12}N, we would expect its excitation energy to be about $0.0165\, m_1 c^2$ (the average of $0.0144\, m_1 c^2$ and $0.0186\, m_1 c^2$), or, taking $m_1 c^2 = 931.5$ MeV, about 15.3 MeV. In fact there is such a state, a spin 1 state of ^{12}C that is 15.11 MeV above the ^{12}C ground state, which decays into the ground state by emission of a photon. This is the $T_3 = 0$ member of the triplet.

Why Isotopic Spin Symmetry?

One may wonder why nuclear forces should obey a symmetry principle that is not obeyed by other forces, such as those of electromagnetism. Indeed, one *should* wonder. An invariance principle that applies only to some phenomena and not others can hardly be regarded as a fundamental physical principle. This puzzle became resolved in the modern theory of strong nuclear forces known as *quantum chromodynamics*.[11] Briefly, in this theory the neutron and proton are composed of two kinds of elementary spin 1/2 particles, the up quark with charge $2e/3$ and the down quark with charge $-e/3$. In close analogy with how ^{3}He and ^{3}H are composed of protons and neutrons, the proton is composed of two up quarks and a down quark, while the neutron consists of one up quark and two down quarks. Nuclear forces in quantum chromodynamics are carried by eight fields like the electromagnetic field, only interacting with a quantum number known whimsically as color instead of charge. At the energies characteristic of nuclear phenomena these forces are much stronger than electromagnetic forces, which is why the composite nature of protons and neutrons is not apparent in most nuclear phenomena and why electromagnetism can be treated as a small perturbation in studying light nuclei. The quarks all carry the same set of colors, so strong nuclear forces do not distinguish up from down quarks, but isotopic spin symmetry is not imposed on the theory. In fact, unlike protons and neutrons, the up and down quarks have quite different masses: according to one estimate, the down quark mass is almost twice the

[11] Quantum chromodynamics is part of our present theory of elementary particles and their interactions, the Standard Model. Formulating and testing this model has been the work of many physicists. For an informal history see Weinberg, "Half a Century of the Standard Model," listed in the bibliography. A more detailed account with references to much of this work can be found in Weinberg, *The Quantum Theory of Fields, Vol. II: Modern Applications* (Cambridge University Press, Cambridge, UK, 1996).

up quark mass. The reason for the isotopic spin symmetry of strong forces is just that there is no room in the theory for any violation of the symmetry other than the quark masses, and the quark masses although unequal are very small. Almost all of the masses of the proton and neutron comes from the strong nuclear forces acting among the quarks within a single proton or neutron, not from the quark masses.

The small mass difference between the proton and the neutron comes both from differences in the quark masses and from electromagnetic forces among the quarks, but the quark mass difference is somewhat more important. This why the neutron is heavier than the proton, even though the electric charges of the quarks in the proton are larger than those in the neutron. It is both the smallness of the quark masses and the relative weakness of electromagnetic effects that makes the neutron–proton mass difference, 1.293 MeV/c^2, so tiny compared with the proton mass, 938 MeV/c^2.

Pions

Isotopic spin symmetry had important implications for the new strongly interacting particles discovered after World War II. The first of these particles was the pi meson, or pion as it is frequently called. In 1947 a group at the University of Bristol,[12] studying photographic plates that had been exposed to cosmic rays at high altitudes in the Pyrenees and Andes, found evidence of a strongly interacting particle with a mass intermediate (hence the name "meson") between the electron and the nucleon. It is today known that these charged pions come with charges $+e$ and $-e$, both with masses 139.570 MeV/c^2. These particles are produced singly in reactions such as $p + p \to p + n + \pi^+$, and so if baryon number is conserved these particles must be supposed to have $B = 0$. Equation (6.2.2) then indicates that the π^+ and π^- have $T_3 = +1$ and $T_3 = -1$, respectively. No doubly charged particles with similar mass have ever been found, so the pions cannot be part of an isotopic spin multiplet with $t \geq 2$, and therefore must be part of a triplet, with $t = 1$. The neutral $T_3 = 0$ member of the triplet, the π^0, was discovered at the Berkeley cyclotron in 1950 – the first particle to be found at an accelerator before it being discovered in cosmic rays. The mass of the π^0 is now known to be 134.977 MeV/c^2.

In quantum chromodynamics, the π^+ and π^- are respectively $u + \overline{d}$ and $d + \overline{u}$, where u and d stand for up and down quarks, and the bar denotes antiquarks. The π^0 is a 50–50 superposition of $u + \overline{u}$ and $d + \overline{d}$. The quark masses contribute equally to all three pions, so the $\simeq$4.6 MeV/c^2 mass difference between charged and neutral pions is entirely due to electromagnetic forces. In fact, this is the one mass difference in an isotopic spin multiplet

[12] C. M. G. Lattes, H. Muirhead, G. P. S. Ochiallini, and C. F. Powell, Nature **159**, 694 (1947).

of elementary particles that has been successfully calculated as a purely electromagnetic effect.

Although the charged and neutral pions are joined in an isotopic spin triplet, their decays occur through interactions that do not respect isotopic spin symmetry and hence they have very different decay rates and decay modes. The neutral pion decays into two photons through purely electromagnetic interactions, with mean lifetime $(8.52 \pm 0.18) \times 10^{-12}$ seconds. The charged pion decays much more slowly through the weak interactions discussed in Section 6.5, with mean lifetime $(2.6033 \pm 0.0005) \times 10^{-8}$ seconds, primarily into a neutrino and a muon, a particle similar to an electron but 210 times heavier, discovered in cosmic rays in 1937.

Appendix: The Three–Three Resonance

There are no clearly identified multiplets of nuclear states larger than triplets, but there is a conspicuous quartet of unstable particles that decay into a nucleon and a pion, with masses all close to 1210 MeV$/c^2$. This is the "three–three resonance" Δ, where "three–three" means that it has $t = 3/2$ and $j = 3/2$, and "resonance" indicates that these are seen as sharp peaks in pion–nucleon scattering, interpreted as the formation of an unstable intermediate state that decays back into a nucleon and a pion. As discussed at the end of the appendix to Section 5.6, the total decay rate of each of these four states is measured as the width of the peak of the cross section as a function of energy, divided by $\hbar$; the rate of decay into any particular pion–nucleon state equals the total decay rate times the branching ratio, the fraction of scattering events at the resonant energy that produce that pion–nucleon state.

Since the formation and decay of the Δ both indicate that it has the same baryon number $B = 1$ as the nucleon, Eq. (6.2.2) indicates that the four states of the quartet with charges $2e$, e, 0, and $-e$ have $T_3 = 3/2$, $T_3 = 1/2$, $T_3 = -1/2$, and $T_3 = -3/2$. Like the proton and neutron the Δ states are interpreted as composites of three quarks: respectively uuu, uud, udd, and ddd.

The three–three resonance provides a good example of the power of symmetry principles such as isotopic spin symmetry to do more than dictate how energy eigenstates are grouped into multiplets. The conservation of isotopic spin tells us that the nucleon and pion produced when a Δ decays must be in a state of total isotopic spin 3/2 rather than a mixture of isotopic spins 3/2 and 1/2. For a three–three resonance Δ with a given value of T_3, the nucleon–pion state has wave function

$$\sum_{\pm} C_{1,1/2}(3/2, T_3; T_3 \mp 1/2, \pm 1/2)\psi^{\pi\,N}_{T_3 \mp 1/2, \pm 1/2} ,$$

where $\psi^{\pi\,N}_{T_3 \mp 1/2, \pm 1/2}$ is the wave function for a pion and a nucleon with their third components of isotopic spin equal respectively to $T_3 \mp 1/2$ and $\pm 1/2$, and

$C_{1,1/2}(3/2, T_3; t, t')$ is the Clebsch–Gordan coefficient discussed in Section 5.4. The rates of decay of a three–three resonance with a given T_3 into various pion–nucleon states are then given by

$$\Gamma(\Delta(T_3) \to \pi(T_3 \mp 1/2) + N(\pm 1/2))$$
$$= \Gamma_\Delta \left| C_{1,1/2}(3/2, T_3; T_3 \mp 1/2, \pm 1/2) \right|^2 ,$$

where Γ_Δ is the total decay rate of a three–three particle of any charge; it is another consequence of isotopic spin symmetry that these total decay rates are the same for all four charges of the Δ. Looking up the Clebsch–Gordan coefficients in Table 5.1 for combining states of spin 1 and 1/2 to form a state of spin 3/2, we see that for the $T_3 = 1/2$ state Δ^+ we have

$$\Gamma(\Delta^+ \to \pi^+ + n) = \Gamma_\Delta/3 , \quad \Gamma(\Delta^+ \to \pi^0 + p) = 2\Gamma_\Delta/3 ,$$

while, for the $T_3 = -1/2$ state Δ^0,

$$\Gamma(\Delta^0 \to \pi^- + p) = \Gamma_\Delta/3 , \quad \Gamma(\Delta^0 \to \pi^0 + n) = 2\Gamma_\Delta/3 .$$

For the Δ^{++} and Δ^- there is only one available decay channel, so without looking up Clebsch–Gordan coefficients we know that

$$\Gamma(\Delta^{++} \to \pi^+ + p) = \Gamma(\Delta^- \to \pi^- + n) = \Gamma_\Delta .$$

These predictions were verified in experiments on pion–nucleon scattering carried out by Fermi's group at Chicago in the early 1950s.

6.3 Shell Structure

In nuclei as in atoms it is a fair approximation to adopt a Hartree approximation, in which each nucleon feels an effective potential due to all the other nucleons. Neutrons and protons are fermions, so their states in nuclei are governed by the Pauli exclusion principle, like the states of electrons in atoms. In particular, there are nuclei in which protons or neutrons or both form closed shells like the electrons in noble gases, and therefore are more tightly bound than other nuclei of similar weight.

The great difference between the closed shells in atoms and nuclei arises from the difference in the form of their effective potentials. Both potentials have approximate spherical symmetry, but in nuclei, unlike atoms, there is nothing special at the center of symmetry that would make the nuclear potential singular there. Since the nuclear potential is a function only of the radial coordinate r, and is expected to be analytic in the Cartesian components of the coordinate vector $\mathbf{x}$, it must be a power series in $r^2 = \mathbf{x}^2$. Within some neighborhood of the origin, it is therefore approximately linear in $\mathbf{x}^2$, a relation we shall write as

$$V(\mathbf{x}) \simeq V_0 + \frac{1}{2} m_N \omega^2 \mathbf{x}^2 \tag{6.3.1}$$

where m_N can be taken as the mean nucleon mass, and ω is a constant with the dimensions of frequency. The total Hamiltonian is then a sum of one-nucleon Hamiltonians, each of the form

$$H = V_0 + \frac{\mathbf{P}^2}{2m_N} + \frac{m_N \omega^2}{2} \mathbf{X}^2 , \tag{6.3.2}$$

with $\mathbf{X}$ the operator that multiplies the wave function with the coordinate argument $\mathbf{x}$, and $\mathbf{P}$ the operator that acts on the wave function as the differential operator $-i\hbar\boldsymbol{\nabla}$. This is the Hamiltonian for an harmonic oscillator with circular frequency ω, the first problem solved using Heisenberg's matrix mechanics at the beginning of quantum mechanics.[13]

To find the spectrum of eigenvalues of this Hamiltonian, we introduce a vector operator

$$\mathbf{a} \equiv \frac{1}{\sqrt{2m_N \omega \hbar}} \mathbf{P} - i \sqrt{\frac{m_N \omega}{2\hbar}} \mathbf{X} . \tag{6.3.3}$$

Recalling the commutation relations (5.3.22),

$$[X_i, P_j] = i\hbar\delta_{ij} , \quad [X_i, X_j] = [P_i, P_j] = 0 , \tag{6.3.4}$$

it is straightforward to calculate that

$$[a_i, a_j^\dagger] = \delta_{ij} , \qquad [a_i, a_j] = [a_i^\dagger, a_j^\dagger] = 0 . \tag{6.3.5}$$

The Hamiltonian (6.3.1) can be expressed as

$$H = V_0 + \frac{\hbar\omega}{2} \big[\mathbf{a} \cdot \mathbf{a}^\dagger + \mathbf{a}^\dagger \cdot \mathbf{a} \big] .$$

Using the commutators (6.3.5), this is

$$H = V_0 + \frac{3\hbar\omega}{2} + \hbar\omega\, \mathbf{a}^\dagger \cdot \mathbf{a} . \tag{6.3.6}$$

The operators $\mathbf{a}^\dagger$ and $\mathbf{a}$ play the role of raising and lowering operators for the energy. Using Eq. (6.3.6) and the commutation relations (6.3.5), we easily see that

$$[\mathbf{a}, H] = \hbar\omega\mathbf{a} , \qquad [\mathbf{a}^\dagger, H] = -\hbar\omega\mathbf{a}^\dagger . \tag{6.3.7}$$

It follows that if $H\psi = E\psi$, then

$$H(\mathbf{a}^\dagger \psi) = (E + \hbar\omega)(\mathbf{a}^\dagger \psi) , \tag{6.3.8}$$

[13] W. Heisenberg, Zeit. Phys. **33**, 879 (1925). This article is reprinted in English in Van der Waerden, *Sources of Quantum Mechanics*, listed in the bibliography.

$$H(\mathbf{a}\psi) = (E - \hbar\omega)(\mathbf{a}\psi) \,. \tag{6.3.9}$$

We assume that there is a one-nucleon state with some minimum energy E_0. In this case the wave function ψ_0 for this state must satisfy

$$\mathbf{a}\psi_0 = 0 \,, \tag{6.3.10}$$

since otherwise according to Eq. (6.3.9) the wave function $\mathbf{a}\psi_0$ would be an energy eigenfunction with an even smaller energy, $E_0 - \hbar\omega$. Using $\mathbf{a}\psi_0 = 0$, Eq. (6.3.6) then gives

$$H\psi_0 = E_0\psi_0 \,, \tag{6.3.11}$$

where E_0 is the minimum energy:

$$E_0 = V_0 + \frac{3\hbar\omega}{2} \,.$$

The energy $\hbar\omega/2$ associated with each of the three coordinate components is known as the *zero-point energy*. The appearance of a zero-point energy for harmonic oscillators is an inevitable feature of quantum mechanics. Inspection of the Hamiltonian (6.3.2) shows that for a state to have energy as low as V_0 its wave function would have to be an eigenfunction of both **P** and **X** with eigenvalues zero, which is impossible since the commutator $[X_i, P_i] = i\hbar$ cannot vanish acting on any wave function.

Equation (6.3.8) shows that acting on any wave function with any component $a_i^\dagger$ raises the energy of the state by $\hbar\omega$, so we can find a complete set of energy eigenfunctions

$$\psi_{n_1 n_2 n_3} \equiv (a_1^\dagger)^{n_1}(a_2^\dagger)^{n_2}(a_3^\dagger)^{n_3}\psi_0 \tag{6.3.12}$$

for which

$$H\psi_{n_1 n_2 n_3} = (E_0 + n\hbar\omega)\psi_{n_1 n_2 n_3} \,, \tag{6.3.13}$$

where $n = n_1 + n_2 + n_3$.

We could just as well construct an eigenfunction of H with eigenvalue $E_0 + n\hbar\omega$ if in place of $(a_1^\dagger)^{n_1}(a_2^\dagger)^{n_2}(a_3^\dagger)^{n_3}$ we operated on ψ_0 with any homogeneous polynomial of order n in the components of $\mathbf{a}^\dagger$ – that is, any sum of terms, each proportional to a product like that in Eq. (6.3.12) of a total of n factors of components of $\mathbf{a}^\dagger$. In order to make clear the angular momentum content of these states, it is much more convenient to use the set of homogeneous polynomials encountered in Eq. (5.2.16). Expressed as a function of any vector **v**, these are

$$\mathcal{Y}_\ell^m(\mathbf{v}) \equiv |\mathbf{v}|^\ell Y_\ell^m(\hat{v}) \,, \tag{6.3.14}$$

where Y_ℓ^m is the spherical harmonic function described in Section 5.2, with ℓ a non-negative integer and m an integer running over the $2\ell+1$ values from $-\ell$ to $+\ell$. For instance, $\mathcal{Y}_0^0(\mathbf{v})$ is a constant, and

$$\mathcal{Y}_1^{\pm 1}(\mathbf{v}) = \mp\sqrt{\frac{3}{8\pi}}(v_1 \pm i v_2)\ , \quad \mathcal{Y}_1^0(\mathbf{v}) = \sqrt{\frac{3}{4\pi}}\ v_3\ .$$

We can find a complete set of states with energy $E_0+n\hbar\omega$ and angular momentum quantum number ℓ for which $n-\ell$ is an even non-negative integer:

$$\psi^m_{n,\ell} = (\mathbf{a}^\dagger \cdot \mathbf{a}^\dagger)^{(n-\ell)/2}\mathcal{Y}_\ell^m(\mathbf{a}^\dagger)\psi_0\ . \tag{6.3.15}$$

For instance, for $n=0$ we have only $\ell=0$, and $\psi^0_{0,0}$ is proportional to the minimum-energy wave function ψ_0. For $n=1$ we have only $\ell=1$, and $\psi^m_{1,1} = \mathcal{Y}_1^m(\mathbf{a}^\dagger)\psi_0$. For $n=2$ we have both $\ell=2$, with $\psi^m_{2,2} = \mathcal{Y}_2^m(\mathbf{a}^\dagger)\psi_0$ and also $\ell=0$, with $\psi^0_{2,0} \propto (\mathbf{a}^\dagger\cdot\mathbf{a}^\dagger)\psi_0$.

All but the lowest energy states are evidently degenerate. As we have seen, for energy levels with $n=1$ and $n=2$ there are respectively three and $5+1=$ six states with energies respectively $E_0+\hbar\omega$ and $E_0+2\hbar\omega$. In general, the number $\#_n$ of states with energy $E_0+n\hbar\omega$ is the sum of $2\ell+1$ for all non-negative integers ℓ with $n-\ell$ an even non-negative integer 2ν. That is,

$$\#_n = \begin{cases} \sum_{\nu=0}^{n/2}(2n-4\nu+1) = (2n+1)(n/2+1) - 2(n/2)(n/2+1) \\ \text{for } n \text{ even} \\ \sum_{\nu=0}^{(n-1)/2}(2n-4\nu+1) = (2n+1)((n-1)/2+1) \\ -2((n-1)/2)((n-1)/2+1) \\ \text{for } n \text{ odd} \end{cases}$$

and so, whether n is even or odd, the degeneracy (apart from spin) is

$$\#_n = (n+1)(n+2)/2\ . \tag{6.3.16}$$

This can be recognized as the number of ways an integer n can be written as a sum of three non-negative integers, so this is also the number of independent wave functions $\psi_{n_1 n_2 n_3}$ with $n = n_1+n_2+n_3$ defined by Eq. (6.3.12). Thus the wave functions (6.3.15) form a complete set of eigenfunctions of H with eigenvalue $E_0+n\hbar\omega$.

It has been possible to work out the energy eigenvalues and their degeneracies here (as Heisenberg did in 1925) without examining the form of these wave functions as functions of the nucleon coordinates, but it will help to make our discussion more concrete if we take a moment to look at these wave functions. By using Eq. (6.3.3), the defining (6.3.10) for the wave function of the state of minimum energy can be written explicitly in a first-order differential equation

$$\left[\sqrt{\frac{\hbar}{2m_N\omega}}\nabla + \sqrt{\frac{m_N\omega}{2\hbar}}\mathbf{x}\right]\psi_0(\mathbf{x}) = 0\,. \tag{6.3.17}$$

The solution (with arbitrary normalization) is

$$\psi_0(\mathbf{x}) = \exp\left[-\frac{m_N\omega}{2\hbar}\mathbf{x}^2\right]\,. \tag{6.3.18}$$

The wave functions $\psi^m_{n,\ell}(\mathbf{x})$ can be found using Eq. (6.3.15), with $\psi_0(\mathbf{x})$ given by Eq. (6.3.18), and with $a^\dagger$ replaced with the differential operator

$$a^\dagger = \frac{-i}{\sqrt{2m_N\omega\hbar}}\nabla + i\sqrt{\frac{m_N\omega}{2\hbar}}\mathbf{x}\,.$$

For instance,

$$\psi^m_{1,1}(\mathbf{x}) \propto |\mathbf{x}|Y^m_1(\hat{x})\exp\left[-\frac{m_N\omega}{2\hbar}\mathbf{x}^2\right]\,.$$

Taking into account the two spin states of a nucleon, the actual degeneracy of the energy level $E = E_0 + n\hbar\omega$ is twice the quantity (6.3.16), or

$$(n+1)(n+2) = 2,\ 6,\ 12,\ 20,\ 30,\ 42,\ \ldots$$

This leads to the expectation that the protons or neutrons in a nucleus would all form closed shells if the number of protons or of neutrons were equal to

$$2,\ 2+6=8,\ 8+12=20,\quad \text{etc.}$$

These are the so-called *magic numbers* of nuclear physics,[14] analogous to the atomic numbers 2, 10, 18, etc., of the noble gases in atomic physics. We expect nuclei with a magic number of protons or neutrons to be more deeply bound and hence more abundant than other nuclei with similar numbers of neutrons and protons. A nucleus is likely to be particularly deeply bound if it is doubly magic, with a magic number of both protons and neutrons. Indeed, the lightest doubly magic nuclei are ^{4}He, ^{16}O, and ^{40}Ca, which are more tightly bound and abundant than other nuclei of similar weight.

One might expect the magic number following 20 to be $20+20 = 40$, but this is not the case. The degenerate multiplets we found for the harmonic oscillator begin, for heavier nuclei, to be split in energy, both by the interaction of the spin and orbital angular momenta of the nucleons and from the breakdown of the harmonic oscillator approximation (6.3.1) as nucleons in high energy levels spend increasing time away from the nuclear center. In particular, there is a term in the Hamiltonian for each nucleon proportional to $\mathbf{S}\cdot\mathbf{L}$ with a large

[14] M. Goeppert-Mayer and J. H. D. Jensen, *Elementary Theory of Nuclear Shell Structure* (Wiley, New York, 1955).

negative coefficient, which for each n lowers the energy of the single-nucleon state with the largest orbital angular momentum $\ell = n$ and largest total angular momentum, $j = \ell + 1/2 = n + 1/2$, below the energies of other single-nucleon states with the same n.

Without these corrections, the $n = 3$ energy level would have 20 degenerate states with $\ell = 3$ and $\ell = 1$, but these corrections lower the eight $f_{7/2}$ states below the other 12 states, so the magic number following 20 is not 40, but $20 + 8 = 28$. The element with 28 protons is nickel, which is known to be produced abundantly by nuclear reactions occurring in core-collapse supernovae. The most abundant isotope of nickel is not the doubly magic ^{56}Ni; this isotope is less abundant than either ^{58}Ni or ^{60}Ni, which have a magic number only of protons. This is because the negative nuclear potential energy of the additional neutrons is needed to compensate for the Coulomb repulsion of the 28 protons. Even so, as noted in Section 3.4, the deep binding of nickel isotopes makes nickel an exception to the rule that atomic weight steadily increases with atomic number.

The same pattern repeats for larger nucleon numbers. The next shell has nucleons in the $20 - 8 = 12$ states with $n = 3$ and $j < 7/2$, and in the 10 $n = 4$ states with $\ell = 4$ and $j = \ell + 1/2 = 9/2$, giving a magic number $28 + 12 + 10 = 50$. The next shell has nucleons in the $30 - 10 = 20$ states with $n = 4$ and $j < 9/2$, and in the 12 $n = 4$ states with $\ell = 5$ and $j = \ell + 1/2 = 11/2$, giving a magic number $50 + 20 + 12 = 82$. Finally, the next shell has nucleons in the $42 - 12 = 30$ states with $n = 5$ and $j < 11/2$, and in the 14 $n = 6$ states with $\ell = 6$ and $j = \ell + 1/2 = 13/2$, giving a magic number $82 + 30 + 14 = 126$. Thus the complete list of magic numbers is

$$2,\ 8,\ 20,\ 28,\ 50,\ 82,\ 126\ .$$

The only stable doubly magic nucleus heavier than calcium 40 is lead 208.

6.4 Alpha Decay

As we saw in Section 3.3, in the first decade of the twentieth century Rutherford and his collaborators were able to distinguish two kinds of radioactivity. One was beta decay, the subject of Section 6.5. The other was alpha decay, the emission of a charged alpha particle, soon identified as a helium 4 nucleus. These alpha particles furnished Rutherford with a probe of atomic structure, with which he discovered the nucleus of the atom.

Alpha decay has the remarkable feature that to get out of the nucleus the alpha particle must pass through a potential barrier that according to classical physics it cannot inhabit, because the potential energy there is greater than the total energy of the alpha particle. Only because of the wave nature of particles in quantum mechanics is it possible for the alpha particle to leak through the barrier. The presence of this barrier gives the rate of alpha decay an extreme sensitivity to the energy of the emitted alpha particle and the radius of the nucleus. Similar Coulomb barriers govern the rate of spontaneous nuclear fission and of nuclear reactions in stars.

We will assume spherical symmetry, and to avoid mathematical complications consider only s-wave ($l = 0$) decays, which are the most common. The Schrödinger (5.2.19) for the radial wave function $R_E(r)$ with alpha particle energy E and $\ell = 0$ takes the form

$$-\frac{\hbar^2}{2m_\alpha}\frac{1}{r^2}\frac{d}{dr}\left[r^2\frac{dR_E(r)}{dr}\right] + V(r)R_E(r) = ER_E(r)\,, \tag{6.4.1}$$

where $V(r)$ is taken to include both the Coulomb repulsion and the nuclear attraction between the alpha particle and the rest of the nucleus. We take $E > 0$, so that it is energetically possible for the alpha particle to exist far from the nucleus. It proves very convenient to write this instead as a differential equation for the *reduced wave function* $u_E(r) \equiv r\,R_E(r)$:

$$-\frac{\hbar^2}{2m_\alpha}\frac{d^2}{dr^2}u_E(r) + V(r)u_E(r) = Eu_E(r)\,. \tag{6.4.2}$$

As we saw in Section 5.2, the boundary condition for general orbital angular momentum ℓ is that, for $r \to 0$, $R_E(r)$ is proportional to r^ℓ and hence $u_E(r)$ is proportional to $r^{\ell+1}$, so for $\ell = 0$ the condition is that $u_E(r) \propto r$ for $r \to 0$.

It is assumed that for r less than the nuclear radius R the potential $V(r)$ is dominated by the nuclear attraction, which gives it negative values. For r greater than R the nuclear attraction is presumed to be ineffective, so $V(r)$ becomes positive:

$$V(r) = \frac{2Ze^2}{r} \quad \text{for } r > R\,, \tag{6.4.3}$$

where Ze is the electric charge of the final nucleus. We assume that for some range of r greater than R, this potential is greater than E. This is the region that classically cannot be inhabited by the alpha particle. (See Fig. 6.1.)

To see how the wave function behaves in this region, it is convenient to rewrite Eq. (6.4.2) for $r > R$ as

$$\frac{d^2}{dr^2}u_E(r) = \kappa_E^2(r)u_E(r)\,, \tag{6.4.4}$$

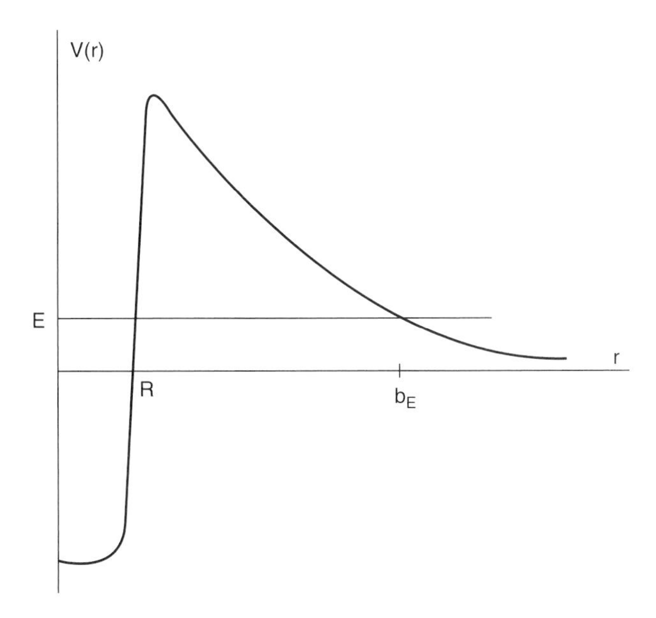

Figure 6.1 An example of the potential $V(r)$ felt by an alpha particle at a distance r from the nuclear center.

where $\kappa_E(r)$ can be taken as the positive square root,

$$\kappa_E(r) = +\sqrt{\frac{2m_\alpha}{\hbar^2}(V(r) - E)} \,. \tag{6.4.5}$$

We note that if $V(r)$ and hence $\kappa_E(r)$ were independent of r, then Eq. (6.4.4) would have solutions proportional to $\exp(\pm\kappa_E\, r)$. It therefore may be guessed that if $\kappa_E(r)$ varies sufficiently slowly with r then the wave function within the barrier takes the approximate form

$$u_E(r) = C_+(E) A_{E+}(r) \exp\left(+\int_R^r \kappa_E(r)\, dr\right) + C_-(E) A_{E-}(r) \exp\left(-\int_R^r \kappa_E(r)\, dr\right) \,, \tag{6.4.6}$$

where the amplitudes $A_{E\pm}(r)$ vary more slowly than the exponentials, and the $C_\pm(E)$ are r-independent factors determined by the conditions that the values and first derivatives of $u_E(r)$ are continuous at the nuclear radius R. The appendix to this section shows that $A_{E+}(r) = A_{E-}(r) = 1/\sqrt{\kappa_E(r)}$, and describes the conditions on $\kappa_E(r)$ under which Eq. (6.4.6) is a good approximation.

Now, if the barrier extended to infinity with $V(r) > E$ then the only allowed values of energy would be those for which the growing exponential term in Eq. (6.4.6) was absent, which would require that E takes a value where $C_+(E) = 0$. These would be the energies of the true bound states of the alpha particle in the nucleus. In fact, $V(r)$ falls to the value E at a radial coordinate $r = b_E$:

$$b_E = 2Ze^2/E \tag{6.4.7}$$

and $V(r) < E$ for $r > b_E$. The condition $C_+(E) = 0$ picks out the energies of unstable states, for which the wave function becomes exponentially small outside the barrier, though not zero.

For instance, if $V(r)$ in the nucleus were a negative constant $-V_0$, then the general solution of Eq. (6.4.2) for $r < R$ would be a linear combination of $\sin qr$ and $\cos qr$, where

$$q \equiv +\frac{1}{\hbar}\sqrt{2m_\alpha(E+V_0)}\ . \tag{6.4.8}$$

The boundary condition that $u_E(r) \propto r$ for $r \to 0$ tells us that (with an arbitrary normalization) the physical solution for $r < R$ is

$$u_E(r) = \sin qr\ . \tag{6.4.9}$$

In this case the continuity at R of the values and first derivatives of the wave functions (6.4.6) and (6.4.9) (with $A_{E\pm}(r) = 1/\sqrt{\kappa_E(r)}$ assumed to vary much more slowly than the exponentials) gives

$$\frac{1}{\sqrt{\kappa_E(R)}}[C_+ + C_-] = \sin qR\ , \qquad \sqrt{\kappa_E(R)}[C_+ - C_-] = q\cos qR\ ,$$

and therefore, for a constant potential in the nucleus,

$$C_\pm(E) \simeq \frac{\sqrt{\kappa_E(R)}}{2}\left[\sin qR \pm \left(\frac{q}{\kappa_E(R)}\right)\cos qR\right]\ . \tag{6.4.10}$$

The condition $C_+(E) = 0$ requires that $\tan qR = -q/\kappa_E(R)$. For a very deep potential well, with $\kappa_E(R)$ much less than $\sqrt{2m_\alpha V_0/\hbar}$ and hence much less than q, the unstable state with lowest energy has q slightly greater than $\pi/2R$.

At a value of E where $C_+(E) = 0$, the wave function outside the barrier is suppressed by a factor $\exp(-G(E))$, where

$$G(E) = \int_R^{b_E} \kappa_E(r)\,dr = \sqrt{\frac{4m_\alpha Ze^2}{\hbar^2}}\int_R^{b_E}\sqrt{\frac{1}{r} - \frac{1}{b_E}}\,dr$$
$$= \sqrt{\frac{4m_\alpha Ze^2 b_E}{\hbar^2}}\,f(R/b_E) = \frac{4Ze^2}{\hbar v_\alpha}f(R/b_E)\ , \tag{6.4.11}$$

where

$$f(x) \equiv \int_x^1 \sqrt{\frac{1}{z} - 1}\, dz = \frac{\pi}{2} - \sqrt{x(1-x)} - \arcsin\sqrt{x}\ , \tag{6.4.12}$$

and $v_\alpha = \sqrt{2E/m_\alpha}$ is the velocity of the alpha particle when it escapes far from the nucleus. At the energy of an unstable state, where $C_+(E) = 0$, the probability density $|R_E(b_E)|^2$ at the outer radius of the barrier is suppressed by a factor of order $\exp(-2G(E))$.

In the earliest successful theory of alpha decay,[15] this factor was interpreted as the probability that an alpha particle coming out of the nucleus would penetrate the Coulomb barrier. That is, the rate Γ_α of alpha decay was presumed to take the form

$$\Gamma_\alpha = \nu \exp(-2G(E))\ , \tag{6.4.13}$$

where ν is some sort of rate factor that reflects conditions within the nucleus. The factor ν is commonly estimated as the rate $\nu \simeq \mathcal{V}/R$ at which alpha particles inside the nucleus classically would strike the nuclear surface, where $\mathcal{V}$ is a typical alpha particle velocity inside the nucleus and R is the nuclear radius. As we have seen, for a very deep potential well the alpha particle wave number inside the nucleus is close to $\pi/2R$, so $\mathcal{V}/R \simeq \hbar\pi/2m_\alpha R^2$, which for a large nucleus with $R \simeq 9 \times 10^{-13}$cm is $3 \times 10^{20}\ \text{sec}^{-1}$. The rate factor ν is usually quoted as $10^{21}\ \text{sec}^{-1}$.

This is sometimes expressed in terms of the spacing of energy levels. For a flat deep nuclear potential with $q \gg \kappa_E(R)$, the energy levels of unstable states where $C_+(E)$ vanishes are at $qR \simeq (n + 1/2)\pi$ with $n = 0, 1, 2, \ldots$, so that their wave numbers are spaced by $\Delta q = \pi/R$. The spacing D in energy is then $D \simeq (dE/dq)\Delta q = \hbar\mathcal{V}/R$, so $\mathcal{V}/R \simeq D/\pi\hbar$.[16]

The appendix to this section gives a thoroughly quantum-mechanical derivation of the decay rate that dispenses with the semi-classical picture of an alpha particle in the nucleus striking the nuclear surface and occasionally leaking through. The rate of decay of an unstable state with energy E_1 is found to be given by Eq. (6.4.54):

$$\Gamma_\alpha = \left|\frac{C_-(E_1)}{\hbar C'_+(E_1)}\right| e^{-2G(E_1)}\ . \tag{6.4.14}$$

The factor multiplying $e^{-2G(E_1)}$ in Eq. (6.4.14) is of the same order of magnitude as the rate factor $\mathcal{V}/R$. For instance, for a flat nuclear potential, Eq. (6.4.10) suggests that the derivative of $C_+(E)$ with respect to wave number is of order

---

[15] G. Gamow, Zeit. f. Physik **52**, 510 (1929); E. U. Condon and R. W. Gurney, Phys. Rev. **33**, 127 (1929).

[16] The rate factor ν multiplying $\exp(-2G)$ is sometimes instead estimated as $\nu \simeq D/2\pi\hbar$; for instance, see J. M. Blatt and V. F. Weisskopf, *Theoretical Nuclear Physics* (John Wiley & Sons, New York, 1952), Section XI.2.

$\kappa_E^{1/2}(R)R$, while $C_-(E_1)$ is of order $\kappa_E^{1/2}(R)$, so $C'_+(E)/C_-(E)$ is of order $(dq/dE)R = R/\hbar\mathcal{V}$ and the factor $C_-(E_1)/\hbar C'_+(E_1)$ in Eq. (6.4.14) is therefore of order $\mathcal{V}/R$.

It must be admitted that taking the rate factor ν as $|C_-(E_1)/\hbar C'_+(E_1)|$ instead of $\mathcal{V}/R$ or $D/\hbar\pi$ is not very important, because none of these estimates take into account the probability that an alpha particle will somehow become detached inside the nucleus from the rest of the nucleus. But at least Eq. (6.4.14) is a precise statement (for thick barriers with a slowly varying potential) of the rate at which an alpha particle that has become detached inside the nucleus will escape, and it does not depend on semi-classical hand-waving.

This theory does correctly describe the extreme sensitivity of alpha particle decay rates to the energy and the nuclear radius, due almost entirely to the barrier penetration factor $\exp(-2G(E))$. In particular, without needing to worry about the rate factor ν we can use the above results for the barrier penetration exponent $G(E)$ to understand the trend of the dependence of the logarithm of the mean lifetime $\tau_\alpha = 1/\Gamma_\alpha$ on energy. Note that for a thick barrier with $b_E \gg R$, the leading and next-to-leading terms in the expansion of Eq. (6.4.11) in powers of R/b_E give

$$G(E) = \frac{4Ze^2}{\hbar v_\alpha}\left[\frac{\pi}{2} - 2\sqrt{\frac{R}{b_E}} + O\left(\frac{R}{b_E}\right)^{3/2}\right]. \tag{6.4.15}$$

Since $v_\alpha \propto \sqrt{E}$ and $b_E \propto 1/E$, we have

$$\ln \tau_\alpha \propto G(E) = \frac{\alpha}{\sqrt{E}} + \beta + O(E)\,, \tag{6.4.16}$$

with α and β constant in energy. This dependence of $\ln \tau_\alpha$ on energy was originally noticed in 1911 as a dependence of the alpha particle range in air on energy, and in that form is known as the *Geiger–Nuttall law*.[17]

For a numerical example let us consider the historically important decay process $^{226}\text{Ra} \to {}^{222}\text{Rn} + {}^4\text{He}$. The nuclei ^{226}Ra, ^{222}Rn, and ^{4}He all have spin zero and even parity, so the alpha particle in this decay has $\ell = 0$, as we assumed in our calculation of $G(E)$. The alpha particles from this decay have a velocity $v_\alpha = 1.519 \times 10^9$ cm/sec, and radon has $Z = 86$, so here the first factor in Eq. (6.4.11) for $G(E)$ is $4Ze^2/\hbar v_\alpha = 49.55$. Also, $b_E = 5.18 \times 10^{-12}$ cm. According to Eq. (6.1.1) the radius of ^{222}Rn is approximately 7.9×10^{-13} cm, to which we should add the radius $\simeq 2 \times 10^{-13}$ of ^{4}He, and so the effective nuclear radius here is $R \simeq 9.9 \times 10^{-13}$ cm, and $R/b_E \simeq 0.19$. The function (6.4.12) is then $f(R/b_E) = 0.72$. Equation (6.4.11) then gives $G(E) = 35.7$, and the barrier penetration probability is $\exp(-2G) \approx 10^{-31}$. If we take $\nu \simeq 10^{21}$ sec^{-1} then Eq. (6.4.13) gives a radium mean life $1/\Gamma_\alpha$ of order 10^{10}

[17] H. Geiger and J. M. Nuttall, Phil. Mag. **22**, 613 (1911); **23**, 439 (1912).

sec. It is the smallness of the factor $\exp(-2G)$ that is responsible for the radium 226 nuclei produced in a chain of radioactive decays from uranium 238 living long enough to be discovered in uranium ores in 1898 by Marie and Pierre Curie. The predicted mean lifetime, of order 10^{10} sec, may be compared with the measured mean life of 2300 years $= 7 \times 10^{10}$ sec. The agreement, such as it is, is somewhat accidental, because the decay rate is so sensitive to the nuclear radius R. For instance, if we had taken the effective nuclear radius as $R = 9.3 \times 10^{-13}$ cm instead of $R = 9.9 \times 10^{-13}$ cm, then with everything else the same we would have found a predicted mean life of 5600 years. Indeed, rather than using known values of R to calculate alpha decay rates of various nuclei, the observed decay rates were historically used to estimate R. For this purpose, it is not important to be precise about the value of the factor ν multiplying $\exp(-2G)$ in Eq. (6.4.13). But it is worth trying to be precise about this in order to make sure that we understand the decay process.

Appendix: Quantum Theory of Barrier Penetration Rates

This appendix presents a thoroughly quantum-mechanical solution of a somewhat artificial problem. We consider a particle in a negative nuclear potential well surrounded by a positive potential barrier, whose wave function is initially confined to the nuclear potential well, and we calculate the rate at which the particle escapes to infinity, without relying on the semi-classical picture of particles in the nucleus continually banging into the potential barrier and occasionally leaking through. The calculations in this appendix do not depend on the detailed form of the potential in the barrier, and so apply also the case where $\ell \neq 0$, where, as in Eq. (5.2.19), we include a centrifugal term $\hbar^2\ell(\ell+1)/2mr^2$ in the potential.

Our strategy will be to assume some initial wave function for the particle, entirely confined in the nuclear potential well, expand it in orthonormalized solutions $u_E^{(N)}(r)$ of the Schrödinger equation for various energies E, give each such solution a time dependence $\exp(-iEt/\hbar)$, and see what happens as the time increases.[18] In the course of this calculation we will be able to give an idea of the conditions under which the approximation (6.4.6) is valid, and find the amplitudes $A_{E\pm}(r)$.

Our first task is to calculate the not-yet-normalized reduced wave function inside the barrier, where it satisfies Eq. (6.4.4). This differential equation has no general analytic solution. We again guess that if the potential varies slowly (in a sense to be determined) then Eq. (6.4.4) has approximate solutions

---

[18] This follows the approach of E. Fermi, *Nuclear Physics*, lecture notes compiled by J. Orear, A. H. Rosenfeld, and R. A. Schluter, revised ed. (University of Chicago Press, Chicago, 1950), Chapter III. The treatment in this appendix is somewhat simplified by working throughout with continuum wave functions, and supplies some justifications skipped over by Fermi.

$$u_E(r) = A_{E\pm}(r) \exp\left[\pm \int \kappa_E(r)\, dr\right] , \qquad (6.4.17)$$

where the amplitudes $A_{E\pm}(r)$ vary more slowly than the exponentials. Before making any approximations, Eq. (6.4.4) can be written as a differential equation for $A_{E\pm}(r)$:

$$2\kappa_E A'_{E\pm} + \kappa'_E A_{E\pm} \pm A''_{E\pm} = 0 . \qquad (6.4.18)$$

We can implement the approximation that $A_{E\pm}(r)$ varies slowly by dropping the second derivative $A''_{E\pm}(r)$, solving Eq. (6.4.18), using the solution to calculate $A''_{E\pm}(r)$, and checking under what conditions it may indeed be neglected. With the term $\pm A''_{E\pm}(r)$ dropped, Eq. (6.4.18) becomes $A'_{E\pm}/A_{E\pm} = -\kappa'_E/2\kappa_E$, which has the easy solution $A_{E\pm}(r) \propto 1/\sqrt{\kappa_E(r)}$. Then

$$\frac{A''_{E\pm}}{\kappa'_E A_{E\pm}} = -\frac{\kappa''_E}{2\kappa'_E \kappa_E} + \frac{3}{4}\frac{\kappa'_E}{\kappa_E^2} .$$

Thus $A''_{E\pm}$ is indeed negligible compared with the term $\kappa'_E A_{E\pm}$ in Eq. (6.4.18) if

$$\frac{1}{\kappa_E}\left|\frac{\kappa''_E}{\kappa'_E}\right| \ll 1 \quad \text{and} \quad \frac{1}{\kappa_E}\left|\frac{\kappa'_E}{\kappa_E}\right| \ll 1 , \qquad (6.4.19)$$

which is to say that both $\kappa'_E(r)$ and $\kappa_E(r)$ undergo only small fractional changes in a distance of order $1/\kappa_E(r)$. Under these conditions, Eq. (6.4.4) has the two independent approximate solutions

$$\frac{1}{\sqrt{\kappa_E(r)}} \exp\left(\pm \int \kappa_E(r)\, dr\right) .$$

This is known as the *WKB approximation.*[19] We can write the general solution of Eq. (6.4.4) inside the barrier as a linear combination of these solutions:

$$u_E(r) = \frac{C_+(E)}{\sqrt{\kappa_E(r)}} \exp\left(+\int_R^r \kappa_E(r)\, dr\right) + \frac{C_-(E)}{\sqrt{\kappa_E(r)}} \exp\left(-\int_R^r \kappa_E(r)\, dr\right) . \qquad (6.4.20)$$

Beyond the barrier, where $r > b_E$ (with $V(b_E) \equiv E$), it is convenient to write the Schrödinger equation (6.4.2) for the reduced wave function as

$$\frac{d^2}{dr^2} u_E(r) = -k_E^2(r) u_E(r) , \qquad (6.4.21)$$

[19] G. Wentzel, Zeit. f. Phys. **38**, 518 (1926); H. A. Kramers, Zeit. f. Phys. **39**, 828 (1926); L. Brillouin, Compt. Rendus Acad. Sci. **183**, 24 (1926).

where

$$k_E(r) = +\frac{1}{\hbar}\sqrt{2m_\alpha(E - V(r))} \; . \tag{6.4.22}$$

Following the same arguments as before, provided that

$$\frac{1}{k_E}\left|\frac{k''_E}{k'_E}\right| \ll 1 \quad \text{and} \quad \frac{1}{k_E}\left|\frac{k'_E}{k_E}\right| \ll 1 \; , \tag{6.4.23}$$

we can use the WKB approximation to find solutions

$$u_E(r) \propto \frac{1}{\sqrt{k_E(r)}} \cos\left(\int k_E(r)\,dr + \vartheta\right) \; , \tag{6.4.24}$$

where ϑ is any angle. We have two independent solutions, given by using Eq. (6.4.24) with ϑ taken as two different angles.

We need to work out how each of the two independent solutions of the Schrödinger equation inside the Coulomb barrier, for $r < b_E$, merges with linear combinations of the two independent solutions beyond the barrier, where $r > b_E$. Unfortunately we cannot do this by equating the value and derivative of the WKB solutions for r just below and just above b_E, because $\kappa_E(r)$ and $k_E(r)$ both vanish at $r = b_E$, and so the conditions (6.4.19) and (6.4.23) for the validity of the WKB approximation break down near b_E. This is a well-known problem in the use of the WKB approximation to calculate bound state energies, but here we will encounter an additional difficulty.

We will make the reasonable assumption that $V(r) - E$ approaches a function proportional to $b_E - r$ for r near b_E. In this case, for $r > b_E$,

$$k_E \to \beta_E\sqrt{r - b_E} \quad \text{for} \quad r \to b_E \; , \tag{6.4.25}$$

with β_E a positive function of E. It is convenient to define a new independent variable

$$\phi \equiv \int_{b_E}^{r} k_E(r')dr' \to \frac{2\beta_E}{3}(r - b_E)^{3/2} \; . \tag{6.4.26}$$

The Schrödinger equation (6.4.21) then takes the form

$$\frac{d^2u}{d\phi^2} + \frac{1}{3\phi}\frac{du}{d\phi} + u = 0 \; , \tag{6.4.27}$$

with two independent solutions

$$u \propto \phi^{1/3} J_{\pm 1/3}(\phi) \; , \tag{6.4.28}$$

where J_ν is the usual Bessel function of order ν. Likewise, for $r < b_E$ we have

$$\kappa_E \to \beta_E\sqrt{b_E - r} \quad \text{for} \quad r \to b_E \; , \tag{6.4.29}$$

with β_E the same positive function of E as in Eq. (6.4.25). Here it is convenient to define

$$\overline{\phi} \equiv \int_r^{b_E} \kappa_E(r')dr' \to \frac{2\beta_E}{3}(b_E - r)^{3/2} . \tag{6.4.30}$$

The Schrödinger equation (6.4.4) then takes the form

$$\frac{d^2u}{d\overline{\phi}^2} + \frac{1}{3\overline{\phi}}\frac{du}{d\overline{\phi}} - u = 0 , \tag{6.4.31}$$

with two independent solutions

$$u \propto \overline{\phi}^{1/3} I_{\pm 1/3}(\overline{\phi}) , \tag{6.4.32}$$

where here $I_\nu(\overline{\phi})$ is the Bessel function of order ν with imaginary argument:

$$I_\nu(\overline{\phi}) \equiv e^{-i\pi\nu/2} J_\nu(e^{+i\pi\nu/2}\overline{\phi}) . \tag{6.4.33}$$

To see how the solutions for $r > b_E$ and $r < b_E$ merge with each other at $r = b_E$, we note that, for $\phi \to 0$,

$$\phi^{1/3} J_{1/3}(\phi) \to \frac{\phi^{2/3}}{2^{1/3}\Gamma(4/3)} , \quad \phi^{1/3} J_{-1/3}(\phi) \to \frac{2^{1/3}}{\Gamma(2/3)} ,$$

while, for $\overline{\phi} \to 0$,

$$\overline{\phi}^{1/3} I_{1/3}(\overline{\phi}) \to \frac{\overline{\phi}^{2/3}}{2^{1/3}\Gamma(4/3)} , \quad \overline{\phi}^{1/3} I_{-1/3}(\overline{\phi}) \to \frac{2^{1/3}}{\Gamma(2/3)} .$$

But $\phi^{2/3} \to (2\beta_E/3)^{2/3}(r - b_E)$ and $\overline{\phi}^{2/3} \to (2\beta_E/3)^{2/3}(b_E - r)$, so

$$\phi^{1/3} J_{\pm 1/3}(\phi) \Longleftrightarrow \mp\overline{\phi}^{1/3} I_{\pm 1/3}(\overline{\phi}) , \tag{6.4.34}$$

where "$\Longleftrightarrow$" means "connects smoothly at $r = b_E$."

To learn from these results about the WKB solutions, we note that the conditions (6.4.19) and (6.4.23) are satisfied if $\overline{\phi} \gg 1$ and $\phi \gg 1$. As long as the approximations (6.4.25) and (6.4.29) are still valid for these large values of $\overline{\phi}$ and ϕ, we can take wave functions in the WKB approximations as the asymptotic limits of the solutions (6.4.28) and (6.4.32):

$$\begin{aligned} \phi^{1/3} J_{\pm 1/3}(\phi) &\to \sqrt{\frac{2}{\pi}}\phi^{-1/6}\cos\left(\phi \mp \frac{\pi}{6} - \frac{\pi}{4}\right) \\ &= \sqrt{\frac{2}{\pi}}\left(\frac{3\beta_E^2}{2}\right)^{1/6} k_E^{-1/2}(r)\cos\left(\int_{b_E}^r k_E(r')dr' \mp \frac{\pi}{6} - \frac{\pi}{4}\right) , \end{aligned} \tag{6.4.35}$$

and

$$\overline{\phi}^{1/3} I_{\pm 1/3}(\overline{\phi}) \to \sqrt{\frac{1}{2\pi}}\overline{\phi}^{-1/6} \exp\left(\overline{\phi}\right)$$
$$= \sqrt{\frac{1}{2\pi}} \left(\frac{3\beta_E^2}{2}\right)^{1/6} \kappa_E^{-1/2}(r) \exp\left(\int_r^{b_E} \kappa_E(r')dr'\right) , \qquad (6.4.36)$$

but

$$\overline{\phi}^{1/3}\left[I_{+1/3}(\overline{\phi}) - I_{-1/3}(\overline{\phi})\right] \to -\sqrt{\frac{3}{\pi}}\overline{\phi}^{-1/6} \exp\left(-\overline{\phi}\right)$$
$$= -\sqrt{\frac{3}{\pi}} \left(\frac{3\beta_E^2}{2}\right)^{1/6} \kappa_E^{-1/2}(r) \exp\left(-\int_r^{b_E} \kappa_E(r')dr'\right) . \qquad (6.4.37)$$

From Eqs. (6.4.37), (6.4.35), and (6.4.34), we see that

$$\kappa_E^{-1/2}(r) \exp\left(-\int_r^{b_E} \kappa_E(r')dr'\right)$$
$$\Longleftrightarrow \sqrt{\frac{2}{3}} k_E^{-1/2}(r) \left[\cos\left(\int_{b_E}^r k_E(r')dr' - \frac{\pi}{6} - \frac{\pi}{4}\right)\right.$$
$$\left. + \cos\left(\int_{b_E}^r k_E(r')dr' + \frac{\pi}{6} - \frac{\pi}{4}\right)\right]$$
$$= 2k_E^{-1/2}(r) \cos\left(\int_{b_E}^r k_E(r')dr' - \frac{\pi}{4}\right) . \qquad (6.4.38)$$

To find the other connection formula we need, we now have a problem. Inspecting Eqs. (6.4.36) and (6.4.35), how do we decide in using Eq. (6.4.34) whether $\overline{\phi}^{-1/6} \exp(\overline{\phi})$ connects smoothly with $-2\phi^{-1/6} \cos(\phi - \pi/6 - \pi/4)$ or with $+2\phi^{1/6} \cos(\phi + \pi/6 - \pi/4)$? This puzzle arises because, lurking under the term proportional to $\overline{\phi}^{-1/6} \exp\left(\overline{\phi}\right)$ in the asymptotic expansion of $\overline{\phi}^{1/3} I_{\pm 1/3}(\overline{\phi})$, there are terms with unknown coefficients that are proportional to $\overline{\phi}^{-1/6} \exp\left(-\overline{\phi}\right)$ and are therefore negligible for $\overline{\phi} \gg 1$ but that nevertheless, as shown in Eq. (6.4.38), connect smoothly with a term proportional to $\phi^{-1/6} \cos(\phi - \pi/4)$. (This is known as the *Stokes phenomenon*.) As we saw in deriving Eq. (6.4.38), the difference between $-\phi^{-1/6} \cos(\phi - \pi/6 - \pi/4)$ and $+\phi^{1/6} \cos(\phi + \pi/6 - \pi/4)$ is proportional to $\phi^{-1/6} \cos(\phi - \pi/4)$, so we can take $\overline{\phi}^{-1/6} \exp(\overline{\phi})$ to connect smoothly with $-\phi^{-1/6} \cos(\phi - \pi/6 - \pi/4)$ or with $+\phi^{1/6} \cos(\phi + \pi/6 - \pi/4)$ or with their average, plus a term proportional to $\phi^{-1/6} \cos(\phi - \pi/4)$ with a coefficient that cannot be calculated within our present approximations. Using the average, we have then from Eqs. (6.4.36), (6.4.35), and (6.4.34):

$$\kappa_E^{-1/2}(r)\exp\left(\int_r^{b_E}\kappa_E(r')dr'\right) \Longleftrightarrow -k_E^{-1/2}(r)$$
$$\times\left[\cos\left(\int_r^{b_E} k_E(r')dr' + \frac{\pi}{4}\right) + \xi(E)\cos\left(\int_r^{b_E} k_E(r')dr' - \frac{\pi}{4}\right)\right], \tag{6.4.39}$$

with $\xi(E)$ an unknown coefficient.[20]

We can write the exponentials in Eq. (6.4.20) as

$$\exp\left(\pm\int_R^r \kappa_E(r')\,dr'\right) = e^{\pm G(E)}\exp\left(\mp\int_r^{b_E}\kappa_E(r')\,dr'\right),$$

where $G(E)$ is the barrier penetration exponent:

$$G(E) \equiv \int_R^{b_E}\kappa_E(r)\,dr, \tag{6.4.40}$$

given by Eq. (6.4.11) for a Coulomb potential with $\ell = 0$. Then the wave function $u_E(r)$ that takes the form (6.4.20) for $R < r < b_E$ takes the following form for $r > b_E$:

$$u_E(r) = -\frac{1}{\sqrt{k_E(r)}}\left[\left(2C_+(E)e^{G(E)} + \xi(E)C_-(E)e^{-G(E)}\right)\right.$$
$$\times\cos\left(\int_{b_E}^r k_E(r')dr' - \frac{\pi}{4}\right)$$
$$\left. + C_-(E)e^{-G(E)}\cos\left(\int_{b_E}^r k_E(r')dr' + \frac{\pi}{4}\right)\right]. \tag{6.4.41}$$

Now we need to consider the normalization of the wave functions. Since the Hamiltonian is Hermitian, and allowed values of energy E form a continuum, we know that wave functions with different energy are orthogonal, in the sense that

$$\int_0^\infty u_E(r)\,u_{E'}(r)\,dr = \int_0^\infty R_E(r)\,R_{E'}(r)\,r^2\,dr = N^2(E)\delta(E - E'). \tag{6.4.42}$$

The only question is, what is the coefficient $N^2(E)$? Once we know this, we can define orthonormalized wave functions

$$u_E^{(N)}(r) \equiv N^{-1}(E)u_E(r), \tag{6.4.43}$$

[20] Without explanation, Fermi in the reference in footnote 18 took $\xi = 0$. This is not justified, but as we shall see it makes no difference in the decay rate.

for which

$$\int_0^\infty u_E^{(N)}(r)\, u_{E'}^{(N)}(r)\, dr = \delta(E - E')\,, \tag{6.4.44}$$

and use these in an expansion of the time-dependent wave function.

To find the coefficient of the delta function in Eq. (6.4.42), we can discard any term in the integral that remains finite as $E \to E'$. The singularity as $E' \to E$ in this integral comes entirely from the infinite range where r is much larger than b_E, so in the integral we use the asymptotic form of Eq. (6.4.41):

$$u_E(r) \to -\frac{1}{\sqrt{k}}\Big[\big(2C_+(E)e^{+G(E)} + \xi(E)C_-(E)e^{-G(E)}\big)\cos(kr - \pi/4) \\ + C_-(E)e^{-G(E)}\cos(kr + \pi/4)\Big]\,, \tag{6.4.45}$$

where now k is the wave number of the free particle,

$$k \equiv \frac{1}{\hbar}\sqrt{2m_\alpha E}\,. \tag{6.4.46}$$

To calculate the singular part of the integral (6.4.42), we insert a convergence factor $\exp(-\epsilon r)$ and consider the limit as $\epsilon \to 0+$ and $E' \to E$. A straightforward calculation gives

$$\int_0^\infty dr\; e^{-\epsilon r}\cos\left(kr \pm \frac{\pi}{4}\right)\cos\left(k'r \pm \frac{\pi}{4}\right) \\ = \frac{1}{2}\left[\frac{k+k'}{\epsilon^2 + (k+k')^2} + \frac{\epsilon}{\epsilon^2 + (k-k')^2}\right],$$

$$\int_0^\infty dr\; e^{-\epsilon r}\cos\left(kr \pm \frac{\pi}{4}\right)\cos\left(k'r \mp \frac{\pi}{4}\right) \\ = \frac{1}{2}\left[\frac{\epsilon}{\epsilon^2 + (k+k')^2} \pm \frac{k-k'}{\epsilon^2 + (k-k')^2}\right].$$

Using a well-known representation of the delta function,

$$\frac{\epsilon}{\epsilon^2 + (k-k')^2} = (\pi/2)\delta(k - k')\,,$$

and discarding any terms that are not singular when we set $k = k'$ and then let ϵ go to zero, we have

$$\int_0^\infty \cos(kr \pm \pi/4)\cos\big(k'r \pm \pi/4\big) = (\pi/4)\delta(k - k') = \frac{\pi\hbar^2 k}{4m_\alpha}\delta(E - E')\,,$$

$$\int_0^\infty \cos(kr \pm \pi/4)\cos\big(k'r \mp \pi/4\big) = 0\,. \tag{6.4.47}$$

Equation (6.4.45) then gives Eq. (6.4.42), with

$$N^2(E) = \frac{\pi\hbar^2}{4m_\alpha}\left[\left(2C_+(E)\,e^{G(E)} + \xi(E)C_-(E)\,e^{-G(E)}\right)^2 + C_-^2(E)\,e^{-2G(E)}\right]. \tag{6.4.48}$$

We have been ruthless here in discarding non-singular terms, but this is a precise result, apart from the WKB approximation, which was used in deriving Eq. (6.4.41).

We now turn to calculating the time-dependent reduced wave function $u(r,t)$, assuming that at $t = 0$ it takes the form

$$u(r,0) = \begin{cases} u_{E_1}(r) & r < R \\ 0 & r > R\,, \end{cases} \tag{6.4.49}$$

where E_1 is the energy of an unstable state for which $C_+(E_1) = 0$. We can expand this in orthonormalized solutions of the Schrödinger equation

$$\begin{aligned} u(r,0) &= \int_0^\infty dE u_E^{(N)}(r) \int_0^\infty dr' u_E^{(N)}(r') u(r',0) \\ &= \int_0^\infty dE u_E^{(N)}(r) \int_0^R dr' u_E^{(N)}(r') u_{E_1}(r')\,. \end{aligned} \tag{6.4.50}$$

The time-dependent Schrödinger equation tells us that to find the wave function at any later time we must insert a factor $e^{-iEt/\hbar}$ in the integrand:

$$\begin{aligned} u(r,t) &= \int_0^\infty dE e^{-iEt/\hbar} u_E^{(N)}(r) \int_0^R dr' u_E^{(N)}(r') u_{E_1}(r') \\ &= \frac{4m_\alpha}{\pi\hbar^2} \int_0^\infty dE e^{-iEt/\hbar}\, u_E(r) \\ &\quad \times \frac{\int_0^R dr' u_E(r') u_{E_1}(r')}{[2C_+(E)\,e^{G(E)} + \xi(E)C_-(E)\,e^{-G(E)}]^2 + C_-^2(E)\,e^{-2G(E)}}. \end{aligned} \tag{6.4.51}$$

For $r' < R$ the wave function $u_E(r')$ is unaffected by the potential barrier, and therefore (as shown for example in Eq. (6.4.9)) varies smoothly with E. On the other hand, the term in the denominator proportional to $e^{2G(E)}$ makes the integrand very small except very near the energies of unstable states, where $C_+(E)$ vanishes. The integral is therefore dominated by values of E very near the energies E_n at which $C_+(E)$ vanishes. These are the energies of nearly stable states, so the wave functions $u_{E_n}(r)$ are approximate eigenfunctions of the Hamiltonian, and therefore are approximately orthogonal, so in Eq. (6.4.51) the integral $\int_0^R dr' u_{E_n}(r') u_{E_1}(r')$ is very small for $n \neq 1$. For E very near E_1, we can approximate $C_+(E) \to C'_+(E_1)(E - E_1)$. Since the contribution of other energies is exponentially suppressed, we can set $E = E_1$ everywhere except in the factor $E - E_1$ and the exponential, and extend the range of energy integration to run over the whole real axis, and for $r < R$ write

$$u(r,t) \simeq \frac{4m_\alpha}{\pi\hbar^2} u_{E_1}(r) \int_0^R dr' u_{E_1}^2(r') \int_{-\infty}^{\infty} dE$$

$$\times \frac{e^{-iEt/\hbar}}{[2C'_+(E_1)\, e^{G(E_1)}(E - E_1) + \xi(E_1)C_-(E_1)e^{-G(E_1)}]^2 + C_-^2(E_1)\, e^{-2G(E_1)}} \,. \tag{6.4.52}$$

For $t > 0$ the contour of integration over E can be closed with a large semicircle in the lower half of the complex plane, on which $e^{-iEt/\hbar}$ is exponentially small. Since this contour is now closed clockwise, the integral is given by $-2i\pi$ times the residue of the pole at $E = \mathcal{E}_1 - i|C_-(E_1)/2C'_+(E_1)|e^{-2G(E_1)}$, where

$$\mathcal{E}_1 = E_1 - \xi(E_1)e^{-2G(E_1)}C_-(E_1)/C_+(E_1) \,.$$

The wave function in the potential well therefore goes as

$$u(r,t) \propto e^{-i\mathcal{E}_1 t/\hbar} e^{-\Gamma_\alpha t/2} u_{E_1}(r) \,, \tag{6.4.53}$$

where

$$\Gamma_\alpha = \left| \frac{C_-(E_1)}{\hbar C'_+(E_1)} \right| e^{-2G(E_1)} \,. \tag{6.4.54}$$

From the square of Eq. (6.4.53), we see that Γ_α is the rate of decay of the probability density $|u(r,t)|^2$ of the alpha particle within the nucleus. The Stokes phenomenon has led to an incalculable but exponentially small shift in the oscillation frequency of the wave function, but has no effect on the decay rate. Equation (6.4.54) justifies the appearance of the suppression factor $e^{-2G(E_1)}$ in Eq. (6.4.14).

6.5 Beta Decay

The earliest studies of radioactivity revealed the existence of a distinct class of radioactive processes, beta decay, in which an electron is emitted in a transition between nuclear states. For instance, ^{234}Th, which is itself a product of the alpha decay of ^{238}U, undergoes beta decay to ^{234}Pa. As we have seen, at first beta decay was taken as evidence for the view that nuclei consist of protons and electrons, but this interpretation was abandoned with the realization in the 1930s that nuclei are composed of protons and neutrons. An electron is created at the moment of beta decay when a neutron turns into a proton – the electron is no more in the nucleus before it is emitted than a photon is in an atom before it is radiated.

But there was a peculiar difference between the observed energies of the photons emitted in atomic transitions and the electrons emitted in beta decay. As we saw in Section 3.4, Bohr had realized in 1913 that a photon emitted in any given atomic transition has a unique energy, given by the difference in energies of the initial and final atomic states. Chadwick discovered in 1914 that the energies of electrons emitted in a beta transition between any specific nuclear states do not have any one value, but occupy a range up to some definite maximum. This might be explained if a photon is emitted along with the electron, with the energy of the nuclear transition shared between the electron and the photon in a proportion that varies from one decay event to another. The electron energy would come close to a maximum value, equal to the energy released in the nuclear transition, only when the photon happens to have very low energy. If beta decay produced a photon along with the electron, then when these decay products are caught in a surrounding medium the heat energy given to the medium would be the same in each decay event, equal to the energy difference of the initial and final nuclear states, and hence equal to the maximum value observed for the electron energy in this decay. But experiments in 1927 by C. D. Ellis (1895–1980) and W. A. Wooster (1903–1984) showed that the average energy deposited in the medium surrounding the decaying nucleus was *not* equal to the maximum energy of the electron, but instead to its average, as if whatever energy was not carried by the electron was simply lost. Bohr was even led by this to speculate that energy might not be conserved in beta decays.

A different explanation was offered in 1930 by Pauli. He proposed that the electron in beta decay is indeed accompanied by another particle that because electrically neutral had escaped detection, but this neutral particle is not a photon. Rather, it is an extremely penetrating particle that is not captured in the surrounding medium. The particle soon became known as a *neutrino*, symbolized ν. The underlying reaction is $n \to p + e^- + \nu$ (where n and p stand for the neutron and proton, and the electron is denoted e^-, for a reason we will come to presently). Among many other examples, this is responsible for the decay of the ground state of boron 12 to carbon in the reaction $^{12}\text{B} \to {}^{12}\text{C} + e^- + \nu$, as well as for the decay of the free neutron, $n \to p + e^- + \nu$. Since neutrons, protons, and electrons have spin 1/2, angular momentum conservation requires the neutrino to have a half-integer spin. It is in fact known to have spin 1/2.

There are also radioactive decays in which instead of an electron there is emitted a positron, e^+, the electron's antiparticle, with the same mass but opposite electric charge.[21] The conservation of energy forbids the process $p \to n + e^+ + \nu$

[21] The existence of the positron was anticipated in 1930 by P. A. M. Dirac, Proc. Roy. Soc. **A126**, 360 (1930). He had developed a relativistic version of the Schrödinger equation, which turned out to have solutions corresponding to states of negatively charged electrons with negative energy as well as states with positive energy. Dirac's interpretation was that these negative-energy states are normally filled, one electron to each negative-energy state in accordance with the Pauli exclusion principle, but that occasionally there

for free protons, but in nuclei this process can produce decays such as the beta decay of the ground state of nitrogen 12, ${}^{12}\mathrm{N} \to {}^{12}\mathrm{C} + e^+ + \nu$.

In 1934 Fermi proposed a detailed theory of beta decay.[22] In Fermi's theory the interaction Hamiltonian takes the form

$$H_\beta = (\hbar c)^3 G_F \eta_{\mu\nu} \int d^3x\, V^\mu \mathcal{V}^\nu + c.c.\,, \tag{6.5.1}$$

where G_F is a constant; V^μ and $\mathcal{V}^\nu$ are operators with the same Lorentz and space inversion transformation properties as the electric current J^μ and with the dimensionality of densities (that is, inverse volumes); V^μ acts to change neutrons to protons; $\mathcal{V}^\nu$ acts to create electrons and neutrinos; and as usual $c.c.$ indicates the adjoint of the foregoing term. The factor $(\hbar c)^3$ is extracted from G_F for later convenience. As we will see in the appendix to Section 7.4 these currents are bilinear functions of Dirac fields, but we will not need that information for our limited purposes here.

Fermi's theory almost immediately needed modification. The three-vector part of the current V^μ is odd under space inversion, so when acting on nuclear states it gives a contribution proportional to nucleon velocities $\mathbf{v}$, and so is suppressed by a factor of order $|\mathbf{v}|/c$, which is small in nuclei, as in atoms. This leaves the time component V^0, which is even under space inversion and is a rotational scalar. For decays that are not suppressed by a centrifugal barrier there is no orbital angular momentum, so in these decays neither the parity nor spin of the nuclear states can change. But many beta decays were observed in which the spin of the nuclear state did change by one unit, and which yet

---

appears a vacancy which we observe as a particle of positive charge and positive energy. At first Dirac identified these holes as protons, but then in 1932 a positively charged particle was unexpectedly found in cosmic rays by C. D. Anderson, Phys. Rev. **43**, 491 (1932). (This article is included in Beyer, *Foundations of Nuclear Physics*, listed in the bibliography.) The cloud chamber tracks of these particles were observed to have the same curvature in a magnetic field as electron tracks, but in the opposite direction, consistent with a particle having the same mass as an electron and a charge of equal magnitude but opposite sign. It was widely supposed that these were Dirac's holes.

The interpretation of positrons as vacancies in a sea of negative-energy electrons has largely been abandoned. Dirac's relativistic wave equation works only for particles of spin 1/2. This at first seemed like a triumph because protons and electrons were known to have spin 1/2, but by now we know of several particles of spin 0 and spin 1 (the H^0, W^+, W^-, and Z^0) that seem every bit as elementary as the electron. Furthermore, the W^+ and W^- are each other's antiparticles, in the same sense as the e^+ and e^-. But these are bosons, which do not obey the exclusion principle and so could not form a stable sea of negative-energy particles. As described in the appendix to Section 7.4, Dirac's equation survives as the field equation satisfied by the quantum field of particles of spin 1/2 but not, as Dirac thought, as a relativistic version of a Schrödinger equation for a probability amplitude.

As explained in Section 7.4, we now understand as a consequence of Lorentz invariance and quantum mechanics that for every species of particle, elementary or not, fermion or boson, there is a corresponding species of antiparticle, with the same mass and spin but opposite electric charge. The only qualification is that a few types of electrically neutral particles like the photon and the Z^0 are their own antiparticles.

22 E. Fermi, Zeit. Phys. **88**, 161 (1934). This article is reprinted in Beyer, *Foundations of Nuclear Physics*, listed in the bibliography. In his article Fermi cited an unpublished suggestion by Pauli that a neutral weakly interacting particle was emitted along with electrons in beta decay.

seemed to be "allowed" in the sense of having rates comparable to typical other beta decays with similar energy. For instance, the ground states of the nuclei ^{12}B and ^{12}N mentioned in Section 6.2 have spin one and even parity, and yet have allowed beta decays into the ground state of ^{12}C, which has spin zero and even parity.

In order to allow such decays, Fermi's theory was modified by adding an additional term to the interaction Hamiltonian:[23]

$$H_\beta = (\hbar c)^3 G_F \eta_{\mu\nu} \int d^3x [V^\mu \mathcal{V}^\nu + A^\mu \mathcal{A}^\nu] + c.c. \,, \tag{6.5.2}$$

where A^μ like V^μ turns neutrons into protons; $\mathcal{A}^\nu$ like $\mathcal{V}^\nu$ creates electrons and neutrinos; and A^μ and $\mathcal{A}^\nu$ are *axial vectors* – that is, like V^μ and $\mathcal{V}^\nu$ they transform as four-vectors under proper Lorentz transformations, but they have opposite properties under space inversion: $\mathbf{A}$ is even and A^0 is odd, and likewise for $\mathcal{A}^\nu$. In consequence $\mathbf{A}$ acting on nuclear states can make contributions proportional to nucleon spin vectors, allowing beta decays in which spin changes by one unit, such as the beta decays of ^{12}B and ^{12}N, without suppression of the rate by factors of order $|\mathbf{v}|/c$ or centrifugal barriers. With the interaction Hamiltonian (6.5.2), the selection rules for "allowed" beta decays are that the nuclear parity does not change and that the nuclear spin can change by at most one unit.

To estimate these "allowed" rates, we note that in order for H_β to have the dimensionality of energy, the constant $(\hbar c)^3 G_F$ must have the dimensions of energy times volume. Hence G_F has the convenient (and conventional) dimensionality of energy^{-2}, which is why the factor $(\hbar c)^3$ was inserted in Eqs. (6.5.1) and (6.5.2). The rate Γ_β of any beta decay process is proportional to G_F^2, and if the energy E released in the decay is much larger than $m_e c^2 = 0.511$ MeV then apart from the factor G_F^2 it can only depend on E, so in order for it to have the dimensionality of a rate it must take the form

$$\Gamma_\beta \approx \frac{1}{\hbar} G_F^2 E^5 \,. \tag{6.5.3}$$

This E^5 dependence is observed for high-energy beta decays that satisfy the selection rules for "allowed" beta decays. The energy $G_F^{-1/2}$ turns out to be very large. For instance, as we saw in Section 6.2, the energy released in the beta decay of ^{12}B to the ground state of ^{12}C is $0.0144\, m_p c^2 = 13.4$ MeV, much larger then $m_e c^2$. The rate is 48.5 sec^{-1}. Using these numbers in Eq. (6.5.3) gives $G_F^{-1/2} \approx 2 \times 10^3$ GeV. (This energy is large in part because weak interactions are transmitted by a heavy particle, the $W^\pm$ particle with

[23] G. Gamow and E. Teller, Phys. Rev. **49**, 895 (1936). There were other possibilities involving scalar and tensor operators that were not finally excluded by experimental data until the 1950s.

mass 80.4 GeV/c^2, and in part because the interactions that emit and absorb $W^{\pm}$ particles are characterized by a small constant, of the same order as the fine-structure constant $e^2/\hbar c \simeq 1/137$ of electrodynamics, which gives $G_F^{-1/2} \approx m_W c^2 \times \sqrt{137} \approx 10^3$ GeV. A more accurate value along with a more precise definition for G_F will be given in the appendix to Section 7.4.)

The extremely low rate at which neutrinos were absorbed by the medium surrounding the radioactive nuclei in the 1927 experiments of Ellis and Wooster is due to the extreme weakness of interactions such as beta decay that involve neutrinos. In general the rates of neutrino interaction processes are characterized by the presence in the rate of the factor G_F^2, which is what makes them so weak. For instance, the cross section for the neutrino reactions $\nu + p \to e^+ + n$ and $\nu + n \to e^- + p$ (whether for a free proton or a proton or neutron inside a nucleus) is proportional to G_F^2, and so, since it has the units of area, dimensional analysis requires that at a neutrino energy E considerably above $m_e c^2$ the cross section takes the form

$$\sigma \approx (\hbar c E)^2 G_F^2 \, .$$

Recalling that $\hbar c = 197$ MeV $\times\, 10^{-13}$ cm, we see that for a relatively high-energy beta decay neutrino with energy $E = 10$ MeV the cross section σ is of order 10^{-44} cm^2. In ordinary matter, with a number density n of nucleons of order 10^{24} cm^{-3}, this gives a mean free path $1/n\sigma \approx 10^{20}$ cm, or about 100 light years. It is no wonder that Ellis and Wooster did not detect energy deposited by neutrinos in their experiment. There never was any hope of detecting neutrinos from ordinary laboratory samples of radioactive material, but nuclear reactors emit such enormous floods of neutrinos from the beta decay of fission products that at last in 1956 Clyde Cohan, Jr. (1919–1974) and Frederick Reines (1918–1998) were able to detect neutrinos produced at the Savannah River reactor by detecting gamma rays from the annihilation of positrons produced in the reaction $\nu + p \to n + e^+$.

All rates for processes involving neutrinos are suppressed by the factor G_F^2, and there are also reactions due to other weak interactions that do not involve neutrinos but are similarly suppressed. Among these are the decays of a particle called the K meson, with a mass of 495 MeV/c^2, into two-pion and three-pion states, decays that are very slow compared with processes such as the decay of the three–three resonance into a pion and a nucleon that occur through the action of strong interactions.

There is another common feature of weak interaction processes, beyond their weakness. They violate some of the symmetry principles obeyed by strong and electromagnetic interactions. It appeared that the charged K meson decayed both into two-pion states that are invariant under the space inversion transformation $\mathbf{x} \to -\mathbf{x}$, and also into three-pion states that change sign under space inversion, which would not be possible if the space inversion operator commutes with the Hamiltonian. It was this that led Tsung-Dao Lee (1926–) and

Chen-Ning Yang (1922–) in 1956[24] to suggest that weak interactions in general do not respect invariance under space inversion, a suggestion that was soon verified in the beta decay[25] of ^{60}Co and in the decays of charged pions.[26] Weak interactions also violate a symmetry between particles and antiparticles, and they violate the conservation of several quantities (collectively known as *flavors*), that are conserved by the strong and electromagnetic interactions.

It used to be thought that neutrinos have zero mass. For massless particles the helicity, the component of angular momentum in the direction of motion, is Lorentz invariant. In 1957 Lee and Yang proposed[27] that the neutrinos emitted with electrons or positrons always have helicities $\hbar/2$ and $-\hbar/2$, respectively. This was only possible if weak interactions violate invariance under space inversion, because space inversion transformations reverse the direction of the neutrino's motion while leaving its spin unchanged, and so reverse the helicity. This proposal was incorporated in another change in the beta decay interaction that in our present schematic notation takes the form

$$H_\beta = \frac{(\hbar c)^3 G_F}{\sqrt{2}} \eta_{\mu\nu} \int d^3x [V^\mu + A^\mu][\mathcal{V}^\nu + \mathcal{A}^\nu] + c.c. \,, \tag{6.5.4}$$

in which the terms $V^\mu \mathcal{A}_\mu$ and $A^\mu \mathcal{V}_\mu$ evidently violate space inversion symmetry.

But Lee's and Yang's proposal regarding neutrino helicity could not be universally and literally true unless neutrinos were massless, because the observed direction of motion of a massive particle is reversed if the observer travels with higher speed in the same direction, which does not affect its spin. It is now known that neutrinos have very small but non-zero mass, much less than the electron mass, and so like any massive particle of spin 1/2 neutrinos exist both in states with angular momentum components $\hbar/2$ and $-\hbar/2$ in any direction. Experiment shows that the emission of an electron or positron in the beta decay of a nucleus *at rest* is accompanied by the emission of a neutrino that is overwhelmingly likely to be in a state with angular momentum component in the direction of motion, respectively $+\hbar/2$ or $-\hbar/2$, as proposed by Lee and Yang, but this would not be the case if the neutrino were viewed by a more rapidly moving observer.

There is a complication regarding neutrinos that I have so far not mentioned. There are two other charged leptons, particles like the electron that have only electromagnetic and weak but not strong interactions. These are the muon, with mass 105.658 MeV$/c^2$, mentioned briefly in Section 4.3, and the more recently

[24] T. D. Lee and C. N. Yang, Phys. Rev. **104**, 254 (1956).
[25] C. S. Wu *et al.*, Phys. Rev. **104**, 254 (1957).
[26] R. Garwin, L. Lederman, and M. Weinrich, Phys. Rev. **105**, 1415 (1957); J. Friedman and V. Telegdi, Phys. Rev. **105**, 1681 (1957).
[27] T. D. Lee and C. N. Yang, Phys. Rev. **105**, 1671 (1957).

discovered tauon, with mass 1776.82 MeV$/c^2$. Both, like electrons, are emitted by strongly interacting particles along with neutrinos, but these neutrinos are not the same as the neutrinos emitted along with electrons or positrons in beta decay. Rather, the neutrinos emitted in beta decay and along with the production of muons and tauons are of three different types. For instance, the neutrino emitted along with a muon in the decay of a charged pion can create another muon in a reaction $\nu + n \to p + \mu^-$, but it cannot create an electron, and the neutrino emitted along with an electron in beta decay can create another electron in a reaction $\nu + n \to p + e^-$, but even if its energy were high enough it could not create a muon.

Except that, in a sense, it can. For years there was a mysterious deficiency in the number of neutrinos observed to be coming from the Sun.[28] These would be electron-type neutrinos, created in reactions such as $p + p \to d + e^+ + \nu$. Bruno Pontecorvo (1919–1993) suggested[29] that this is because neutrinos have mass but the states with definite mass are not electron-type or muon-type or tauon-type neutrinos. Rather, each of these is a superposition of neutrino states of definite mass. According to this idea, the electron-type neutrinos emitted by the Sun are superpositions of states of definite mass, which oscillate at different rates on their way to the Earth, arriving as incoherent mixtures of neutrinos of all three types. In the search for solar neutrinos the detectors were looking for the reaction $\nu + {}^{37}\text{Cl} \to e^- + {}^{37}\text{Ar}$, and were therefore sensitive only to electron-type neutrinos, which according to Pontecorvo is why fewer neutrinos were detected than would have been the case if neutrinos were massless, the undetected neutrinos arriving as muon-type or tauon-type. This hypothesis was confirmed when it became possible to detect solar neutrinos in the reaction $\nu + d \to \nu + p + n$, which is equally sensitive to neutrinos of all three types, and the number seen was just what was expected. The existence of neutrino masses has by now been convincingly confirmed in numerous terrestrial experiments, which, although they have not yielded values for individual neutrino masses, indicate that they are in the range of 0.01 to 0.1 eV$/c^2$.

When neutrinos were thought to have zero mass it was common to call the particle emitted along with an electron an *antineutrino*, reserving the term neutrino for the particle emitted along with a positron. This was to preserve a widely accepted conservation law, of a quantity known as *lepton number*, analogous to baryon number. Electrons and neutrinos were supposed to have lepton number $+1$; positrons and antineutrinos would have lepton number -1, while protons and neutrons would have lepton number zero, so that lepton number would be conserved in both kinds of beta decay. But it is not possible to attribute different values for lepton number or any other conserved quantity to the neutral particles

[28] J. N. Bahcall, Phys. Rev. Lett. **12**, 300 (1964); Phys. Rev. **135**, B137 (1964); R. Davis, Jr., Phys. Rev. Lett. **12**, 303 (1964); R. Davis, Jr., D. S. Harmer, and K. C. Hoffmann, Phys. Rev. Lett. **26**, 1205 (1968).

[29] B. Pontecorvo, JETP **53**, 1717 (1967).

emitted with electrons or positrons in beta decay if they are just different spin states of the same particle.

Now that we know that neutrinos have mass, there are two widely considered points of view regarding the nature of neutrinos and of lepton number and the origin of neutrino masses.

First, in order to preserve the exact conservation of lepton number it would be necessary to suppose that the neutrino fields of electron-type, muon-type, and tauon-type with lepton number -1 each have distinct adjoints with lepton number $+1$. In this view it is the states of helicity $+\hbar/2$ of the field with lepton number -1 that have been observed to be emitted with electrons in beta decay, and it is the states of helicity $-\hbar/2$ of the adjoint field with lepton number $+1$ that have been observed to be emitted with positrons, while the other helicity states of the two fields of each type exist but are so far unobserved. Neutrinos of this description are often called Dirac neutrinos, because their fields are described in the same way as in the description of electrons by Dirac, discussed in the appendix to Section 6.4.

The other possibility, often associated with the name of Ettore Majorana (1906–1938), is that lepton number is not conserved, and the three types of neutral particles emitted with negative and positive leptons are states of the same three spin 1/2 particles, which as a consequence of Eq. (6.5.4) are overwhelmingly likely to be emitted with helicity $\hbar/2$ when emitted with e^-, μ^-, τ^- and with helicity $-\hbar/2$ when emitted with e^+, μ^+, τ^+. We can then regard these three neutral particles as their own antiparticles, like the photon or the π^0.

For what it is worth, the Majorana alternative seems to me a more economical and plausible view, which is why in this section I have not distinguished neutrinos and antineutrinos. In the Dirac case neutrinos get masses in much the same way as the other leptons and the quarks of the Standard Model, so it is mysterious why they are so light compared with other elementary particles. On the other hand, the masses of Majorana neutrinos can only arise from effects at very high energy, and are naturally in the observed range.

Fermi's theory correctly described the probability distribution for the energy of the electron or positron emitted in beta decay, a distribution that was unaffected by the subsequent modifications in the interaction Hamiltonian described above. With these modifications Fermi's theory has survived as a correct approximate theory for nuclear beta decay. It was in fact the first successful application of quantum field theory outside the context of electrodynamics.

7
Quantum Field Theory

Chapter 5 described quantum mechanics in the context of particles moving in a potential. This application of quantum mechanics led to great advances in the 1920s and 1930s in our understanding of atoms, molecules, and much else. But starting around 1930, and increasingly since then, theoretical physicists have become aware of a deeper description of matter, in terms of *fields*. Just as Einstein and others had much earlier recognized that the energy and momentum of the electromagnetic field is packaged in bundles, the particles later called photons, so also there is an electron field whose energy and momentum are packaged in particles, observed as electrons, and likewise for every other sort of elementary particle. Indeed, in practice this is what we now mean by an elementary particle: it is the quantum of some field, which appears as an ingredient in whatever seem to be the fundamental equations of physics at any stage in our progress.

This is a good place to warn of an old misunderstanding. It used to be thought by some theorists (perhaps de Broglie) that the wave function of a particle is a field, something like the electromagnetic field. Just as the creation and annihilation of photons was seen as a consequence of the application of quantum mechanics to the electromagnetic field, some theorists came to think that the creation and annihilation of electrons and other particles could be understood through the application of quantum mechanics to the wave function itself, a process known as *second quantization*. This does not work. The electromagnetic field cannot be interpreted like a wave function as a probability amplitude, and the Schrödinger wave function does not have the Lorentz transformation property of a scalar field. The wave function is not a field – it is a representation of a physical state. As discussed in Section 5.10, it is the component of the state vector in some basis, such as one labeled by the possible positions of a particle. Even though it is not generally useful to do so, we can also introduce wave functions for fields – they are functionals of the field, quantities that depend on the value taken by the field at every point in space, equal to the component of the state vector in a basis labeled by these field values. One still sometimes hears talk of second quantization, but this idea is an obsolete historical relic.

7.1 Canonical Formalism for Fields

We begin by restating the canonical formalism described in Section 5.7, now in the context of fields. Here the $q_N(t)$ are fields $\varphi_n(\mathbf{x},t)$, with sums over the label N now comprising both sums over the discrete label n which distinguishes one type of field from another and integrals over the spatial argument $\mathbf{x}$. In order to have any chance of a Lorentz-invariant theory, the action here must take the form of an integral over spacetime of a function of space derivatives as well as time derivatives of the fields

$$I[\varphi] = \int d^3x \int_{-\infty}^{+\infty} dt\; \mathcal{L}(\varphi_n(\mathbf{x},t), \nabla\varphi_n(\mathbf{x},t), \dot{\varphi}_n(\mathbf{x},t)) \;. \tag{7.1.1}$$

The function $\mathcal{L}$ is known as the *Lagrangian density*. Comparing this with Eq. (5.7.11), we see that the Lagrangian here is

$$L[\varphi(t), \dot{\varphi}(t)] = \int d^3x\; \mathcal{L}(\varphi_n(\mathbf{x},t), \nabla\varphi_n(\mathbf{x},t), \dot{\varphi}_n(\mathbf{x},t)) \;. \tag{7.1.2}$$

This Lagrangian is a functional rather than a function of $\varphi_n(\mathbf{x},t)$ and $\dot{\varphi}_n(\mathbf{x},t)$; that is, it depends on the values of $\varphi_n(\mathbf{x},t)$ and $\dot{\varphi}_n(\mathbf{x},t)$ for all n and $\mathbf{x}$ at a given time t. Therefore, where derivatives of L appear in the canonical formalism as described in Section 5.7, they should now be interpreted as *functional derivatives*. In general, the functional derivatives $\delta F/\delta\varphi$ and $\delta F/\delta\dot{\varphi}$ of any functional F of $\varphi_n(\mathbf{x},t)$ and $\dot{\varphi}_n(\mathbf{x},t)$ at a fixed time t are defined by the prescription that the effect of independent infinitesimal variations in the arguments of the functional is given by

$$\begin{aligned} &F[\varphi(t)+\delta\varphi(t), \dot{\varphi}(t)+\delta\dot{\varphi}(t)] \\ &\quad \equiv F[\varphi(t),\dot{\varphi}(t)] + \sum_n \int d^3x \left[\frac{\delta F[\varphi(t),\dot{\varphi}(t)]}{\delta\varphi_n(\mathbf{x},t)}\delta\varphi_n(\mathbf{x},t) + \frac{\delta F[\varphi(t),\dot{\varphi}(t)]}{\delta\dot{\varphi}_n(\mathbf{x},t)}\delta\dot{\varphi}_n(\mathbf{x},t)\right] . \end{aligned} \tag{7.1.3}$$

For the particular functional (7.1.2), we have

$$\begin{aligned} &L[\varphi(t)+\delta\varphi(t), \dot{\varphi}(t)+\delta\dot{\varphi}(t)] \\ &\quad = \int d^3x\; \mathcal{L}(\varphi_n(\mathbf{x},t)+\delta\varphi_n(\mathbf{x},t), \nabla\varphi_n(\mathbf{x},t)+\nabla\delta\varphi_n(\mathbf{x},t), \dot{\varphi}_n(\mathbf{x},t)+\delta\dot{\varphi}_n(\mathbf{x},t)) \end{aligned}$$

$$= L[\varphi(t), \dot{\varphi}(t)]$$
$$+ \sum_n \int d^3x g \left[\frac{\partial \mathcal{L}(\varphi_n(\mathbf{x},t), \nabla\varphi_n(\mathbf{x},t), \dot{\varphi}_n(\mathbf{x},t))}{\partial \varphi_n(\mathbf{x},t)} \delta\varphi_n(\mathbf{x},t) \right.$$
$$+ \frac{\partial \mathcal{L}(\varphi_n(\mathbf{x},t), \nabla\varphi_n(\mathbf{x},t), \dot{\varphi}_n(\mathbf{x},t))}{\partial(\partial\varphi_n(\mathbf{x},t)/\partial x_i)} \delta(\partial\varphi_n(\mathbf{x},t)/\partial x_i)$$
$$\left. + \frac{\partial \mathcal{L}(\varphi_n(\mathbf{x},t), \nabla\varphi_n(\mathbf{x},t), \dot{\varphi}_n(\mathbf{x},t))}{\partial \dot{\varphi}_n(\mathbf{x},t)} \delta\dot{\varphi}_n(\mathbf{x},t) g \right]$$
$$= L[\varphi(t), \dot{\varphi}(t)]$$
$$+ \sum_n \int d^3x g \left[\frac{\partial \mathcal{L}(\varphi_n(\mathbf{x},t), \nabla\varphi_n(\mathbf{x},t), \dot{\varphi}_n(\mathbf{x},t))}{\partial \varphi_n(\mathbf{x},t)} \delta\varphi_n(\mathbf{x},t) \right.$$
$$- \frac{\partial}{\partial x_i} \left[\frac{\partial \mathcal{L}(\varphi_n(\mathbf{x},t), \nabla\varphi_n(\mathbf{x},t), \dot{\varphi}_n(\mathbf{x},t))}{\partial(\partial\varphi_n(\mathbf{x},t)/\partial x_i)} \right] \delta\varphi_n(\mathbf{x},t)$$
$$\left. + \frac{\partial \mathcal{L}(\varphi_n(\mathbf{x},t), \nabla\varphi_n(\mathbf{x},t), \dot{\varphi}_n(\mathbf{x},t))}{\partial \dot{\varphi}_n(\mathbf{x},t)} \delta\dot{\varphi}_n(\mathbf{x},t) g \right]$$

where as usual a repeated index i is summed over the values 1, 2, 3. Comparing this with the definition (7.1.3), we have

$$\frac{\delta L}{\delta \varphi_n(\mathbf{x},t)} = \frac{\partial \mathcal{L}}{\partial \varphi_n(\mathbf{x},t)} - \frac{\partial}{\partial x_i} \left[\frac{\partial \mathcal{L}}{\partial(\partial\varphi_n(\mathbf{x},t)/\partial x_i)} \right], \tag{7.1.4}$$

$$\frac{\delta L}{\delta \dot{\varphi}_n(\mathbf{x},t)} = \frac{\partial \mathcal{L}}{\partial \dot{\varphi}_n(\mathbf{x},t)}, \tag{7.1.5}$$

in which to save writing we have dropped the arguments of L and $\mathcal{L}$.

Field Equations

We take the derivatives of the Lagrangian in the equations of motion (5.7.12) to be functional derivatives:

$$\frac{\partial}{\partial t} \left[\frac{\delta L}{\delta \dot{\varphi}_n(\mathbf{x},t)} \right] = \frac{\delta L}{\delta \varphi_n(\mathbf{x},t)}. \tag{7.1.6}$$

Equations (7.1.4) and (7.1.5) then give the field equations

$$\frac{\partial}{\partial t} \left[\frac{\partial \mathcal{L}}{\partial \dot{\varphi}_n(\mathbf{x},t)} \right] = \frac{\partial \mathcal{L}}{\partial \varphi_n(\mathbf{x},t)} - \frac{\partial}{\partial x_i} \left[\frac{\partial \mathcal{L}}{\partial(\partial\varphi_n(\mathbf{x},t)/\partial x_i)} \right]. \tag{7.1.7}$$

These are known as the *Euler–Lagrange equations*. We can put Eq. (7.1.7) into a form that appears more consistent with Lorentz invariance:

$$\frac{\partial \mathcal{L}}{\partial \varphi_n(\mathbf{x},t)} = \frac{\partial}{\partial x^\mu} \left[\frac{\partial \mathcal{L}}{\partial(\partial\varphi_n(\mathbf{x},t)/\partial x^\mu)} \right], \tag{7.1.8}$$

in which as usual the repeated index μ is summed over the values $\mu = 1, 2, 3, 0$, again with $x^0 = ct$.

Commutation Relations

The field equations (7.1.8) could have been derived more easily by directly requiring that the action (7.1.1) must be stationary with respect to arbitrary infinitesimal variations of the $\varphi_n(\mathbf{x}, t)$ that vanish when $|\mathbf{x}| \to \infty$ or when $|t| \to \infty$. The calculation of functional derivatives is however important in finding the commutation relations of the fields. The canonical conjugate $\pi_n(\mathbf{x}, t)$ to $\psi_n(\mathbf{x}, t)$ is defined as in Eq. (5.7.13) but with a functional derivative of L in place of an ordinary derivative:

$$\pi_n(\mathbf{x}, t) = \frac{\delta L}{\delta \dot{\varphi}_n(\mathbf{x}, t)} = \frac{\partial \mathcal{L}}{\partial \dot{\varphi}_n(\mathbf{x}, t)} . \tag{7.1.9}$$

The canonical commutation relations (5.7.5) here read

$$\begin{aligned} &[\varphi_n(\mathbf{x}, t), \pi_m(\mathbf{y}, t)] = i\hbar \delta_{nm} \delta^3(\mathbf{x} - \mathbf{y}) , \\ &[\varphi_n(\mathbf{x}, t), \varphi_m(\mathbf{y}, t)] = [\pi_n(\mathbf{x}, t), \pi_m(\mathbf{y}, t)] = 0 . \end{aligned} \tag{7.1.10}$$

We will explore the consequences of these relations in the next section.

Energy and Momentum

In order to calculate the energies of the various states in a quantum field theory, we need to know the Hamiltonian. Returning to Eq. (5.7.14) and again replacing derivatives with functional derivatives and sums with sums and integrals, we have

$$H = \sum_n \int d^3x \left[\frac{\delta L}{\delta \dot{\varphi}_n(\mathbf{x}, t)} \dot{\varphi}_n(\mathbf{x}, t) \right] - L = \int d^3x \left[\sum_n \pi_n(\mathbf{x}, t) \dot{\varphi}_n(\mathbf{x}, t) - \mathcal{L} \right] , \tag{7.1.11}$$

evaluated at any time t. As explained in Section 5.7, the momentum operator of any system is the generator of space translations. Under an infinitesimal space translation $\mathbf{x} \to \mathbf{x} + \boldsymbol{\epsilon}$, the fields are changed by

$$\varphi_n(\mathbf{x}, t) \to \varphi_n(\mathbf{x} + \boldsymbol{\epsilon}, t) = \varphi_n(\mathbf{x}, t) + \boldsymbol{\epsilon} \cdot \nabla \varphi_n(\mathbf{x}, t) ;$$

so, according to the general rule Eq. (5.7.16), the momentum operator is

$$\mathbf{P} = \sum_n \int d^3x \frac{\delta L}{\delta \dot{\varphi}_n(\mathbf{x}, t)} \nabla \varphi_n(\mathbf{x}, t) = \sum_n \int d^3x \; \pi_n(\mathbf{x}, t) \nabla \varphi_n(\mathbf{x}, t) . \tag{7.1.12}$$

7.2 Free Real Scalar Field

We next consider the simplest example of a quantum field theory, with a single real scalar field $\varphi(x)$, "free" in the sense that the field equations are linear. Of course, we are really interested in what happens when fields interact, but, as we will see in the next section, the first step in dealing with interacting fields is to understand the content of the free-field theory.

We will take the Lagrangian density to have the form

$$\mathcal{L}_0(x) = -\frac{1}{2}\eta^{\mu\nu}\frac{\partial\varphi(x)}{\partial x^\mu}\frac{\partial\varphi(x)}{\partial x^\nu} - \frac{m^2c^2}{2\hbar^2}\varphi^2 , \tag{7.2.1}$$

the justification being that, as we shall see, this gives a sensible theory of free spinless particles of mass m. (We are using the conventions described in Chapter 4, with $x^0 = ct$; repeated indices are summed, with $\eta^{\mu\nu} = +1$ for $\mu = \nu = 1, 2, 3$, $\eta^{\mu\nu} = -1$ for $\mu = \nu = 0$, and $\eta^{\mu\nu} = 0$ otherwise. This makes the action $\int d^4x\, \mathcal{L}_0$ Lorentz invariant.) The subscript 0 on $\mathcal{L}$ is to remind us that this is just the part of the Lagrangian density that would describe free fields if nothing else were added. We will have to add additional terms in the following section to include interactions.

The field equations (7.1.8) here are

$$-(m^2c^2/\hbar^2)\varphi = -\frac{\partial}{\partial x^\mu}\left[\eta^{\mu\nu}\frac{\partial\varphi(x)}{\partial x^\nu}\right] ,$$

or more simply

$$(\Box - m^2c^2/\hbar^2)\varphi = 0 , \tag{7.2.2}$$

where $\Box$ is the d'Alembertian operator:

$$\Box \equiv \eta^{\mu\nu}\frac{\partial}{\partial x^\mu}\frac{\partial}{\partial x^\nu} = \nabla^2 - \frac{1}{c^2}\frac{\partial^2}{\partial t^2} . \tag{7.2.3}$$

The general real solutions of Eq. (7.2.2) are of the form

$$\varphi(\mathbf{x}, t) = \int d^3p \left[A(\mathbf{p}) \exp\left(\frac{i}{\hbar}(\mathbf{p}\cdot\mathbf{x} - E(\mathbf{p})t)\right) + A^\dagger(\mathbf{p}) \exp\left(\frac{-i}{\hbar}(\mathbf{p}\cdot\mathbf{x} - E(\mathbf{p})t)\right)\right] \tag{7.2.4}$$

where $E(\mathbf{p}) = \sqrt{c^2\mathbf{p}^2 + m^2c^4}$, and the coefficients $A(\mathbf{p})$ and $A^\dagger(\mathbf{p})$ are spacetime-independent operators whose properties are to be determined from the canonical commutation relations.

The canonical conjugate to φ is here

$$\pi(\mathbf{x},t) = \frac{\partial \mathcal{L}_0}{\partial \dot{\varphi}(\mathbf{x},t)} = \frac{1}{c^2}\dot{\varphi}(\mathbf{x},t) . \tag{7.2.5}$$

Then

$$\begin{aligned}
[\varphi(\mathbf{x},t), \pi(\mathbf{y},t)] &= \int d^3p \int d^3p' \; (-iE(\mathbf{p}')/c^2\hbar) \\
&\times \left[\left\{ A(\mathbf{p}) \exp\left(\frac{i}{\hbar}(\mathbf{p}\cdot\mathbf{x} - E(\mathbf{p})t)\right) + A^\dagger(\mathbf{p}) \exp\left(\frac{-i}{\hbar}(\mathbf{p}\cdot\mathbf{x} - E(\mathbf{p})t)\right) \right\} , \right. \\
&\left. \left\{ A(\mathbf{p}') \exp\left(\frac{i}{\hbar}(\mathbf{p}'\cdot\mathbf{y} - E(\mathbf{p}')t)\right) - A^\dagger(\mathbf{p}') \exp\left(\frac{-i}{\hbar}(\mathbf{p}'\cdot\mathbf{y} - E(\mathbf{p}')t)\right) \right\} \right] .
\end{aligned}$$

Terms in the integrand that are proportional to the product $\exp(-iE(\mathbf{p})t/\hbar) \times \exp(-iE(\mathbf{p}')t/\hbar)$ or to the product $\exp(iE(\mathbf{p})t/\hbar)\exp(iE(\mathbf{p}')t/\hbar)$ would make different time-dependent contributions to the integral for any values of $\mathbf{p}$ and $\mathbf{p}'$, so since the canonical commutation rules give a time-independent commutator, we must have

$$[A(\mathbf{p}), A(\mathbf{p}')] = [A^\dagger(\mathbf{p}), A^\dagger(\mathbf{p}')] = 0 . \tag{7.2.6}$$

The commutator is then

$$\begin{aligned}
[\varphi(\mathbf{x},t), \pi(\mathbf{y},t)] &= \int d^3p \int d^3p' \; (-iE(\mathbf{p}')/c^2\hbar) \\
&\times \left[-[A(\mathbf{p}), A^\dagger(\mathbf{p}')] \exp\left(\frac{i}{\hbar}(\mathbf{p}\cdot\mathbf{x} - E(\mathbf{p})t)\right) \exp\left(\frac{-i}{\hbar}(\mathbf{p}'\cdot\mathbf{y} - E(\mathbf{p}')t)\right) \right. \\
&\left. - [A(\mathbf{p}'), A^\dagger(\mathbf{p})] \exp\left(\frac{i}{\hbar}(\mathbf{p}'\cdot\mathbf{y} - E(\mathbf{p}')t)\right) \exp\left(\frac{-i}{\hbar}(\mathbf{p}\cdot\mathbf{x} - E(\mathbf{p})t)\right) \right] .
\end{aligned}$$

The commutator must be proportional to $\delta^3(\mathbf{x}-\mathbf{y})$, and in particular must be a function only of $\mathbf{x}-\mathbf{y}$, so the commutator of $A(\mathbf{p})$ with $A^\dagger(\mathbf{p}')$ must be proportional to $\delta^3(\mathbf{p}-\mathbf{p}')$,

$$[A(\mathbf{p}), A^\dagger(\mathbf{p}')] = f(\mathbf{p})\delta^3(\mathbf{p}-\mathbf{p}') ,$$

which also ensures the cancellation of time-dependent factors. The commutator is then

$$[\varphi(\mathbf{x},t), \pi(\mathbf{y},t)] = \int d^3p \; (iE(\mathbf{p})/c^2\hbar)[f(\mathbf{p}) + f(-\mathbf{p})]e^{i\mathbf{p}\cdot(\mathbf{x}-\mathbf{y})/\hbar} .$$

The canonical commutation relations require that

$$[\varphi(\mathbf{x},t),\pi(\mathbf{y},t)] = i\hbar\delta^3(\mathbf{x}-\mathbf{y}) = \frac{i\hbar}{(2\pi\hbar)^3}\int d^3p\; e^{i\mathbf{p}\cdot(\mathbf{x}-\mathbf{y})/\hbar}$$

and therefore

$$f(\mathbf{p}) + f(-\mathbf{p}) = \frac{c^2\hbar^2}{(2\pi\hbar)^3 E(\mathbf{p})} .$$

At this point, we have to look at the commutator of the field with itself. Using what we have already learned about $[A(\mathbf{p}), A(\mathbf{p}')]$ and $[A(\mathbf{p}), A^\dagger(\mathbf{p}')]$, we see that

$$[\varphi(\mathbf{x},t),\varphi(\mathbf{y},t)] = \int d^3p\; [f(\mathbf{p}) - f(-\mathbf{p})]e^{i\mathbf{p}\cdot(\mathbf{x}-\mathbf{y})/\hbar} .$$

Since this commutator has to vanish, we must have $f(\mathbf{p}) = f(-\mathbf{p})$, so we must take

$$f(\mathbf{p}) = f(-\mathbf{p}) = \frac{c^2\hbar^2}{(2\pi\hbar)^3\, 2E(\mathbf{p})} .$$

It is therefore convenient to define $a(\mathbf{p}) \equiv A(\mathbf{p})/\sqrt{f(\mathbf{p})}$, so that

$$[a(\mathbf{p}), a^\dagger(\mathbf{p}')] = \delta^3(\mathbf{p}-\mathbf{p}') \tag{7.2.7}$$

and

$$\varphi(\mathbf{x},t) = \hbar c \int \frac{d^3p}{\sqrt{2E(\mathbf{p})}(2\pi\hbar)^{3/2}} \times \left[a(\mathbf{p})\exp\left(\frac{i}{\hbar}(\mathbf{p}\cdot\mathbf{x} - E(\mathbf{p})t)\right) + a^\dagger(\mathbf{p})\exp\left(\frac{-i}{\hbar}(\mathbf{p}\cdot\mathbf{x} - E(\mathbf{p})t)\right)\right] . \tag{7.2.8}$$

The operators $a(\mathbf{p})$ and $a^\dagger(\mathbf{p})$ are analogous to the operators a_i and $a_i^\dagger$ introduced in our discussion of the harmonic oscillator Hamiltonian in Section 6.3 but with a continuum momentum argument in place of the three-valued index i and a delta function instead of a Kronecker delta.

The Hamiltonian for the free scalar field is given by

$$H_0 = \int d^3x[\pi\dot{\varphi} - \mathcal{L}_0] = \frac{1}{2}\int d^3x \left[\frac{1}{c^2}\dot{\varphi}^2 + (\nabla\varphi)^2 + \frac{m^2c^2}{\hbar^2}\varphi^2\right] .$$

Since this is quadratic in the field φ, when we insert the expression (7.2.8) for φ in H_0, we encounter a double integral over momentum. The integral over $\mathbf{x}$ yields $(2\pi\hbar)^3$ factors times momentum delta functions that reduce this to a single momentum integral. The time-dependent terms in the integrand proportional to $\exp(-2iE(\mathbf{p})t/\hbar)$ or $\exp(2iE(\mathbf{p})t/\hbar)$ are also proportional to $-E^2(\mathbf{p})/c^2 + \mathbf{p}^2 + m^2c^2$, which vanishes. This leaves the time-independent terms

$$H_0 = \frac{1}{2}\int d^3p\ E(\mathbf{p})\left[a^\dagger(\mathbf{p})a(\mathbf{p}) + a(\mathbf{p})a^\dagger(\mathbf{p})\right]. \tag{7.2.9}$$

In the same way, we find the momentum operator

$$\mathbf{P} = \frac{1}{2}\int d^3p\ \mathbf{p}\left[a^\dagger(\mathbf{p})a(\mathbf{p}) + a(\mathbf{p})a^\dagger(\mathbf{p})\right]. \tag{7.2.10}$$

We can check that Eq. (7.2.9) is consistent with what we know to be the time dependence of the field. The canonical commutation relations have been constructed so that

$$\begin{aligned} i\hbar\dot{\varphi}(x) = [\varphi(x), H_0] = \hbar c\int & \frac{d^3p}{\sqrt{2E(\mathbf{p})}(2\pi\hbar)^{3/2}} \\ & \times\Bigg[[a(\mathbf{p}), H_0]\exp\left(\frac{i}{\hbar}(\mathbf{p}\cdot\mathbf{x} - E(\mathbf{p})t)\right) \\ & + [a^\dagger(\mathbf{p}), H_0]\exp\left(\frac{-i}{\hbar}(\mathbf{p}\cdot\mathbf{x} - E(\mathbf{p})t)\right)\Bigg]. \end{aligned}$$

From Eq. (7.2.9) we have

$$[a(\mathbf{p}), H_0] = E(\mathbf{p})a(\mathbf{p}), \qquad [a^\dagger(\mathbf{p}), H_0] = -E(\mathbf{p})a^\dagger(\mathbf{p}), \tag{7.2.11}$$

so

$$\begin{aligned} i\dot{\varphi}(\mathbf{x}, t) = c\int & \frac{d^3p}{\sqrt{2E(\mathbf{p})}(2\pi\hbar)^{3/2}}E(\mathbf{p}) \\ & \times\left[a(\mathbf{p})\exp\left(\frac{i}{\hbar}(\mathbf{p}\cdot\mathbf{x} - E(\mathbf{p})t)\right) - a^\dagger(\mathbf{p})\exp\left(\frac{-i}{\hbar}(\mathbf{p}\cdot\mathbf{x} - E(\mathbf{p})t)\right)\right] \end{aligned}$$

which is the same as would be given directly by taking the time derivative of Eq. (7.2.8).

Likewise, from Eq. (7.2.10), we have

$$[a(\mathbf{p}), \mathbf{P}] = \mathbf{p}a(\mathbf{p}), \qquad [a^\dagger(\mathbf{p}), \mathbf{P}] = -\mathbf{p}a^\dagger(\mathbf{p}), \tag{7.2.12}$$

from which we can see that, as expected, $[\varphi(x), \mathbf{P}] = -i\hbar\nabla\varphi(x)$.

Equations (7.2.11) and (7.2.12) show that $a(\mathbf{p})$ and $a^\dagger(\mathbf{p})$ act as annihilation and creation operators, analogous to the lowering and raising operators for energy a_i and $a_i^\dagger$ in Section 6.3 and to the operators $J_1 - iJ_2$ and $J_1 + iJ_2$ that we used in working out the content of angular momentum multiplets in Section 5.4. Suppose a state represented by a wave function ψ has definite values E_ψ and $\mathbf{p}_\psi$ for the total energy and total momentum. That is,

$$H_0\psi = E_\psi\psi, \qquad \mathbf{P}\psi = \mathbf{p}_\psi\psi.$$

Then

$$H_0\, a(\mathbf{p})\psi = [H_0, a(\mathbf{p})]\psi + a(\mathbf{p})H_0\psi = (E_\psi - E(\mathbf{p}))a(\mathbf{p})\psi\ ;$$

so if $a(\mathbf{p})\psi$ does not vanish, it is the wave function for a state with energy $E_\psi - E(\mathbf{p})$. Likewise, this is a state with momentum $\mathbf{p}_\psi - \mathbf{p}$, while $a^\dagger(\mathbf{p})\psi$ is the wave function for a state with energy $E_\psi + E(\mathbf{p})$ and total momentum $\mathbf{p}_\psi + \mathbf{p}$. In other words, $a(\mathbf{p})$ and $a^\dagger(\mathbf{p})$ respectively annihilate and create a particle of momentum $\mathbf{p}$. This is what we mean when refer to elementary particles being bundles of the energy and momentum in some field.

At this point, and for the rest of this chapter, we will abandon the language of wave mechanics, and instead employ the more abstract language of state vectors and scalar products that was outlined in Section 5.10. In quantum field theory the wave function of any state such as the vacuum is a complicated functional of the fields, and the action of operators like $a(\mathbf{p})$ on these wave functions involves functional derivatives with respect to these fields. None of these complications plays a role in most calculations. What we use instead are the properties of operators, such as the field equations and the canonical commutation relations, and limited assumptions about physical states.

In particular, it is a plausible physical assumption that there should exist a physical state, the vacuum Ψ_{vac}, with the lowest possible energy. Then $a(\mathbf{p})\Psi_{\text{vac}}$ must vanish,

$$a(\mathbf{p})\Psi_{\text{vac}} = 0\ , \tag{7.2.13}$$

since otherwise it would be a state with energy less by an amount $E(\mathbf{p})$.

To calculate the energy and momentum of the vacuum, it is convenient to use the commutator of a with $a^\dagger$ to rewrite Eqs. (7.2.9) and (7.2.10) as

$$H_0 = \int d^3p\ E(\mathbf{p})\ \Big[a^\dagger(\mathbf{p})a(\mathbf{p}) + E_{\text{vac}}\Big]\ , \tag{7.2.14}$$

$$\mathbf{P} = \int d^3p\ \mathbf{p}\ a^\dagger(\mathbf{p})a(\mathbf{p})\ , \tag{7.2.15}$$

where E_{vac} is an infinite constant:

$$E_{\text{vac}} = \frac{1}{2}\int d^3p\ E(\mathbf{p})\,\delta^3(0) = \frac{1}{2(2\pi\hbar)^3}\int d^3p\ E(\mathbf{p})\int d^3x\ . \tag{7.2.16}$$

From Eqs. (7.2.14) and (7.2.13), we see that this is the energy of the vacuum:

$$H_0\Psi_{\text{vac}} = E_{\text{vac}}\Psi_{\text{vac}}\ , \tag{7.2.17}$$

while Eqs. (7.2.15) and (7.2.13) show that the momentum of the vacuum is zero.

For most purposes a constant term in the energy such as E_{vac} makes no difference, because the same constant appears in the energies of all states and therefore has no effect in applications of the conservation of energy. The one phenomenon that is affected by such a constant is gravitation, which is coupled to all forms of energy. In a finite volume, Eq. (7.3.16) corresponds to an infinite vacuum energy density

$$\rho_{\text{vac}} = \frac{1}{2(2\pi\hbar)^3}\int d^3p\ E(\mathbf{p})\ .$$

But Einstein's general theory of relativity allows a term in the field equations of gravitation, known as the cosmological constant, that has just the same effects as ρ_{vac}. There is no reason why the cosmological constant should not include an infinite negative term that simply cancels ρ_{vac}, possibly leaving over a finite remaining energy density. Observations of an accelerated expansion of the universe have shown that this remaining energy density is not zero, though it is tiny compared with the energy densities encountered in atomic and nuclear physics.[1]

The quantum states of this free field can be constructed by acting on the vacuum with any number of creation operators. If we define

$$\Psi_{\mathbf{p}_1,\mathbf{p}_2,\mathbf{p}_3,\ldots} \equiv a^\dagger(\mathbf{p}_1)\,a^\dagger(\mathbf{p}_2)\,a^\dagger(\mathbf{p}_3)\ \cdots\ \Psi_{\text{vac}} \tag{7.2.18}$$

then from Eqs. (7.2.11), (7.2.12), and (7.2.17) we see that

$$H_0\Psi_{\mathbf{p}_1,\mathbf{p}_2,\mathbf{p}_3,\ldots} = [E_{\text{vac}} + E(\mathbf{p}_1) + E(\mathbf{p}_2) + E(\mathbf{p}_3) + \cdots]\,\Psi_{\mathbf{p}_1,\mathbf{p}_2,\mathbf{p}_3,\ldots} \tag{7.2.19}$$

and

$$\mathbf{P}\Psi_{\mathbf{p}_1,\mathbf{p}_2,\mathbf{p}_3,\ldots} = [\mathbf{p}_1 + \mathbf{p}_2 + \mathbf{p}_3 + \cdots]\Psi_{\mathbf{p}_1,\mathbf{p}_2,\mathbf{p}_3,\cdots}\ . \tag{7.2.20}$$

These are states with any number of particles. The superpositions of all such states make up what is called *Fock space*, named after Vladimir Fock (1898–1974). Because the operators $a^\dagger(\mathbf{p}_1)$, $a^\dagger(\mathbf{p}_2)$, etc. all commute with one another, the states (7.2.18) are symmetric in the momenta of the particles, and hence these spinless particles are bosons.

The states $\Psi_{\mathbf{p}_1,\mathbf{p}_2,\mathbf{p}_3,\ldots}$ are no longer eigenstates of the Hamiltonian if we add higher-order terms such as φ^3, φ^4, etc. to the Lagrangian density. Such terms drive transitions between these states, corresponding to the creation and annihilation of particles. We will discuss this further in the next section. As we will see there, knowledge of the free-field theory is an essential ingredient in these calculations, which is why we have gone into it here.

---

[1] For a textbook discussion and references to the original literature, see S. Weinberg, *Cosmology* (Oxford University Press, Oxford, 2008), Sections 1.4 and 1.5.

7.3 Interactions

We shall now consider how to calculate transition rates in theories of interacting fields. Most (though not all) useful calculations in quantum field theory rely on perturbation theory. We write the Hamiltonian as $H = H_0 + H'$, where H_0 is the Hamiltonian of a free-field theory, like the Hamiltonian discussed in the previous section, and H' is an interaction term that is considered to be small enough to allow physical quantities to be calculated as power series in H'.

In Section 5.9 we saw how perturbation theory is used to calculate shifts in energy levels in quantum mechanics. In second and higher orders in a perturbation, we encounter energy denominators, such as that shown in Eq. (5.9.27). Similar energy denominators occur in perturbative calculations of scattering amplitudes using the Lippmann–Schwinger equation (5.6.29) and its iterations. The fact that denominators involve energy but not momentum differences makes it obvious that they are not Lorentz invariant, so they make it difficult to keep track of Lorentz invariance in relativistic theories. This sort of perturbation theory, which is now known as "old-fashioned perturbation theory," was all that was available for calculations in quantum field theory in the 1930s, making progress difficult. In particular, it was not clear how to deal with the divergent integrals occurring in these calculations without losing the Lorentz invariance of the underlying theory.

In the late 1940s, independently, Richard Feynman[2] (1918–1988), Julian Schwinger[3] (1918–1994), and Sin-Itiro Tomonaga (1906–1979) and his collaborators[4] were able to carry out manifestly relativistic perturbative calculations in quantum electrodynamics. The equivalence of their methods was shown by Freeman Dyson[5] (1923–2020), who gave a systematic account of a method of calculation that would maintain manifest Lorentz invariance to all orders of perturbation theory. We shall now describe this method. Here and for the balance of this chapter, as is usual in work on quantum field theory, we shall use what are called "natural units," in which $\hbar = c = 1$, and we shall continue to represent physical states as vectors in Hilbert space, as described in Section 5.10.

[2] R. P. Feynman, *Rev. Mod. Phys.* **20**, 367 (1948); *Phys. Rev.* **74**, 939, 1430 (1948); *ibid.*, **76**, 749, 769 (1949); *ibid* **80**, 440 (1950).

[3] J. Schwinger, *Phys. Rev.* **74**, 1439 (1948); *ibid.*, **75**, 651 (1949); *ibid.*, **76**, 790 (1949); *ibid.*, **82**, 664, 914 (1951); *ibid.*, **91**, 713 (1953); *Proc. Nat. Acad. Sci.* **37**, 452 (1951).

[4] S. Tomonaga, *Prog. Theor. Phys. Rev. Mod. Phys.* **1**, 27 (1946); Z. Koba, T. Tati, and S. Tomonaga, *ibid.*, **2**, 101 (1947); S. Kanesawa and S. Tomonaga, *ibid.*, **3**, 1, 101 (1948); S. Tomonaga, *Phys. Rev.* **74**, 224 (1948); D. Ito, Z. Koba, and S. Tomonaga, *Prog. Theor. Phys.* **3**, 276 (1948); Z. Koba and S. Tomonaga, *ibid.*, **3**, 290 (1948).

[5] F. J. Dyson, *Phys. Rev.* **75**, 486, 1736 (1949).

Time-Ordered Perturbation Theory

We saw in the appendix to Section 5.6 how the rate for any transition $\alpha \to \beta$ between free-particle states α and β can be calculated from a knowledge of the S-matrix, $S_{\beta\alpha}$. Our first task now is to see how to express $S_{\beta\alpha}$ in a form that allows its calculation in modern perturbation theory.

In the appendix to Section 5.6 we showed how to construct an eigenstate Ψ_α of the full Hamiltonian, with $H\Psi_\alpha = E_\alpha\Psi_\alpha$, with the special property that at very early times it looks like the eigenstate Φ_α of the free-particle Hamiltonian, with $H_0\Phi_\alpha = E_\alpha\Phi_\alpha$, in the sense of Eq. (5.6.34): for $t \to -\infty$,

$$\int g(\alpha)e^{-iHt}\Psi_\alpha\, d\alpha \to \int g(\alpha)e^{-iH_0t}\Phi_\alpha\, d\alpha \;,$$

where $g(\alpha)$ is a smooth function of the momenta of all the particles in state α, introduced to give meaning to the limit for $t \to -\infty$. (Recall that the label α is intended to include the momenta, spin 3-components or helicities, and species labels of all the particles in the state α, and an integral over α is intended to include integrals over all momenta and sums over all spin 3-components or helicities and species labels.) We also showed that at very late times the same state Ψ_α looks like the superposition $\int d\beta S_{\beta\alpha}\Phi_\beta$, in the sense of Eq. (3.6.35): for $t \to +\infty$,

$$\int g(\alpha)e^{-iHt}\Psi_\alpha\, d\alpha \to \int g(\alpha)e^{-iH_0t}\, d\alpha \int d\beta\; S_{\beta\alpha}\Phi_\beta \;.$$

With considerable loss of mathematical rigor, we multiply both sides by the operator e^{iHt} and equate coefficients of $g(\alpha)$ on both sides, so these formulas yield

$$\Psi_\alpha = \Omega(-\infty)\Phi_\alpha = \Omega(+\infty)\int d\beta\; S_{\beta\alpha}\Phi_\beta \;,$$

where

$$\Omega(t) \equiv e^{iHt}e^{-iH_0t} \;.$$

From the two equalities we can conclude that

$$S_{\beta\alpha} = (\Phi_\beta, \Omega^{-1}(+\infty)\Omega(-\infty)\Phi_\alpha) = (\Phi_\beta, U(+\infty,-\infty)\Phi_\alpha) \;, \qquad (7.3.1)$$

where

$$U(t,t_0) \equiv \Omega^{-1}(t)\Omega(t_0) = e^{iH_0t}e^{-iH(t-t_0)}e^{-iH_0t_0} \;. \qquad (7.3.2)$$

The justification (such as it is) for treating $\Omega(t)$ as if it had well-defined limits for $t \to \pm\infty$ is that at very early and very late times the incoming and outgoing particles are so far apart that the interaction $H - H_0$ is ineffective. As we shall see, at least in perturbation theory the limits $t \to \pm\infty$ do lead to well-defined probability amplitudes.

To construct a perturbation series for U in powers of the interaction $H - H_0$, we first take the derivative of Eq. (7.3.2) with respect to t:

$$\frac{d}{dt}U(t,t_0) = -i\exp(iH_0t)[H - H_0]\exp(-iH(t-t_0))\exp(-iH_0t_0)$$
$$= -iH'(t)U(t,t_0)\,, \tag{7.3.3}$$

where $H'(t)$ is the interaction in what is called the *interaction picture*, in which the time dependence is governed by the free-particle Hamiltonian H_0:

$$H'(t) \equiv \exp(iH_0t)[H - H_0]\exp(-iH_0t)\,. \tag{7.3.4}$$

The time dependence of any operator in the interaction picture is given by its commutator with H_0, which is one reason why we need to understand the free-field theory before taking interactions into account.

The differential (7.3.3) together with the initial condition $U(t_0,t_0) = 1$ is incorporated in the integral equation

$$U(t,t_0) = 1 - i\int_{t_0}^{t} H'(t_1)U(t_1,t_0)dt_1\,. \tag{7.3.5}$$

We can solve this at least formally by iteration:

$$U(t,t_0) = 1 - i\int_{t_0}^{t} dt_1\; H'(t_1) + (-i)^2\int_{t_0}^{t} dt_1\int_{t_0}^{t_1} dt_2\; H'(t_1)\,H'(t_2) + \cdots\,. \tag{7.3.6}$$

Instead of using limits on the integrals to impose an ordering of the integration variables t_1, t_2, etc., we can integrate all these variables over the whole range from t_0 to t, which with n integrals includes $n!$ permutations of the order of the integration variables; we then correct for this multiplicity of permutations by dividing by $n!$ and reimpose the ordering of time variables by changing the product of H' operators to a time-ordered product denoted $T\{\cdots\}$, in which operator factors appear in order of decreasing time arguments. For instance

$$\int_{t_0}^{t} dt_1\int_{t_0}^{t_1} dt_2\; H'(t_1)\,H'(t_2) = \frac{1}{2}\int_{t_0}^{t} dt_1\int_{t_0}^{t} dt_2\; T\{H'(t_1)\,H'(t_2)\}$$

where

$$T\{H'(t_1)\,H'(t_2)\} = \begin{cases} H'(t_1)\,H'(t_2) & t_1 > t_2 \\ H'(t_2)\,H'(t_1) & t_2 > t_1\,. \end{cases}$$

The complete sum is then

$$U(t,t_0) = 1 + \sum_{n=1}^{\infty}\frac{(-i)^n}{n!}\int_{t_0}^{t} dt_1\cdots\int_{t_0}^{t} dt_n\; T\{H'(t_1)\cdots H'(t_n)\}\,. \tag{7.3.7}$$

This begins to look Lorentz invariant if we take the limits $t \to +\infty$ and $t_0 \to -\infty$ and suppose that $H'(t)$ is the integral over all space of a scalar density $\mathcal{H}'(x)$, such as a polynomial function of the field $\varphi(x)$ discussed in the previous section:

$$H'(t) = \int d^3x \; \mathcal{H}'(\mathbf{x}, t) \tag{7.3.8}$$

in which case Eq. (7.3.7) becomes

$$U(\infty, -\infty) = 1 + \sum_{n=1}^{\infty} \frac{(-i)^n}{n!} \int d^4x_1 \cdots \int d^4x_n \; T\{\mathcal{H}'(x_1) \cdots \mathcal{H}'(x_n)\} \,, \tag{7.3.9}$$

which can be used along with Eq. (7.3.1) to calculate the S-matrix.

Lorentz Invariance

The remaining problem with Lorentz invariance is that the integrand is still time-ordered. As we saw in Section 4.7, the ordering in time of two events at spacetime positions x_1 and x_2 is Lorentz invariant if the separation $x_1 - x_2$ is time-like or light-like – that is (using units with $c = 1$), if

$$\eta_{\mu\nu}(x_1 - x_2)^\mu (x_1 - x_2)^\nu = (\mathbf{x}_1 - \mathbf{x}_2)^2 - (t_1 - t_2)^2 \leq 0 \,.$$

Thus to make the scattering operator Lorentz invariant we need the densities to commute at space-like separations:

$$\left[\mathcal{H}'(x_1), \mathcal{H}'(x_2)\right] = 0 \quad \text{for} \quad \eta_{\mu\nu}(x_1 - x_2)^\mu (x_1 - x_2)^\nu > 0 \,. \tag{7.3.10}$$

The vanishing of this commutator tells us that there is no obstacle to finding states that are eigenstates of both $\mathcal{H}'(x_1)$ and $\mathcal{H}'(x_2)$, which can also be justified on grounds of causality since for $x_1 - x_2$ space-like no signal could travel from a measurement of $\mathcal{H}$ at x_1 to interfere with a measurement of $\mathcal{H}$ at x_2.

Any space-like separation $x_1 - x_2$ can be obtained from a purely spatial separation with $t_1 = t_2$ by a Lorentz transformation, so as long as $\mathcal{H}'(x)$ is a scalar, the necessary and sufficient condition for (7.3.10) is that the commutator should vanish at equal times:

$$\left[\mathcal{H}'(\mathbf{x}_1, t), \mathcal{H}'(\mathbf{x}_2, t)\right] = 0 \,. \tag{7.3.11}$$

The scalar field $\varphi(\mathbf{x}, t)$ introduced in the former section satisfies the commutation relation $[\varphi(\mathbf{x}_1, t), \varphi(\mathbf{x}_2, t)] = 0$ for any positions $\mathbf{x}_1$ and $\mathbf{x}_2$, so an interaction Hamiltonian density $\mathcal{H}'$ constructed as any polynomial function of φ will satisfy Eq. (7.3.11). As we shall see in the next section, the condition (7.3.11) is not so easy to satisfy in more general theories, and this leads to the necessity of antiparticles.

Example: Scattering

To make all this concrete, let us calculate the lowest-order amplitude for scattering of a pair of particles in the theory of real scalar fields, with H_0 the free-particle Hamiltonian described in the previous section, and with the simple interaction Hamiltonian density

$$\mathcal{H}' = \frac{g}{6}\varphi^3 , \tag{7.3.12}$$

with g a constant taken small enough to justify the use of perturbation theory.[6] (The factor 1/6 is inserted for later convenience.) To lowest order in g, the S-matrix element for particles with momenta $\mathbf{p}_1$ and $\mathbf{p}_2$ to scatter, with momenta changing to $\mathbf{p}'_1$ and $\mathbf{p}'_2$, is

$$S_{\mathbf{p}'_1\mathbf{p}'_2,\mathbf{p}_1\mathbf{p}_2} = -\frac{1}{2}\left(\frac{g}{6}\right)^2 \int d^4x\, d^4y\, (\Phi_{\mathbf{p}'_1\mathbf{p}'_2}, T\{\varphi^3(x), \varphi^3(y)\}\Phi_{\mathbf{p}_1\mathbf{p}_2}) , \tag{7.3.13}$$

where

$$\Phi_{\mathbf{p}_1\mathbf{p}_2} \equiv \frac{1}{\sqrt{2}} a^\dagger(\mathbf{p}_1)\, a^\dagger(\mathbf{p}_2)\Phi_0 , \tag{7.3.14}$$

with Φ_0 the free-particle vacuum state. The factor $1/\sqrt{2}$ is included to compensate for the sum of two delta functions in the scalar product; using Eqs. (7.2.7) and (7.2.13),

$$(\Phi_{\mathbf{p}'_1\mathbf{p}'_2}, \Phi_{\mathbf{p}_1\mathbf{p}_2}) = \frac{1}{2}\left[\delta^3(\mathbf{p}'_1 - \mathbf{p}_1)\delta^3(\mathbf{p}'_2 - \mathbf{p}_2) + \delta^3(\mathbf{p}'_1 - \mathbf{p}_2)\delta^3(\mathbf{p}'_2 - \mathbf{p}_1)\right] . \tag{7.3.15}$$

There is no term in this S-matrix element that is of first order in g, because there are not enough creation and annihilation operators in a single φ^3 operator to destroy the two initial particles and create the two final particles.

Our strategy in calculating the scattering amplitude (7.3.13) will be to move a pair of the annihilation operators in $\varphi^3(x)$ and/or $\varphi^3(y)$ past the creation operators in $\Phi_{\mathbf{p}_1\mathbf{p}_2}$, which gives a pair of commutators of annihilation with creation operators, and use the fact that annihilation operators give zero when acting on Φ_0; also, to move a pair of the creation operators in $\varphi^3(x)$ and/or $\varphi^3(y)$ to the left side of the scalar product, so that their adjoints act as annihilation operators on $\Phi_{\mathbf{p}'_1\mathbf{p}'_2}$, and then move them past the creation operators in $\Phi_{\mathbf{p}'_1\mathbf{p}'_2}$, giving another pair of commutators of annihilation with creation operators and again using the fact that annihilation operators give zero when acting on Φ_0.[7]

[6] This theory is actually unphysical, because $\mathcal{H}' \to -\infty$ if $g > 0$ and $\varphi \to -\infty$ or if $g < 0$ and $\varphi \to +\infty$. This problem does not emerge in perturbation theory, and in any case can be dealt with by adding higher even powers of φ with positive coefficients in $\mathcal{H}'$.

[7] In moving these annihilation operators out of the time-ordered product to the right or their adjoints to the left, we are ignoring their commutators with the other fields in $\varphi^3(x)$ and $\varphi^3(y)$, because these terms involve momentum delta functions that vanish if we assume that neither $\mathbf{p}'_1$ nor $\mathbf{p}'_2$ equal either $\mathbf{p}_1$ or $\mathbf{p}_2$.

If we separate the annihilation and creation parts of $\varphi(x)$, so that

$$\varphi(x) = \varphi_{\rm an}(x) + \varphi^\dagger_{\rm an}(x)\ , \tag{7.3.16}$$

$$\varphi_{\rm an}(\mathbf{x},t) = \int \frac{d^3p}{\sqrt{2E(\mathbf{p})}(2\pi)^{3/2}} a(\mathbf{p}) \exp\big(i(\mathbf{p}\cdot\mathbf{x} - E(\mathbf{p})t)\big)\ , \tag{7.3.17}$$

then

$$[\varphi_{\rm an}(x)\ ,\ a^\dagger(\mathbf{p})] = \frac{1}{\sqrt{2E(\mathbf{p})}(2\pi)^{3/2}} e^{ip\cdot x}\ , \tag{7.3.18}$$

where

$$p\cdot x \equiv \eta_{\mu\nu}p^\mu x^\nu = \mathbf{p}\cdot\mathbf{x} - E(\mathbf{p})t\ .$$

Following this strategy, we encounter three terms in the S-matrix element:

a. The annihilation operator that destroys the particle with momentum $\mathbf{p}_1$ and the creation operator that creates the particle with momentum $\mathbf{p}'_1$ come from the same φ^3 operator in Eq. (7.3.13), while the annihilation operator that destroys the particle with momentum $\mathbf{p}_2$ and the creation operator that creates the particle with momentum $\mathbf{p}'_2$ come from the other φ^3 operator.
b. The annihilation operator that destroys the particle with momentum $\mathbf{p}_1$ and the creation operator that creates the particle with momentum $\mathbf{p}'_2$ come from the same φ^3 operator in Eq. (7.3.13), while the annihilation operator that destroys the particle with momentum $\mathbf{p}_2$ and the creation operator that creates the particle with momentum $\mathbf{p}'_1$ come from the other φ^3 operator.
c. The annihilation operators that destroy both initial particles come from the same φ^3 operator, while the creation operators that create both final particles come from the other φ^3 operator.

In each case, one of the $\varphi(x)$ and one of the $\varphi(y)$ fields is left over in the time-ordered product. Also, since the time-ordered product in Eq. (7.3.13) is symmetric in the spacetime arguments x and y, each of the above contributions is a sum of two equal terms with x and y interchanged, so we can make an arbitrary choice of which of the φ^3 operators in the three cases above is $\varphi^3(x)$ and which is $\varphi^3(y)$, and drop the factor 1/2 in (7.3.13). Instead, a factor 1/2 appears owing to the factors $1/\sqrt{2}$ given by Eq. (7.3.14) in the initial and final states. The factor 1/6 in Eq. (7.3.12) is cancelled by the 3! ways of choosing each of the fields in φ^3.

Here in turn are these three contributions:

$$S^{(a)}_{\mathbf{p}'_1\mathbf{p}'_2,\mathbf{p}_1\mathbf{p}_2} = -\frac{g^2}{2}\int d^4x\ d^4y\ g([\varphi_{\rm an}(x), a^\dagger(\mathbf{p}'_1)][\varphi_{\rm an}(y), a^\dagger(\mathbf{p}'_2)]\Phi_0\ ,$$

$$T\{\varphi(x),\varphi(y)\} \times [\varphi_{\rm an}(x), a^\dagger(\mathbf{p}_1)][\varphi_{\rm an}(y), a^\dagger(\mathbf{p}_2)]\Phi_0 g)$$

$$= \frac{-g^2}{2(2\pi)^6\sqrt{2E(\mathbf{p}_1)\cdot 2E(\mathbf{p}_2)\cdot 2E(\mathbf{p}'_1)\cdot 2E(\mathbf{p}'_2)}}$$

$$\times \int d^4x\, d^4y e^{-ip'_1\cdot x}e^{-ip'_2\cdot y}e^{ip_1\cdot x}e^{ip_2\cdot y}(\Phi_0, T\{\varphi(x),\varphi(y)\}\Phi_0)\,. \tag{7.3.19}$$

$$S^{(b)}_{\mathbf{p}'_1\mathbf{p}'_2,\mathbf{p}_1\mathbf{p}_2} = -\frac{g^2}{2}\int d^4x\, d^4y\; g([\varphi_{\rm an}(y),a^\dagger(\mathbf{p}'_1)][\varphi_{\rm an}(x),a^\dagger(\mathbf{p}'_2)]\Phi_0\,,$$

$$T\{\varphi(x),\varphi(y)\}\times[\varphi_{\rm an}(x),a^\dagger(\mathbf{p}_1)][\varphi_{\rm an}(y),a^\dagger(\mathbf{p}_2)]\Phi_0 g)$$

$$= \frac{-g^2}{2(2\pi)^6\sqrt{2E(\mathbf{p}_1)\cdot 2E(\mathbf{p}_2)\cdot 2E(\mathbf{p}'_1)\cdot 2E(\mathbf{p}'_2)}}$$

$$\times \int d^4x\, d^4y e^{-ip'_1\cdot y}e^{-ip'_2\cdot x}e^{ip_1\cdot x}e^{ip_2\cdot y}(\Phi_0, T\{\varphi(x),\varphi(y)\}\Phi_0)\,. \tag{7.3.20}$$

$$S^{(c)}_{\mathbf{p}'_1\mathbf{p}'_2,\mathbf{p}_1\mathbf{p}_2} = -\frac{g^2}{2}\int d^4x\, d^4y\; g([\varphi_{\rm an}(y),a^\dagger(\mathbf{p}'_1)][\varphi_{\rm an}(y),a^\dagger(\mathbf{p}'_2)]\Phi_0\,,$$

$$T\{\varphi(x),\varphi(x)\}\times[\varphi_{\rm an}(x),a^\dagger(\mathbf{p}_1)][\varphi_{\rm an}(y),a^\dagger(\mathbf{p}_2)]\Phi_0 g)$$

$$= \frac{-g^2}{2(2\pi)^6\sqrt{2E(\mathbf{p}_1)\cdot 2E(\mathbf{p}_2)\cdot 2E(\mathbf{p}'_1)\cdot 2E(\mathbf{p}'_2)}}$$

$$\times \int d^4x\, d^4y e^{-ip'_1\cdot y}e^{-ip'_2\cdot y}e^{ip_1\cdot x}e^{ip_2\cdot x}(\Phi_0, T\{\varphi(x),\varphi(y)\}\Phi_0)\,. \tag{7.3.21}$$

These three contributions are symbolized in three of what are known as *Feynman diagrams*, shown here in Figure 7.1.

Calculation of the Propagator

Evidently, we need to calculate the vacuum expectation value

$$\big(\Phi_0, T\{\varphi(x),\varphi(y)\}\Phi_0\big),$$

which is known as the *propagator* of the field φ. For this purpose, we again write the scalar field as in Eq. (7.3.16) and use the fact that $\varphi_{\rm an}$ acting to the right on Φ_0 and $\varphi^\dagger_{\rm an}$ acting to the left, where its adjoint acts on Φ_0 as $\varphi_{\rm an}$, both vanish. This gives

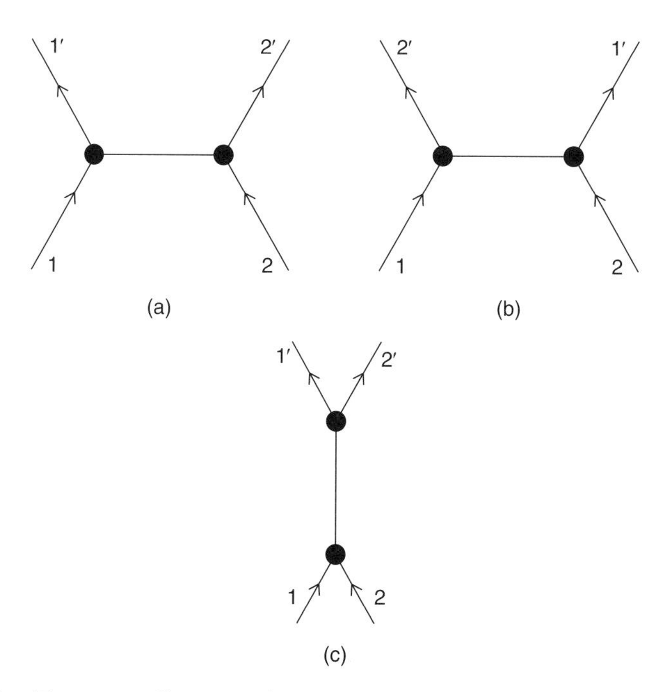

Figure 7.1 Feynman diagrams for the scattering of neutral scalar particles. Here the lines coming into the diagrams from below or going out from the diagrams above represent particles in initial and final states, respectively; the vertices represent an interaction and are proportional to φ^3; the line connecting vertices represents the propagator.

$$\begin{aligned}(\Phi_0, \varphi(x)\varphi(y)\Phi_0) &= (\Phi_0, \varphi_{\text{an}}(x)\varphi^\dagger_{\text{an}}(y)\Phi_0)\\ &= (\Phi_0, [\varphi_{\text{an}}(x), \varphi^\dagger_{\text{an}}(y)]\Phi_0) = \Delta_+(x-y)\end{aligned} \tag{7.3.22}$$

where Δ_+ is the function

$$\Delta_+(z) \equiv \int \frac{d^3p}{(2\pi)^3 2E(\mathbf{p})} \exp(i\mathbf{p}\cdot\mathbf{z} - iE(\mathbf{p})z^0)\,. \tag{7.3.23}$$

The propagator is then

$$(\Phi_0, T\{\varphi(x), \varphi(y)\}\Phi_0) = \theta(x-y)\Delta_+(x-y) + \theta(y-x)\Delta_+(y-x)\,, \tag{7.3.24}$$

where θ is the step function

$$\theta(z) \equiv \begin{cases} 1 & z^0 > 0 \\ 0 & z^0 < 0\,. \end{cases} \tag{7.3.25}$$

What we need in Eqs. (7.3.19)–(7.3.21) is the Fourier transform of the propagator:

$$\Delta(q) \equiv \int e^{-iq\cdot z}\left[\theta(z)\Delta_+(z) + \theta(-z)\Delta_+(-z)\right] d^4z$$

$$= \frac{1}{2E(\mathbf{q})}\Bigg[\int_0^\infty dz^0 \, \exp[(iq^0 - iE(\mathbf{q}))z^0]$$
$$+ \int_{-\infty}^0 dz^0 \, \exp[(iq^0 + iE(\mathbf{q}))z^0]\Bigg] .$$

We can give meaning to these integrals by inserting convergence factors $\exp(-\epsilon z^0)$ in the first integral and $\exp(+\epsilon z^0)$ in the second integral, where ϵ is a positive infinitesimal. The integrals are then elementary:

$$\Delta(q) = \frac{1}{2E(\mathbf{q})}\left[\frac{1}{\epsilon - iq^0 + iE(\mathbf{q})} + \frac{1}{\epsilon + iq^0 + iE(\mathbf{q})}\right] ;$$

so, for $\epsilon \to 0$,

$$\Delta(q) \to \frac{-i}{q^2 + m^2 - 2i\epsilon E(\mathbf{p})} , \qquad (7.3.26)$$

where $q^2 \equiv \eta_{\mu\nu} q^\mu q^\nu = \mathbf{q}^2 - (q^0)^2$. The term $-2i\epsilon E(\mathbf{p})$ in the denominator, though infinitesimal, is important in more complicated calculations where we have to integrate $\Delta(q)$ over a range of its argument in which $q^2 + m^2$ can vanish. (For this purpose, it is only important that it is a negative imaginary infinitesimal, and so is usually written simply as $-i\epsilon$.) In our calculation the integrals over x and y fix the argument of Δ, and we can drop the term $-2i\epsilon E(\mathbf{p})$ in the denominator.[8]

To do the integrals over x and y in Eqs. (7.3.19)–(7.3.21) we set $x = (x - y) + y$ and integrate separately over $x - y$ and y. In each term the integral over y then simply gives a factor $(2\pi)^4 \delta^4(p_1' + p_2' - p_1 - p_2)$, which guarantees the conservation of energy and momentum. The integrals over $x - y$ are given by Eq. (7.3.26), and the sum of (7.3.19), (7.3.20), and (7.3.21) then gives the total second-order scattering amplitude:

$$S_{\mathbf{p}_1'\mathbf{p}_2', \mathbf{p}_1\mathbf{p}_2} = \frac{ig^2 (2\pi)^4 \delta^4(p_1' + p_2' - p_1 - p_2)}{2(2\pi)^6 \sqrt{2E(\mathbf{p}_1) \cdot 2E(\mathbf{p}_2) \cdot 2E(\mathbf{p}_1') \cdot 2E(\mathbf{p}_2')}}$$
$$\times \left[\frac{1}{(p_1 - p_1')^2 + m^2} + \frac{1}{(p_1 - p_2')^2 + m^2} + \frac{1}{(p_1 + p_2)^2 + m^2}\right] . \qquad (7.3.27)$$

[8] The circumstance that all four-momenta are fixed by the delta functions generated by integrals over spacetime coordinates is true of all *tree* diagrams – that is, diagrams like Figure 7.1 that can be disconnected by cutting any single internal line. The contributions to the S-matrix of diagrams with L loops, whose disconnection requires the cutting of a minimum of $L + 1$ internal lines, involve integrals over L four-momenta.

The appearance here of the term $1/[(p_1 - p_1')^2 + m^2]$ may evoke the recollection of an earlier result. In Eq. (5.6.23) we found in the Born approximation that a potential proportional to $\exp(-\kappa r)/r$ gives a scattering amplitude proportional to $1/[(\mathbf{k} - \mathbf{k}')^2 + \kappa^2]$, where $\mathbf{k}$ and $\mathbf{k}'$ are the initial and final wave numbers of the scattered particle, or, in units with $\hbar = 1$, the initial and final momenta. There is no energy term in the denominator in Eq. (5.6.23) because the scattering was supposed there to be due to an external potential that can transfer momentum but not energy. Aside from that, the comparison shows that the exchange of a scalar particle of mass m creates effects, like those of a Yukawa potential, proportional to $\exp(-\kappa r)/r$, with $\kappa = m$ in natural units or, in cgs units, with $\kappa = mc/\hbar$. This was the point made by Yukawa[9] in 1935, which led him to the prediction of a "meson," with mass intermediate between the electron and the proton, to carry the nuclear force.

7.4 Antiparticles, Spin, Statistics

The real scalar field discussed in the previous two sections could not describe a particle that carries any conserved quantity, such as electric charge. If the annihilation part $\varphi_{\text{an}}(x)$ of the field given by Eq. (7.3.17) destroys a certain amount of charge then its adjoint $\varphi_{\text{an}}^\dagger(x)$ would create the same quantity of charge, and no interaction such as φ^3 constructed from the real field $\varphi = \varphi_{\text{an}} + \varphi_{\text{an}}^\dagger$ could possibly conserve this quantity. We could construct interactions that conserve charge by separating φ_{an} and $\varphi_{\text{an}}^\dagger$ and taking the interaction to include equal numbers of factors of each, such as $\varphi_{\text{an}}^2 \varphi_{\text{an}}^{\dagger 2}$, but then we would not be able to preserve Lorentz invariance. The commutator of $\varphi_{\text{an}}(x)$ and $\varphi_{\text{an}}^\dagger(y)$ is the function $\Delta_+(x - y)$ given by Eq. (7.3.23), which does not vanish for $x - y$ space-like, so an interaction such as $\varphi_{\text{an}}^2 \varphi_{\text{an}}^{\dagger 2}$ that treats φ_{an} and $\varphi_{\text{an}}^\dagger$ separately would not satisfy the condition (7.3.10), which we have seen is necessary for Lorentz invariance.

So, what to do? The only known way of restoring Lorentz invariance for charged particles while preserving charge conservation is to take the free field to be complex, the sum of a term that annihilates a particle and another term that creates its *antiparticle*, a particle with the opposite value of electric charge (and of all other conserved quantities) but the same spin and mass. For spinless particles this field takes the form

9 H. Yukawa, Proc. Phys.-Math. Soc. Japan **17**, 48 (1935). This article is reprinted in Beyer, *Foundations of Nuclear Physics*, listed in the bibliography.

$$\varphi(\mathbf{x},t) = \int \frac{d^3p}{\sqrt{2E(\mathbf{p})}(2\pi)^{3/2}} \big[a(\mathbf{p}) \exp(i\mathbf{p}\cdot\mathbf{x} - iE(\mathbf{p})t) + b^\dagger(\mathbf{p}) \exp(-i\mathbf{p}\cdot\mathbf{x} + iE(\mathbf{p})t)\big] \tag{7.4.1}$$

where

$$[a(\mathbf{p}), a^\dagger(\mathbf{p}')] = [b(\mathbf{p}), b^\dagger(\mathbf{p}')] = \delta^3(\mathbf{p} - \mathbf{p}')\,, \tag{7.4.2}$$

$$[a(\mathbf{p}), a(\mathbf{p}')] = [b(\mathbf{p}), b(\mathbf{p}')] = [a(\mathbf{p}), b(\mathbf{p}')] = 0\,, \tag{7.4.3}$$

$$[a^\dagger(\mathbf{p}), b(\mathbf{p}')] = [b^\dagger(\mathbf{p}), a(\mathbf{p}')] = 0\,. \tag{7.4.4}$$

In particular, the commutator of $\varphi(\mathbf{x},t)$ with $\varphi^\dagger(\mathbf{y},t)$ vanishes for space-like $x-y$ because of the same sort of cancellation that we encountered for real scalar fields. Both terms in φ change the electric charge (or any other conserved quantity) by the same amount, so interactions conserve this charge if they contain an equal number of factors of φ and $\varphi^\dagger$. This theory was presented in 1934 by Pauli and Weisskopf,[10] in order to contradict Dirac's view that antiparticles arise as holes in a sea of negative-energy particles. Antiparticles are indispensable for Lorentz invariance in any quantum field theory of particles that carry a conserved quantity, such as electric charge, even if the particles are bosons, which do not satisfy the Pauli exclusion principle and so could not form a stable sea of negative-energy particles. Where particles carry no conserved quantity, as in the previous sections, these particles can be said to be their own antiparticles. This is the case for the neutral pion and for the Z^0 particle.

This sort of free complex field theory can be derived as a consequence of a free-field Lagrangian density, of the form

$$\mathcal{L} = -\eta^{\mu\nu} \frac{\partial \varphi^\dagger}{\partial x^\mu} \frac{\partial \varphi}{\partial x^\nu} - m^2 \varphi^\dagger \varphi\,. \tag{7.4.5}$$

The Euler–Lagrange field equations are again

$$(\Box - m^2)\varphi = 0 \tag{7.4.6}$$

whose general complex solution is Eq. (7.4.1), with spacetime-independent operator coefficients a and $b^\dagger$. But now the canonical conjugate to φ is the time derivative of an independent canonical variable, the adjoint:

$$\pi(\mathbf{x},t) = \frac{\partial}{\partial t} \varphi^\dagger(\mathbf{x},t)\,, \tag{7.4.7}$$

while $\partial\varphi/\partial t$ is the canonical conjugate to $\varphi^\dagger$. The canonical commutation relations (7.1.10) then yield the commutation relations (7.4.2)–(7.4.4).

[10] W. Pauli and V. F. Weisskopf, Helv. Phys. Acta **7**, 709 (1934).

The particles described so far are bosons. The multi-particle state vectors here are

$$a^\dagger(\mathbf{p}_1)a^\dagger(\mathbf{p}_2)a^\dagger(\mathbf{p}_3)\cdots b^\dagger(\mathbf{p}'_1)b^\dagger(\mathbf{p}'_2)b^\dagger(\mathbf{p}'_3)\cdots\Psi_0\,, \tag{7.4.8}$$

where Ψ_0 is the vacuum, satisfying $a(\mathbf{p})\Psi_0 = b(\mathbf{p})\Psi_0 = 0$. By taking the adjoints of the commutation relations (7.4.3), we see that these particles are bosons; the states (7.4.8) are completely symmetric under interchanges of the labels $\mathbf{p}_1$, $\mathbf{p}_2$, etc. of the particles and under interchanges of the labels $\mathbf{p}'_1$, $\mathbf{p}'_2$, etc. of the antiparticles.

Suppose we wanted to construct a theory of spinless neutral *fermions*. We could suppose that all the commutators $[A,B] \equiv [A,B]_- \equiv AB - BA$ that we previously derived from the canonical commutation relations are now replaced with anticommutators, $[A,B]_+ \equiv AB + BA$. For instance, we can try introducing a real scalar field like (7.3.16):

$$\varphi(\mathbf{x},t) = \varphi_{\rm an}(\mathbf{x},t) + \varphi^\dagger_{\rm an}(\mathbf{x},t)\,,$$

$$\varphi_{\rm an}(\mathbf{x},t) = \int \frac{d^3p}{(2\pi)^{3/2}\sqrt{2E(\mathbf{p})}}\exp(i\mathbf{p}\cdot\mathbf{x} - iE(\mathbf{p})t)\;a(\mathbf{p})$$

but we now suppose that the annihilation and creation operators satisfy the anticommutation rules:

$$[a(\mathbf{p}), a^\dagger(\mathbf{p}')]_+ = \delta^3(\mathbf{p}-\mathbf{p}')\,, \tag{7.4.9}$$

$$[a(\mathbf{p}), a(\mathbf{p}')]_+ = [a^\dagger(\mathbf{p}), a^\dagger(\mathbf{p}')]_+ = 0\,. \tag{7.4.10}$$

The anticommutation relations (7.4.10) imply the complete antisymmetry of the multi-particle state vector

$$a^\dagger(\mathbf{p}_1)a^\dagger(\mathbf{p}_2)a^\dagger(\mathbf{p}_3)\cdots\Psi_0 \tag{7.4.11}$$

under interchange of the labels $\mathbf{p}_1$, $\mathbf{p}_2$, etc. of the particles, as required for fermions.

In place of the vanishing of the equal-time commutator $[\varphi(\mathbf{x},t),\varphi(\mathbf{y},t)]$ we now have

$$\begin{aligned}[\varphi(\mathbf{x},t),\varphi(\mathbf{y},t)]_+ &= [\varphi_{\rm an}(\mathbf{x},t),\varphi^\dagger_{\rm an}(\mathbf{y},t)]_+ + [\varphi^\dagger_{\rm an}(\mathbf{x},t),\varphi_{\rm an}(\mathbf{y},t)]_+ \\ &= \Delta_+(\mathbf{x}-\mathbf{y},0) + \Delta_+(\mathbf{y}-\mathbf{x},0)\end{aligned}$$

where $\Delta_+(\mathbf{x}-\mathbf{y}, x^0 - y^0)$ is the function (7.3.23), which at equal times is non-zero and even:

$$\Delta_+(\mathbf{x}-\mathbf{y},0) = \int \frac{d^3p}{2E(\mathbf{p})(2\pi)^3}e^{i\mathbf{p}\cdot(\mathbf{x}-\mathbf{y})} = \frac{m}{4\pi^2}\frac{K_1(m|\mathbf{x}-\mathbf{y}|)}{|\mathbf{x}-\mathbf{y}|}\,.$$

We see that here the two terms in $[\varphi(\mathbf{x},t),\varphi(\mathbf{y},t)]_+$ do not cancel but add, unlike the bosonic case. It is in fact impossible to construct scalar fields that anticommute at equal times for spinless fermions.

Though it is not possible here to go into so much detail, the same sort of analysis leads to the general conclusion cited in Section 5.5, that integer-spin particles (including spinless particles) must be bosons, while particles with half-odd-integer spin must be fermions. The free fields for spinning particles take the general form

$$\varphi_n(\mathbf{x},t) = \varphi_{n,\mathrm{an}}(\mathbf{x},t) + \varphi_{n,\mathrm{cr}}(\mathbf{x},t)\ , \tag{7.4.12}$$

$$\varphi_{n,\mathrm{an}}(\mathbf{x},t) = \sum_\sigma \int \frac{d^3p}{(2\pi)^{3/2}\sqrt{2E(\mathbf{p})}} u_n(\mathbf{p},\sigma)\exp(i\mathbf{p}\cdot\mathbf{x} - iE(\mathbf{p})t)\ a(\mathbf{p},\sigma)\ , \tag{7.4.13}$$

$$\varphi_{n,\mathrm{cr}}(\mathbf{x},t) = \sum_\sigma \int \frac{d^3p}{(2\pi)^{3/2}\sqrt{2E(\mathbf{p})}} v_n(\mathbf{p},\sigma)\exp(-i\mathbf{p}\cdot\mathbf{x} + iE(\mathbf{p})t)\ b^\dagger(\mathbf{p},\sigma)\ . \tag{7.4.14}$$

Here $a(\mathbf{p},\sigma)$ is the operator that annihilates a particle of momentum $\mathbf{p}$ and spin 3-component σ; $b^\dagger(\mathbf{p},\sigma)$ is the operator that creates its antiparticle (so that both terms in φ have the same effect on the charge and all other conserved quantities); and $u_n(\mathbf{p},\sigma)$ and $v_n(\mathbf{p},\sigma)$ are functions about which more later. For neutral particles that are their own antiparticles, $a(\mathbf{p},\sigma) = b(\mathbf{p},\sigma)$. For bosons or fermions, the operators a and b satisfy the commutation or anticommutation relations

$$[a(\mathbf{p},\sigma), a^\dagger(\mathbf{p}',\sigma')]_\mp = [b(\mathbf{p},\sigma), b^\dagger(\mathbf{p}',\sigma')]_\mp = \delta_{\sigma'\sigma}\delta^3(\mathbf{p}-\mathbf{p}')\ , \tag{7.4.15}$$

$$[a(\mathbf{p},\sigma), a(\mathbf{p}',\sigma')]_\mp = [b(\mathbf{p},\sigma), b(\mathbf{p}',\sigma')]_\mp = [a(\mathbf{p},\sigma), b(\mathbf{p}',\sigma')]_\mp = 0\ , \tag{7.4.16}$$

$$[a^\dagger(\mathbf{p},\sigma), b(\mathbf{p}',\sigma')]_\mp = [b^\dagger(\mathbf{p},\sigma), a(\mathbf{p}',\sigma')]_\mp = 0\ , \tag{7.4.17}$$

the $\mp$ signs being minus, denoting commutators, for bosons or plus, denoting anticommutators, for fermions.

The functions $u_n(\mathbf{p},\sigma)$ and $v_n(\mathbf{p},\sigma)$ are governed by what is assumed for the Lorentz-transformation property of the fields. Under a Lorentz transformation $x^\mu \to x'^\mu = \Lambda^\mu{}_\nu x^\nu$, the various fields $\varphi_n(x)$ undergo various matrix transformations[11]

$$\varphi_n(x) \to \varphi_n^\Lambda(\Lambda x) = \sum_m D_{nm}(\Lambda)\varphi_m(x)\ , \tag{7.4.18}$$

[11] These transformations are often written as actions of a quantum-mechanical operator $U(\Lambda)$, as $U(\Lambda)\varphi_n(x)U^{-1}(\Lambda) = \sum_m D^{-1}_{nm}(\Lambda)\varphi_m(\Lambda x)$. This is the same as Eq. (7.4.18) if we identify $\varphi_n^\Lambda(\Lambda x) = U^{-1}(\Lambda)\varphi_n(x)U(\Lambda)$.

so that for an observer who uses coordinates $x'^\mu = \Lambda^\mu{}_\nu x^\nu$ the field is related by a matrix D to the field for an observer who uses coordinates x^μ at the same spacetime point, a point that is thus given different coordinates by the two observers. When we perform two successive Lorentz transformations Λ_1 and then Λ_2, the effect on the fields is

$$\varphi_n(x) \to \varphi_n^{\Lambda_1}(\Lambda_1 x) = \sum_l D_{nl}(\Lambda_1)\varphi_l(x)$$

$$\to \sum_l D_{nl}(\Lambda_1)\varphi_l^{\Lambda_2}(\Lambda_2 x) = \sum_{m,l} D_{nl}(\Lambda_1)D_{lm}(\Lambda_2)\varphi_m(x) ,$$

while the effect of the compound Lorentz transformation $(\Lambda_1\Lambda_2)^\mu{}_\nu = \Lambda_1^\mu{}_\rho \Lambda_2^\rho{}_\nu$ is

$$\varphi_n(x) \to \varphi_n^{\Lambda_1\Lambda_2}(\Lambda_1\Lambda_2 x) = \sum_m D_{nm}(\Lambda_1\Lambda_2)\varphi_m(x) .$$

These transformations must be the same, so

$$D(\Lambda_1)D(\Lambda_2) = D(\Lambda_1\Lambda_2) , \tag{7.4.19}$$

where $[D(\Lambda_1)D(\Lambda_2)]_{nm}$ is the usual matrix product:

$$[D(\Lambda_1)D(\Lambda_2)]_{nm} = \sum_l D_{nl}(\Lambda_1)D_{lm}(\Lambda_2) .$$

Such matrices are said to form a *representation* of the group of Lorentz transformations. We classify the various kinds of field according to the representation they furnish of the Lorentz group.

It is always possible to write the Lagrangian density in terms of fields that are *irreducible*, in the sense that their components cannot be divided into sets that, under Lorentz (and perhaps space inversion) transformations, transform only into linear combinations of the field components in the same set. Among these irreducible fields are a single scalar field for spin zero, for which $D(\Lambda)$ is the unit matrix, or a single four-vector field for spin one, for which $D(\Lambda)$ is Λ itself. For spin 1/2 there is the four-component Dirac field, briefly described in the appendix to this section. For our present purposes, the important thing about irreducible fields is that the coefficient functions $u_n(\mathbf{p}, \sigma)$ and $v_n(\mathbf{p}, \sigma)$ are uniquely determined up to constant factors by what is assumed for the Lorentz-transformation properties of the fields and the spin of the particles.

As discussed in the previous section, for the Lorentz invariance of the theory it is not enough that the interaction Hamiltonian density $\mathcal{H}'$ should be a scalar; it also has to satisfy the condition (7.3.11), that $\mathcal{H}'(x)$ should commute with $\mathcal{H}'(y)$ at equal times $x^0 = y^0$. For this, it is necessary that $\mathcal{H}'$ should be formed from bosonic fields that all commute with each other at equal times, plus some even number (perhaps zero) of fermionic fields that anticommute with each other at equal times (and commute with the bosonic fields at equal times).

For particles that are not their own antiparticles, the commutators or anticommutators of the φ_n with each other and of the $\varphi_n^\dagger$ with each other trivially vanish. On the other hand, the equal-time commutator or anticommutator of any field with its adjoint is

$$[\varphi_n(\mathbf{x},t), \varphi_m^\dagger(\mathbf{y},t)]_\mp = \Delta_{nm}(\mathbf{x}-\mathbf{y}) \mp \overline{\Delta}_{nm}(\mathbf{y}-\mathbf{x}) \tag{7.4.20}$$

where

$$\Delta_{nm}(\mathbf{x}-\mathbf{y}) \equiv \sum_\sigma \int \frac{d^3p}{2E(\mathbf{p})(2\pi)^3} u_n(\mathbf{p},\sigma) u_m^*(\mathbf{p},\sigma) e^{i\mathbf{p}\cdot(\mathbf{x}-\mathbf{y})}\ , \tag{7.4.21}$$

$$\overline{\Delta}_{nm}(\mathbf{x}-\mathbf{y}) \equiv \sum_\sigma \int \frac{d^3p}{2E(\mathbf{p})(2\pi)^{3/2}} v_n(\mathbf{p},\sigma) v_m^*(\mathbf{p},\sigma) e^{i\mathbf{p}\cdot(\mathbf{x}-\mathbf{y})}\ . \tag{7.4.22}$$

The first and second terms on the right of Eq. (7.4.20) come respectively from the commutator or anticommutator of the annihilation part of $\varphi_n(\mathbf{x},t)$ with the creation part of $\varphi_m^\dagger(\mathbf{y},t)$ and from the commutator or anticommutator of the creation part of $\varphi_n(\mathbf{x},t)$ with the annihilation part of $\varphi_m^\dagger(\mathbf{y},t)$. (The crucial $\mp$ sign that distinguishes bosons from fermions appears in the second term of Eq. (7.4.20) because this term comes from the part of the commutator or anticommutator of φ_n with $\varphi_m^\dagger$ in which $b^\dagger$ appears to the left of b.) Detailed calculations beyond the scope of this book show that[12]

$$\overline{\Delta}_{nm}(\mathbf{y}-\mathbf{x}) = (-1)^{2j}|\lambda|^2\Delta_{nm}(\mathbf{x}-\mathbf{y}) \tag{7.4.23}$$

where j is the particle spin and λ depends on how the u_n and v_n are normalized. (If we multiply u_n and v_n by factors α and β then λ is changed by a factor β/α.) For equal-time commutators or anticommutators of fields and their adjoints to vanish, the two terms in Eq. (7.4.20) must cancel. For this we need

$$|\lambda|^2(-1)^{2j} = \pm 1\ ,$$

with the top sign for bosons and the bottom sign for fermions. This requires that $|\lambda| = 1$, which can always be arranged by adjusting the relative normalization of u_n and v_n, and thereby imposes a relation between the strengths of interactions of particles and antiparticles. But with $|\lambda| = 1$, we also need

$$(-1)^{2j} = \pm 1\ . \tag{7.4.24}$$

This is the famous connection between spin and statistics:[13] particles with j an integer are bosons, and particles with j a half odd integer are fermions.

[12] For a textbook treatment, see e.g. S. Weinberg, *The Quantum Theory of Fields*, Vol. I (Cambridge University Press, Cambridge, UK, 1995), Section 5.7.

[13] M. Fierz, Helv. Phys. Acta **12**, 3 (1939); W. Pauli, Phys. Rev. **58**, 716 (1940).

Appendix: Dirac Fields

In 1928 Dirac introduced a relativistic wave equation[14] that he thought would provide the basis for a formulation of quantum mechanics consistent with special relativity. About this program he was wrong; the successful relativistic formulation of quantum mechanics turned out to take the form of quantum field theory. But his equation survives as the field equation of the quantum fields for particles of spin 1/2, and their antiparticles, and leads to some of the same consequences, such as formulas for the fine structure of atomic spectra. This appendix provides just a sketch of Dirac's formalism, skipping most proofs.

The Dirac field is a set of four operators $\psi_n(x)$, characterized by their Lorentz transformations: for $x \to \Lambda x$,

$$\psi_n(x) \to \psi_n^\Lambda(\Lambda x) = \sum_m D_{nm}(\Lambda)\psi_m(x) \,, \tag{7.4.25}$$

with the matrix $D(\Lambda)$ furnishing a representation of the Lorentz group with the special property that

$$D^{-1}(\Lambda)\gamma^\mu D(\Lambda) = \Lambda^\mu{}_\nu \gamma^\nu \tag{7.4.26}$$

where the γ^μ are a set of four 4×4 matrices satisfying the anticommutation relations

$$\gamma^\mu\gamma^\nu + \gamma^\nu\gamma^\mu = 2\eta^{\mu\nu} \equiv 2 \times \begin{cases} +1 & \mu = \nu = 1,2,3 \\ -1 & \mu = \nu = 0 \\ 0 & \mu \neq \nu \,. \end{cases} \tag{7.4.27}$$

This allows a Lorentz-invariant first-order free-field equation for mass m:

$$\left(\gamma^\mu \frac{\partial}{\partial x^\mu} + m\right)\psi(x) = 0 \,. \tag{7.4.28}$$

Using the commutativity of partial derivatives and the anticommutation rules (7.4.27), we see that Eq. (7.4.28) has the consequence

$$0 = \left(\gamma^\mu \frac{\partial}{\partial x^\mu} - m\right)\left(\gamma^\mu \frac{\partial}{\partial x^\mu} + m\right)\psi = (\Box - m^2)\psi \,.$$

For this reason, Dirac thought of his equation as a sort of square root of the relativistic Schrödinger (or Klein–Gordon) free-particle equation $(\Box - m^2)\psi = 0$.

The general solution of Eq. (7.4.28) is

$$\psi_n(x) = \int \frac{d^3p}{(2\pi)^{3/2}\sqrt{2E(\mathbf{p})}}\left[e^{ip\cdot x}u_n(\mathbf{p},\sigma)a(\mathbf{p},\sigma) + e^{-ip\cdot x}v_n(\mathbf{p},\sigma)b^\dagger(\mathbf{p},\sigma)\right] \tag{7.4.29}$$

[14] P. A. M. Dirac, Proc. Roy. Soc. (London) **A117**, 610 (1928).

where $p \cdot x \equiv \eta_{\mu\nu} p^\mu x^\nu$; $p^0 = E(\mathbf{p}) = +(\mathbf{p}^2 + m^2)^{1/2}$; the $u_n(\mathbf{p}, \sigma)$ and $v_n(\mathbf{p}, \sigma)$ are independent solutions of the equations

$$\left(i\eta_{\mu\nu}\gamma^\mu p^\nu + m\right) u = 0 \,, \tag{7.4.30}$$

$$\left(-i\eta_{\mu\nu}\gamma^\mu p^\nu + m\right) v = 0 \,; \tag{7.4.31}$$

and $a(\mathbf{p}, \sigma)$ and $b(\mathbf{p}, \sigma)$ are operator coefficients, with σ labeling the independent solutions of Eqs. (7.4.30) and (7.4.31).

We can count the number of independent solutions, noting that any column w_n can be decomposed as

$$w = w_+ + w_- \,, \quad \left(i\eta_{\mu\nu}\gamma^\mu p^\nu \pm m\right) w_\pm = 0 \,,$$

by taking

$$w_\pm = \frac{\left(\mp i\eta_{\mu\nu}\gamma^\mu p^\nu + m\right)}{2m} w \,.$$

Thus, with a total of four components, there must be just two independent $u_n(\mathbf{p}, \sigma)$ satisfying Eq. (7.4.30) and two independent $v_n(\mathbf{p}, \sigma)$ satisfying Eq. (7.4.31). The index σ therefore takes just two values, corresponding to the two values of the third component of spin for a particle of spin 1/2. Dirac thought that the solutions $e^{-ip\cdot x} v_n(\mathbf{p}, \sigma)$ were the wave functions for a free negatively charged electron with negative energy; instead, just as we saw for scalar fields, they are the coefficients of the creation operator $b^\dagger$ for a positively charged antielectron, or positron, of positive energy.

In forming a Lagrangian density $\mathcal{L}(x)$, we need to include both fields and their adjoints in such a way that $\mathcal{L}(x)$ is a scalar. Here $\sum_n \psi_n^\dagger \psi_n$ is not a scalar, but here and more generally there is always a matrix β_{nm} for which $\sum_{n,m} \psi_n^\dagger \beta_{nm} \psi_m$ does transform as a scalar. This is because for any matrices A and B, we have $(AB)^{\dagger -1} = (B^\dagger A^\dagger)^{-1} = A^{\dagger -1} B^{\dagger -1}$, so the inverse of the adjoint $D^\dagger_{nm} \equiv D^*_{mn}$ satisfies the same multiplication rule (7.4.19) as D itself:

$$D^{\dagger -1}(\Lambda_1) D^{\dagger -1}(\Lambda_2) = D^{\dagger -1}(\Lambda_1 \Lambda_2) \,. \tag{7.4.32}$$

This does not mean that $D^{\dagger -1}(\Lambda) = D(\Lambda)$, but for irreducible representations they are equal up to a similarity transformation; there is a matrix β for which

$$D^{\dagger -1}(\Lambda) = \beta D(\Lambda) \beta^{-1} \,,$$

or, multiplying on the left with $D^\dagger$ and on the right with β,

$$D^\dagger(\Lambda) \beta D(\Lambda) = \beta \,. \tag{7.4.33}$$

It follows then that we can define a covariant adjoint

$$\overline{\psi}_m(x) \equiv \sum_m \psi_n^\dagger(x) \beta_{nm} \,, \tag{7.4.34}$$

such that the effect of a Lorentz transformation $x \to \Lambda x$ is

$$\overline{\psi}_m(x) \to \overline{\psi^\Lambda}_m(\Lambda x) = \sum_n \psi_n^\dagger(x)[D^\dagger(\Lambda)\beta]_{nm} = \sum_n \overline{\psi}_n(x)[D^{-1}(\Lambda)]_{nm} \,. \tag{7.4.35}$$

Thus not only is $\overline{\psi}(x)\psi(x) \equiv \sum_m \overline{\psi}_m(x)\psi_m(x)$ a scalar, but also (in the same abbreviated notation) $\overline{\psi}(x)\gamma^\mu\psi(x)$ is a four-vector.

It is now easy to construct a Lorentz-invariant free-field Lagrangian density from which follows the field equation (7.4.28):

$$\mathcal{L}_0 = -\overline{\psi}\left(\gamma^\mu \frac{\partial}{\partial x^\mu} + m\right)\psi \,. \tag{7.4.36}$$

Without going into details, the canonical anticommutation relations here give

$$[a(\mathbf{p},\sigma), a^\dagger(\mathbf{p}',\sigma')]_+ = [b(\mathbf{p},\sigma), b^\dagger(\mathbf{p}',\sigma')]_+ = \delta_{\sigma\sigma'}\delta^3(\mathbf{p}-\mathbf{p}') \tag{7.4.37}$$

$$\begin{aligned}[a(\mathbf{p},\sigma), a(\mathbf{p}',\sigma')]_+ &= [b(\mathbf{p},\sigma), b(\mathbf{p}',\sigma')]_+ \\ &= [a(\mathbf{p},\sigma), b(\mathbf{p}',\sigma')]_+ = [a(\mathbf{p},\sigma), b^\dagger(\mathbf{p}',\sigma')]_+ = 0 \,,\end{aligned} \tag{7.4.38}$$

provided the solutions of Eqs. (7.4.30) and (7.4.31) are normalized so that

$$\begin{aligned}\sum_\sigma u_n(\mathbf{p},\sigma)\overline{u}_m(\mathbf{p},\sigma) &= [-ip^\mu\gamma_\mu + m]_{nm} \,, \\ \sum_\sigma v_n(\mathbf{p},\sigma)\overline{v}_m(\mathbf{p},\sigma) &= [-ip^\mu\gamma_\mu - m]_{nm} \,.\end{aligned} \tag{7.4.39}$$

(As usual, $[A, B]_+$ is defined as $AB + BA$.) The anticommutator of the Dirac field with its adjoint is given by

$$[\psi_n(x), \overline{\psi}_m(y)]_+ = \left[-\gamma^\mu \frac{\partial}{\partial x^\mu} + m\right]_{nm} \int \frac{d^3p}{2p^0(2\pi)^3}\left[e^{ip\cdot(x-y)} - e^{-ip\cdot(x-y)}\right] , \tag{7.4.40}$$

which obviously vanishes for $x^0 = y^0$ and hence for all space-like $x - y$, as required by Lorentz invariance.

We can include the interaction of the Dirac field of the electron with the electromagnetic vector potential using the prescription given in the following section. Replacing $\partial\psi/\partial x^\mu$ in the free-field Lagrangian density with $\partial\psi/\partial x^\mu + ieA_\mu\psi$ gives the Lagrangian for electrons and positrons and their interaction with electromagnetism:

$$\mathcal{L}_{\text{Dirac}} = -\overline{\psi}\left(\gamma^\mu\left[\frac{\partial}{\partial x^\mu} + ieA_\mu\right] + m_e\right)\psi \,. \tag{7.4.41}$$

This yields the Euler–Lagrange field equation for a Dirac field interacting with any electromagnetic field:

$$\left(\gamma^\mu\left[\frac{\partial}{\partial x^\mu}+ieA_\mu\right]+m_e\right)\psi=0\,. \tag{7.4.42}$$

The Dirac wave function used in Dirac's calculations was not the quantum field ψ, but its matrix elements:

$$\psi_{An}(x)\equiv(\Phi_{\text{vac}},\psi_n(x)\Phi_A)\,,\quad \psi_{Bn}(x)\equiv(\Phi_B,\psi_n(x)\Phi_{\text{vac}}) \tag{7.4.43}$$

where Φ_A and Φ_B are states of charge $-e$ and $+e$, respectively, such as states of an electron and a positron in the electromagnetic field of an atom. These wave functions satisfy the same equation as the field;

$$\left(\gamma^\mu\left[\frac{\partial}{\partial x^\mu}+ieA_\mu\right]+m_e\right)\psi_A(x)=\left(\gamma^\mu\left[\frac{\partial}{\partial x^\mu}+ieA_\mu\right]+m_e\right)\psi_B(x)=0\,. \tag{7.4.44}$$

For a time-independent electromagnetic field, the time dependence of the Dirac field is governed by a time-independent Hamiltonian H in the Heisenberg picture, so, for states Φ_A and Φ_B with energy E_A and E_B, the wave functions have the time dependences

$$\psi_{An}(\mathbf{x},t)\propto e^{-iE_At}\,,\qquad \psi_{Bn}(\mathbf{x},t)\propto e^{+iE_Bt}\,. \tag{7.4.45}$$

The different sign of the argument of e^{+iE_Bt} does not arise because the state Φ_B has negative energy, but because it appears to the left of the Dirac field in the definition (7.4.43) of the wave function $\psi_{Bn}(\mathbf{x},t)$. From solutions of the wave (7.4.44) for $\psi_{An}(\mathbf{x},t)$ with time dependence given by (7.4.45) and a pure Coulomb field $A^0=Ze^2/r$, $\mathbf{A}=0$, Dirac was able to calculate the energies of the states of hydrogenic atoms, including their fine structure:

$$E(nj)=m_e\left[1-\frac{Ze^4}{2n^2}+\frac{Z^2e^8}{n^4}\left(\frac{3}{8}-\frac{n}{2j+1}\right)+\cdots\right] \tag{7.4.46}$$

with no dependence on ℓ.

As discussed in Section 6.5, Fermi in his 1934 theory of beta decay proposed an interaction Hamiltonian of the form (6.5.1), proportional to the scalar product of two vector currents. This then had to be modified, first by the introduction of axial vector currents and then by including terms that violate invariance under space inversion, resulting in an interaction of the form (6.5.4). Expressed explicitly in terms of Dirac fields for the proton, neutron, electron, and neutrino, Fermi's original proposed interaction (in units with $\hbar=c=1$) was

$$\mathcal{H}_\beta=G_F(\overline{\psi}_e\gamma^\mu\psi_\nu)(\overline{\psi}_p\gamma^\mu\psi_n)+G_F(\overline{\psi}_\nu\gamma^\mu\psi_e)(\overline{\psi}_n\gamma^\mu\psi_p)\,, \tag{7.4.47}$$

and after 30 years of experiments on nuclear beta decay and other weak interaction processes, this was finally modified to

$$\mathcal{H}_\beta = \frac{G_F}{\sqrt{2}}(\overline{\psi}_e\gamma^\mu(1+\gamma_5)\psi_\nu)(\overline{\psi}_p\gamma^\mu(1+\gamma_5)\psi_n)$$
$$+\frac{G_F}{\sqrt{2}}(\overline{\psi}_\nu\gamma^\mu(1+\gamma_5)\psi_e)(\overline{\psi}_n\gamma^\mu(1+\gamma_5)\psi_p)\,, \qquad (7.4.48)$$

where $G_F = 1.16\times 10^{-5}\,\text{GeV}^{-2}$ and $\gamma_5 \equiv i\gamma_1\gamma_2\gamma_3\gamma_0$. It can be shown from the anticommutation relations that γ_5 is Lorentz invariant, in the sense of commuting with $D(\Lambda)$, so (7.4.48) like (7.4.47) transforms as a scalar under any proper Lorentz transformation. It is the presence of the matrix $1+\gamma_5$ in $\mathcal{H}_\beta$ that produces the violations of invariance under space inversion discussed in Section 6.5, including the fact that if neutrinos were massless, the neutrinos created along with electrons by the first term in Eq. (7.4.48) or along with positrons in the second term in Eq. (7.7.48) would have a component of angular momentum in the direction of motion respectively equal to $\hbar/2$ or $-\hbar/2$. For the very small known masses of neutrinos, these helicities are overwhelmingly likely.

7.5 Quantum Theory of Electromagnetism

We end our treatment of quantum mechanics where we began, with the quantum theory of radiation. We will first present the Lagrangian densities both for the free electromagnetic field and for the fields' interactions with matter, then work out in detail the theory of the free field, which as shown in Section 7.3 is needed to provide the interaction in the interaction picture in perturbation theory and then to apply what we have learned to a classic problem, calculation of the rate of emission of photons in transitions between atomic or molecular states. We close with an account of the interaction of electromagnetism with general matter fields.

Lagrangian Density

It is easy to think of a possible Lagrangian density for the electromagnetic field that is quadratic in the fields, like all free-field Lagrangians and is Lorentz invariant:

$$\mathcal{L}_0 = -\frac{1}{16\pi}\eta_{\mu\rho}\eta_{\nu,\sigma}F^{\mu\nu}F^{\rho\sigma}\,, \qquad (7.5.1)$$

where $F^{\mu\nu}$ is the field strength tensor, given by Eqs. (4.6.7) and (4.6.8):

$$E_1 = F^{01} = -F^{10},\quad E_2 = F^{02} = -F^{20},\quad E_3 = F^{03} = -F^{30}\,, \qquad (7.5.2)$$

$$B_1 = F^{23} = -F^{32}, \quad B_2 = F^{31} = -F^{13}, \quad B_3 = F^{12} = -F^{21} \, . \tag{7.5.3}$$

(The factor $-1/16\pi$ is irrelevant now but will be convenient later, when we consider the coupling of these fields to matter.) This is manifestly Lorentz invariant, but otherwise appears absurd. If we assume that $\int d^4x \, \mathcal{L}$ is stationary under arbitrary infinitesimal variations of the fields $F^{\mu\nu}$, we find Euler–Lagrange equations of the form $F^{\mu\nu} = 0$, which certainly do not describe actual free electromagnetic fields. The error made in deriving this wrong result is that we must not impose conditions for arbitrary variations of $F^{\mu\nu}$, because the field-strength tensor is constrained by the homogeneous Maxwell equations (4.6.15), (4.6.17):

$$0 = \frac{\partial F_{\mu\nu}}{\partial x^\lambda} + \frac{\partial F_{\lambda\mu}}{\partial x^\nu} + \frac{\partial F_{\nu\lambda}}{\partial x^\mu} \tag{7.5.4}$$

where

$$F_{\mu\nu} \equiv \eta_{\mu\rho}\eta_{\nu\sigma} F^{\rho\sigma} \, . \tag{7.5.5}$$

We should only demand that the action is stationary for variations in the fields that preserve the constraint (7.5.4).

It is easy to see that this requirement leads to the remaining free-field Maxwell equations $\partial F^{\mu\nu}/\partial x^\nu = 0$, but in deriving the canonical commutation relations it is awkward to work with functional derivatives with respect to constrained fields like $F^{\mu\nu}$. In electrodynamics it is much easier to express the field-strength tensor in terms of an unconstrained vector potential A_μ, in such a way that the constraint (7.5.4) is automatically respected,

$$F_{\mu\nu} = \frac{\partial A_\nu}{\partial x^\mu} - \frac{\partial A_\nu}{\partial x^\mu} \tag{7.5.6}$$

and take all functional derivatives with respect to the A_μ. As shown in Section 5.8, the introduction of a vector potential is essential anyway in formulating the quantum theory of charged particles in an electromagnetic field.

For the present we will introduce a general Lagrangian density $\mathcal{L}_{\text{mat}}$ for matter and its interaction with the electromagnetic field, and define the electric current four-vector J^μ as the functional derivative with respect to $A_\mu(x)$ of the corresponding term in the action:

$$J^\mu(x) \equiv \frac{\delta}{\delta A_\mu(x)} \int d^4y \, \mathcal{L}_{\text{mat}}(y) \, . \tag{7.5.7}$$

Under an infinitesimal shift in A_μ, the change in the total action is now

$$\delta \int d^4x (\mathcal{L}_0 + \mathcal{L}_{\text{mat}}) = \int d^4x \left[-\frac{1}{4\pi} \frac{\partial F^{\mu\nu}(x)}{\partial x^\nu} + J^\mu(x) \right] \delta A_\mu(x) \, ,$$

and the Euler–Lagrange equations here are

$$\frac{\partial F^{\mu\nu}(x)}{\partial x^\nu} = 4\pi J^\mu(x) , \tag{7.5.8}$$

which we recognize as the inhomogeneous Maxwell equations (4.6.9) (except that there is no factor $1/c$, because we are using natural units with $\hbar = c = 1$).

Gauge Transformations

Now we have a problem. We cannot satisfy the canonical commutation relations for the field A^0, because since $F_{00} = 0$ the Lagrangian density does not contain a time derivative of A_0. To deal with this, we note that the action is invariant under a *gauge transformation*

$$A_\mu(x) \to A_\mu(x) + \frac{\partial \xi(x)}{\partial x^\mu} \tag{7.5.9}$$

with $\xi(x)$ an arbitrary function of the spacetime coordinate. This has no effect on the field-strength tensor (7.5.6), and the consistency of the Maxwell equations requires that the current J^μ is conserved in the sense that $\partial J^\mu(x)/\partial x^\mu = 0$, so that according to Eq. (7.5.7) the change produced in the matter action by the gauge transformation (7.5.9) is

$$\delta \int d^4x\; \mathcal{L}_{\text{mat}} = \int d^4x\; J^\mu(x) \frac{\partial \xi(x)}{\partial x^\mu} = -\int d^4x\; \frac{\partial J^\mu(x)}{\partial x^\mu}\, \xi(x) = 0 \,. \tag{7.5.10}$$

Coulomb Gauge

We can always choose $\xi(x)$ so as to adopt what is known as the Coulomb gauge, for which

$$\boldsymbol{\nabla} \cdot \mathbf{A} = 0 \tag{7.5.11}$$

because if $\boldsymbol{\nabla} \cdot \mathbf{A} \neq 0$, we can make it vanish by performing a gauge transformation with $\nabla^2 \xi = -\boldsymbol{\nabla} \cdot \mathbf{A}$. This is called the Coulomb gauge because the $\mu = 0$ component of the inhomogeneous Maxwell equations (7.5.8) is here

$$4\pi\, J^0 = \frac{\partial F^{0i}}{\partial x^i} = -\nabla^2 A^0$$

with solution given by the familiar Coulomb field

$$A^0(\mathbf{x}, t) = \int d^3y \frac{J^0(\mathbf{y}, t)}{|\mathbf{x} - \mathbf{y}|} \,. \tag{7.5.12}$$

Since A^0 is a functional of the matter fields in J^0 at the same time, it is not to be regarded as an independent canonical variable. The canonical variables of electrodynamics in Coulomb gauge are the spatial components A^i, but subject to the constraint (7.5.11).

The condition (7.5.11) for Coulomb gauge is obviously not Lorentz invariant. Given a vector potential $A^\mu(x)$ that satisfies this condition, the Lorentz-transformed vector potential $\Lambda^\mu{}_\nu A^\nu$ will in general not satisfy Eq. (7.5.11) if Λ is anything but a pure rotation. However, we can always combine any Lorentz transformation with a gauge transformation that takes the vector potential back to Coulomb gauge. Since the action is presumed to be gauge invariant, the physical consequences of the theory calculated in Coulomb gauge turn out to be Lorentz invariant.

The virtue of Coulomb gauge, which here makes up for its lack of manifest Lorentz invariance, is that it displays the physical degrees of freedom of electrodynamics. Even though A^μ has four components, as we have seen in Coulomb gauge A^0 is a functional of matter fields, and $\nabla \cdot \mathbf{A}$ vanishes. We shall see that the two remaining degrees of freedom are the two independent states of photon polarization. It must be admitted, however, that, as a practical matter, in carrying out calculations in quantum electrodynamics more complicated than those essayed here, it is necessary to use techniques that preserve manifest Lorentz invariance, such as the path integral approach of Feynman.[15]

Now we have to consider what is the canonical conjugate to $\mathbf{A}$. According to the usual definition of a functional derivative, if we make an infinitesimal variation $\dot{\mathbf{A}} \to \dot{\mathbf{A}} + \delta\dot{\mathbf{A}}$, then

$$\delta L(t) = \int d^3x \frac{\delta L(t)}{\delta \dot{A}_i(\mathbf{x},t)} \delta \dot{A}_i(\mathbf{x},t) ,$$

but, since $\dot{A}_i$ is constrained by (7.5.11), we are only allowed to consider variations satisfying $\partial\delta\dot{A}_i/\partial x^i = 0$, so $\delta L(t)/\delta\dot{A}_i(\mathbf{x},t)$ is only defined up to gradient terms, of the form $\partial f/\partial x^i$. A direct calculation gives

$$\frac{\partial \mathcal{L}}{\partial \dot{A}_i} = \frac{1}{4\pi}\left[\dot{A}_i + \frac{\partial A^0}{\partial x^i}\right]$$

but we need to take advantage of our freedom to shift this functional derivative by the gradient $-(\partial A^0/\partial x^i)/4\pi$, and take the canonical conjugate to A_i as

$$\pi_i = \dot{A}_i/4\pi , \tag{7.5.13}$$

so that π_i satisfies the same constraint as A_i:

$$\nabla \cdot \boldsymbol{\pi} = 0 . \tag{7.5.14}$$

[15] R. P. Feynman, Ph.D. thesis, *The Principle of Least Action in Quantum Mechanics* (Princeton University, 1942; University Microfilms Publication No. 2948, Ann Arbor).

The usual canonical commutation relations must here be modified to take account of the conditions (7.5.11) and (7.5.14). We use the formula

$$\nabla^2 \frac{1}{|\mathbf{x}-\mathbf{y}|} = -4\pi\delta^3(\mathbf{x}-\mathbf{y}) \, .$$

(This can be derived by showing directly that the left-hand side vanishes for $\mathbf{x} \neq \mathbf{y}$, and using Gauss's theorem to show that its integral over all space is -4π.) Then we have consistency with conditions (7.5.11) and (7.5.14) if we take

$$[A_i(\mathbf{x},t), \pi_j(\mathbf{y},t)] = i\delta_{ij}\delta^3(\mathbf{x}-\mathbf{y}) + i\frac{\partial^2}{\partial x^i \partial x^j}\left(\frac{1}{4\pi|\mathbf{x}-\mathbf{y}|}\right) \tag{7.5.15}$$

and also

$$[A_i(\mathbf{x},t), A_j(\mathbf{y},t)] = [\pi_i(\mathbf{x},t), \pi_j(\mathbf{y},t)] = 0 \, . \tag{7.5.16}$$

Free Fields

As emphasized in Section 7.3, the first step in using time-ordered perturbation theory to calculate processes involving interacting particles is to write explicit formulas for the free fields. With zero current and charge densities, and hence $A^0 = 0$, the field equations (7.5.8) for A^i in Coulomb gauge are

$$0 = \frac{\partial F^{\mu i}}{\partial x^\mu} = \Box A^i - \frac{\partial}{\partial x_i}\frac{\partial A^\mu}{\partial x^\mu} = \Box A^i \, . \tag{7.5.17}$$

The general real solution of Eqs. (7.5.11) and (7.5.17) is conveniently written

$$\mathbf{A}(\mathbf{x},t) = \sqrt{4\pi}\sum_\lambda \int \frac{d^3q}{(2\pi)^{3/2}\sqrt{2|\mathbf{q}|}}\Big[\mathbf{e}(\mathbf{q},\lambda)a(\mathbf{q},\lambda)e^{i\mathbf{q}\cdot\mathbf{x}-i|\mathbf{q}|t} + \mathbf{e}^*(\mathbf{q},\lambda)a^\dagger(\mathbf{q},\lambda)e^{-i\mathbf{q}\cdot\mathbf{x}+i|\mathbf{q}|t}\Big] \tag{7.5.18}$$

where $a(\mathbf{q},\lambda)$ is an operator coefficient whose properties will be found from the canonical commutation relations, and $\mathbf{e}(\mathbf{q},\lambda)$ are any two independent three-vectors normal to $\mathbf{q}$,

$$\mathbf{q}\cdot\mathbf{e}(\mathbf{q},\lambda) = 0 \tag{7.5.19}$$

with λ a two-valued index distinguishing the two solutions of (7.5.19). By a suitable normalization of $a(\mathbf{q},\lambda)$, we can always normalize these vectors so that

$$\sum_\lambda e_i(\mathbf{q},\lambda)e_j^*(\mathbf{q},\lambda) = \delta_{ij} - q_iq_j/|\mathbf{q}|^2 \, . \tag{7.5.20}$$

For instance, for $\mathbf{q}$ in the 3-direction, we can take $\mathbf{e} = (1,i,0)/\sqrt{2}$ for $\lambda = 1$, and $\mathbf{e} = (1,-i,0)/\sqrt{2}$ for $\lambda = -1$, and, for $\mathbf{q}$ in a direction defined by some choice of rotation from the 3-direction, apply the same rotation to **e**. These are the same as the polarization vectors for left- and right-handed circular polarization that appear in the Fourier expansion of an electromagnetic wave.

With this normalization of the polarization vectors, the field (7.5.18) satisfies the canonical commutation relations (7.5.15)–(7.5.16) if we take

$$[a(\mathbf{q},\lambda), a^\dagger(\mathbf{q}',\lambda')] = \delta_{\lambda\lambda'}\delta^3(\mathbf{q}-\mathbf{q}') \ , \tag{7.5.21}$$

and

$$[a(\mathbf{q},\lambda), a(\mathbf{q}',\lambda')] = 0 \ . \tag{7.5.22}$$

Then, just as we saw for a real scalar field in Section 7.2, the operator $a^\dagger(\mathbf{q},\lambda)$ creates a photon of momentum $\mathbf{q}$ and polarization vector $e(\mathbf{q},\lambda)$ in any state vector on which it acts, while if there already is such a photon in the state, the operator $a(\mathbf{q},\lambda)$ removes it.

To see the physical significance of λ, note that for $\mathbf{q}$ in the 3-direction, if we perform a rotation by angle θ around the 3-axis,

$$e_1 \to e_1\cos\theta + e_2\sin\theta \ , \qquad e_2 \to -e_1\sin\theta + e_2\cos\theta \ ,$$

then the polarization vectors change by phases as follows:

$$e(\mathbf{q},\pm 1) \to e^{\mp i\theta} e(\mathbf{q},\pm 1) \ .$$

Since there is nothing special about the 3-direction, this is the effect of rotation by angle θ around the direction of motion for a photon moving in any direction. In accordance with the general discussion of angular momentum in Section 5.4, this means that a photon created by $a^\dagger(\mathbf{q},\lambda)$ has a component of angular momentum around the direction of motion, that is a *helicity*, equal to $\hbar\lambda$ in cgs units.

To calculate the free-field Hamiltonian, we first note that, since $A^0 = 0$ for free fields, the free-field Hamiltonian density is

$$\mathcal{H}_0 = \pi_j \dot{A}_j - \mathcal{L}_0 = \frac{1}{8\pi}\dot{A}_j\dot{A}_j + \frac{1}{16\pi}(\partial_i A_j - \partial_j A_i)(\partial_i A_j - \partial_j A_i) \ ,$$

where as usual i and j run over the values 1, 2, 3, and repeated indices are summed. Using integration by parts and the Coulomb gauge condition (7.5.11) we find the free-field Hamiltonian

$$H_0 = \int d^3x \, \mathcal{H}_0 = \frac{1}{8\pi}\int d^3x \left[\dot{A}_i\dot{A}_i + \partial_i A_j \partial_i A_j\right] \ .$$

Inserting the field (7.5.18) and following just the same steps as in calculating the free-field Hamiltonian for a scalar field in Section 7.2, we find the free-field Hamiltonian for electromagnetism

$$
\begin{aligned}
H_0 &= \frac{1}{2}\sum_\lambda \int d^3q|\mathbf{q}|(a^\dagger(\mathbf{q},\lambda)a(\mathbf{q},\lambda) + a(\mathbf{q},\lambda)a^\dagger(\mathbf{q},\lambda)) \\
&= \sum_\lambda \int d^3q|\mathbf{q}|a^\dagger(\mathbf{q},\lambda)a(\mathbf{q},\lambda) + E_{\rm vac} \,, \qquad (7.5.23)
\end{aligned}
$$

where

$$
E_{\rm vac} = \delta^3(0)\int d^3q|\mathbf{q}| = (2\pi)^{-3}\int d^3x \int d^3q|\mathbf{q}| \,. \qquad (7.5.24)
$$

As in the case of the real scalar field treated in Section 7.2, the vacuum $\Phi_{\rm vac}$, defined as the state of lowest energy, must satisfy the condition

$$
a(\mathbf{q},\lambda)\Phi_{\rm vac} = 0 \,, \qquad (7.5.25)
$$

since otherwise there would be a state $a(\mathbf{q},\lambda)\Phi_{\rm vac}$ with a lower energy than $\Phi_{\rm vac}$. Thus

$$
H_0\Phi_{\rm vac} = E_{\rm vac}\Phi_{\rm vac} \,. \qquad (7.5.26)
$$

The energy (7.5.24) is a contribution to the total vacuum energy that must be added to the contributions of all other fields, such as (7.2.16). The state consisting of a photon with momentum $\mathbf{q}_1$ and helicity λ_1, another photon with momentum $\mathbf{q}_2$ and helicity λ_2, and so on, may be expressed as

$$
\Phi_{\mathbf{q}_1,\lambda_1;\mathbf{q}_2,\lambda_2;\ldots} \propto a^\dagger(\mathbf{q}_1,\lambda_1)a^\dagger(\mathbf{q}_2,\lambda_2)\cdots\Phi_{\rm vac} \,, \qquad (7.5.27)
$$

and has energy $E_{\rm vac} + |\mathbf{q}_1| + |\mathbf{q}_2| + \cdots$. The term $E_{\rm vac}$ appears in the energy of all states, and so aside from gravitational phenomena may be ignored, as we shall do here.

Radiative Decay

We now consider the rate at which an excited atom[16] will drop into a state of lower energy, emitting a photon. We shall neglect relativistic effects and the interaction of the electromagnetic field with the electron spin, so that the Hamiltonian for the atom interacting with the electromagnetic field is given by a sum over the particles in the atom of terms of form (5.8.3). Since we are interested in the emission only of a single photon, the relevant interaction term is the part of this sum linear in **A**:

$$
V = -\sum_n \frac{e_n}{2m_n}[\mathbf{A}(\mathbf{X}_n)\cdot\mathbf{P}_n + \mathbf{P}_n\cdot\mathbf{A}(\mathbf{X}_n)] \,,
$$

where e_n and m_n are the charge and mass of the nth particle (electron or nucleus) while $\mathbf{X}_n$ and $\mathbf{P}_n$ are the position and momentum operators of the nth

[16] The calculations here of radiative decay rates apply to molecules as well as to atoms, but to avoid repeating "or molecules" again and again, I will just refer below to transitions in atoms.

particle and $\mathbf{A}(\mathbf{X})$ is the quantum vector potential in the Schrödinger picture. Because we are using Coulomb gauge, in which $\mathbf{A}$ satisfies Eq. (7.5.11), it makes no difference in what order we write the operators in V, and we can just as well write

$$V = -\sum_n \frac{e_n}{m_n} \mathbf{A}(\mathbf{X}_n) \cdot \mathbf{P}_n \, . \tag{7.5.28}$$

We take the initial and final states of the atom to be eigenstates $\Phi_{i,\mathbf{p}_i}$ and $\Phi_{f,\mathbf{p}_f}$ of the Hamiltonian of the atom, with energies E_i and E_f, respectively, and with total momenta $\mathbf{p}_i$ and $\mathbf{p}_f$, respectively. (Because atomic nuclei are heavy the kinetic energies of the states of the whole atom are always much less than $E_i - E_f$, and so will be neglected.) The atomic state vectors are assumed to be normalized so that

$$(\Phi_{a',\mathbf{p}'}\, , \, \Phi_{a,\mathbf{p}}) = \delta_{a'a}\delta^3(\mathbf{p}' - \mathbf{p}) \, . \tag{7.5.29}$$

Each of these states is a vacuum as far as photons are concerned, so, for any photon momentum $\mathbf{q}$ and helicity λ,

$$a(\mathbf{q}, \lambda)\, \Phi_{i,\mathbf{p}_i} = a(\mathbf{q}, \lambda)\, \Phi_{f,\mathbf{p}_f} = 0 \, . \tag{7.5.30}$$

The initial state of the radiative decay process is then $\Phi_{i,\mathbf{p}_i}$, and the final state is $a^\dagger(\mathbf{q}, \lambda)\Phi_{f,\mathbf{p}_f}$, with $\mathbf{q}$ and λ the momentum and helicity of the emitted photon. To first order in V we can treat $\mathbf{A}$ in Eq. (7.5.28) as a free field, so to this order the S-matrix element (5.6.36) for the decay process is

$$\begin{aligned}
&S[i(\mathbf{p}_i) \to f(\mathbf{p}_f) + \gamma(\mathbf{q}, \lambda)] \\
&\quad = -2\pi i \delta(E_f + |\mathbf{q}| - E_i)(a^\dagger(\mathbf{q}, \lambda)\Phi_{f,\mathbf{p}_f}\, , V\, \Phi_{i,\mathbf{p}_i}) \\
&\quad = -2\pi i \delta(E_f + |\mathbf{q}| - E_i)(\Phi_{f,\mathbf{p}_f}\, , a(\mathbf{q}, \lambda) V\, \Phi_{i,\mathbf{p}_i}) \\
&\quad = 2\pi i \delta(E_f + |\mathbf{q}| - E_i)\sqrt{4\pi} \sum_{\lambda'} \int \frac{d^3q'}{(2\pi)^{3/2}\sqrt{2|\mathbf{q}'|}} \\
&\qquad \times \sum_n \frac{e_n}{m_n} (\Phi_{f\mathbf{p}_f}, a(\mathbf{q}, \lambda)\mathbf{e}^*(\mathbf{q}', \lambda') \cdot \mathbf{P}_n e^{-i\mathbf{q}'\cdot\mathbf{X}_n} a^\dagger(\mathbf{q}', \lambda')\Phi_{i,\mathbf{p}_i}) \, .
\end{aligned} \tag{7.5.31}$$

Using the photon vacuum condition (7.5.30) and the commutation relation (7.5.21), we can replace the product $a(\mathbf{q}, \lambda)a^\dagger(\mathbf{q}', \lambda')$ with $\delta^3(\mathbf{q} - \mathbf{q}')\delta_{\lambda\lambda'}$, and do the integral over $\mathbf{q}'$ and the sum over λ' by just setting $\mathbf{q}' = \mathbf{q}$ and $\lambda' = \lambda$, so

$$\begin{aligned}
S[i(\mathbf{p}_i) \to f(\mathbf{p}_f) + \gamma(\mathbf{q}, \lambda)] &= \frac{2\pi i \sqrt{4\pi}\, \delta(E_f + |\mathbf{q}| - E_i)}{(2\pi)^{3/2}\sqrt{2|\mathbf{q}|}} \\
&\quad \times \sum_n \frac{e_n}{m_n} \left(\Phi_{f\mathbf{p}_f}\, , \, \mathbf{e}^*(\mathbf{q}, \lambda) \cdot \mathbf{P}_n e^{-i\mathbf{q}\cdot\mathbf{X}_n} \Phi_{i,\mathbf{p}_i}\right) .
\end{aligned} \tag{7.5.32}$$

At this point we make a further approximation, known as the *electric dipole approximation*. The wavelength $2\pi/|\mathbf{q}|$ of the emitted photon is typically at least hundreds or thousands of angstroms, while the mean separations of electrons from the center of mass of the atom are typically a few angstroms. It is therefore usually a good approximation (as long as selection rules to be discussed below do not require the result to vanish) to replace each particle position $\mathbf{X}_n$ in the exponent in Eq. (7.5.32) with the center-of-mass coordinate vector

$$\overline{\mathbf{X}} \equiv \frac{1}{M}\sum_n m_n \mathbf{X}_n \,, \qquad M \equiv \sum_n m_n \,. \tag{7.5.33}$$

Now, using the commutators of the momentum and position operators,

$$\left[\sum_n \mathbf{P}_n \,,\, \exp(i\mathbf{q}\cdot\overline{\mathbf{X}})\right] = \mathbf{q}\exp(i\mathbf{q}\cdot\overline{\mathbf{X}}) \,, \tag{7.5.34}$$

so[17]

$$\exp(i\mathbf{q}\cdot\overline{\mathbf{X}})\Phi_{f,\mathbf{p}_i} = \Phi_{f,\mathbf{p}_f+\mathbf{q}} \,. \tag{7.5.35}$$

Hence, replacing all $\mathbf{X}_n$ in the exponent in Eq. (7.5.32) with $\overline{\mathbf{X}}$, and letting the adjoint of this exponential act on the final state, we have

$$S\big[i(\mathbf{p}_i) \to f(\mathbf{p}_f) + \gamma(\mathbf{q},\lambda)\big] = \frac{-2\pi i\sqrt{4\pi}\,\delta(E_f + |\mathbf{q}| - E_i)}{(2\pi)^{3/2}\sqrt{2|\mathbf{q}|}} \times \sum_n \frac{e_n}{m_n}\big(\Phi_{f,\mathbf{p}_f+\mathbf{q}}, \mathbf{e}^*(\mathbf{q},\lambda)\cdot\mathbf{P}_n\Phi_{i,\mathbf{p}_i}\big) \,. \tag{7.5.36}$$

The operators $\mathbf{P}_n$ all commute with the total momentum, so we can write their matrix elements as

$$\big(\Phi_{f,\mathbf{p}_f+\mathbf{q}}, \mathbf{P}_n\Phi_{i,\mathbf{p}_i}\big) = \delta^3(\mathbf{p}_f - \mathbf{p}_i + \mathbf{q})(\mathbf{P}_n)_{fi} \tag{7.5.37}$$

and so

$$S\,[i(\mathbf{p}_i) \to f(\mathbf{p}_f) + \gamma(\mathbf{q},\lambda)] = -2\pi i\,\delta(E_f - E_i + |\mathbf{q}|)\delta^3(\mathbf{p}_f - \mathbf{p}_i + \mathbf{q}) \times M[i(\mathbf{p}_i) \to f(\mathbf{p}_f) + \gamma(\mathbf{q},\lambda)] \,, \tag{7.5.38}$$

[17] This argument does not rule out the possible presence of a numerical factor multiplying the right-hand side of Eq. (7.5.35). Any such factor of proportionality would have to have absolute magnitude unity, because $[\exp(i\mathbf{q}\cdot\overline{\mathbf{X}})]^\dagger\exp(i\mathbf{q}\cdot\overline{\mathbf{X}}) = 1$, and we define both $\Phi_{f,\mathbf{p}_f}$ and $\Phi_{f,\mathbf{p}_f+\mathbf{q}}$ to be normalized in accordance with Eq. (7.5.29). Such a phase factor would depend on our arbitrary choice of the phase of the state $\Phi_{f,\mathbf{p}_f}$ as a function of $\mathbf{p}_f$ and can be defined to be unity, but in any case it cannot affect the radiative transition rate, which is proportional to the absolute value squared of the matrix element for the transition. So this possible phase factor will be ignored here.

where

$$M\,[i(\mathbf{p}_i) \to f(\mathbf{p}_f) + \gamma(\mathbf{q}, \lambda)] = \frac{\sqrt{4\pi}}{(2\pi)^{3/2}\sqrt{2|\mathbf{q}|}} \sum_n \frac{e_n}{m_n} (\mathbf{P}_n)_{fi} \cdot \mathbf{e}^*(\mathbf{q}, \lambda) \,. \tag{7.5.39}$$

To see how this is calculated in wave mechanics, note for example that in hydrogen the initial and final atomic wave functions take the form

$$\psi_{i,\mathbf{p}_i}(\overline{\mathbf{x}}, \mathbf{x}) = \frac{\exp(i\mathbf{p}_i \cdot \overline{\mathbf{x}})}{(2\pi)^{3/2}} \psi_i(\mathbf{x}) \,,$$

$$\psi_{f,\mathbf{p}_f+\mathbf{q}}(\overline{\mathbf{x}}, \mathbf{x}) = \frac{\exp(i[\mathbf{p}_f + \mathbf{q}] \cdot \overline{\mathbf{x}})}{(2\pi)^{3/2}} \psi_f(\mathbf{x}) \,,$$

where $\mathbf{x}$ is the vector separation of the electron and proton and $\overline{\mathbf{x}}$ is the coordinate vector of the center of mass. With $m_e \ll m_p$, the matrix element of the electron momentum operator $\mathbf{P}_e$ is

$$\left(\Phi_{f,\mathbf{p}_f+\mathbf{q}}, \mathbf{P}_e \Phi_{i,\mathbf{p}_i}\right) = \int d^3\mathbf{x} \int d^3\overline{\mathbf{x}}\; \psi^*_{f,\mathbf{p}_f+\mathbf{q}}(\overline{\mathbf{x}}, \mathbf{x}) \mathbf{P}_e \psi_{i,\mathbf{p}_i}(\overline{\mathbf{x}}, \mathbf{x}) \,,$$

which has the same form as Eq. (7.5.37), with

$$(\mathbf{P}_e)_{fi} = -i \int d^3x\; \psi^*_f(\mathbf{x}) \boldsymbol{\nabla} \psi_i(\mathbf{x}) \,.$$

Using Eq. (7.5.39) in Eq. (5.6.45) (with the number N_α of particles in the initial state equal to one), the differential decay rate is

$$\begin{aligned} d\Gamma(i \to f + \mathbf{q}, \lambda) &= 2\pi \left| M[i(\mathbf{p}_i) \to f(\mathbf{p}_f) + \gamma(\mathbf{q}, \lambda)] \right|^2 \\ &\quad \times \delta(E_f + |\mathbf{q}| - E_i)\delta^3(\mathbf{p}_f + \mathbf{q} - \mathbf{p}_i)\, d^3q\, d^3p_f \\ &= \frac{1}{2\pi|\mathbf{q}|} \left| \sum_n \frac{e_n}{m_n} (\mathbf{P}_n)_{fi} \cdot e^*(\mathbf{q}, \lambda) \right|^2 \\ &\quad \times \delta(E_f + |\mathbf{q}| - E_i)\delta^3(\mathbf{p}_f + \mathbf{q} - \mathbf{p}_i) d^3q d^3p_f \,. \end{aligned} \tag{7.5.40}$$

The momentum-conservation delta function just goes to fix the recoil momentum $\mathbf{p}_f = \mathbf{p}_i - \mathbf{q}$, and the energy-conservation delta function fixes the photon energy $|\mathbf{q}| = E_i - E_f$. Writing $d^3q = |\mathbf{q}|^2\, d|\mathbf{q}| d\Omega_\gamma$, we are left with the rate for emission of a photon with helicity λ into a small solid angle $d\Omega_\gamma$:

$$d\Gamma(i \to f + \mathbf{q}, \lambda) = \frac{|q|}{2\pi} \left| \sum_n \frac{e_n}{m_n} (\mathbf{P}_n)_{fi} \cdot e^*(\mathbf{q}, \lambda) \right|^2 d\Omega_\gamma \,. \tag{7.5.41}$$

In the common case where the photon helicity is not measured, the observed rate is given by a sum over helicities. This sum can be calculated using Eq. (7.5.20):

$$\sum_\lambda e_j(\mathbf{q},\lambda)e_k^*(\mathbf{q},\lambda) = \delta_{jk} - \hat{q}_j\hat{q}_k ,$$

where $\hat{q}$ is the unit vector $\mathbf{q}/|\mathbf{q}|$. The observed differential decay rate for emission of a photon with momentum $\mathbf{q}$ into a small solid angle $d\Omega_\gamma$ is

$$\sum_\lambda d\Gamma(i \to f + \mathbf{q},\lambda)$$
$$= \frac{|q|}{2\pi}\left(\sum_n \frac{e_n}{m_n}(P_{nj})_{fi}\right)^*\left(\sum_n \frac{e_n}{m_n}(P_{nk})_{fi}\right)(\delta_{jk} - \hat{q}_j\hat{q}_k)d\Omega_\gamma . \quad (7.5.42)$$

We can now easily integrate over the photon direction, using

$$\int d\Omega_\gamma\left[\delta_{jk} - \hat{q}_j\hat{q}_k\right] = 4\pi\,\delta_{jk}\left[1 - \frac{1}{3}\right] = \frac{8\pi}{3}\delta_{jk} .$$

The total decay rate for emission of a photon in any direction with any helicity is then, in Einstein's notation,

$$A_i^f = \int d\Omega_\gamma \sum_\lambda d\Gamma(i \to f + \mathbf{q},\lambda) = \frac{4|q|}{3}\left|\sum_n \frac{e_n}{m_n}(\mathbf{P}_n)_{fi}\right|^2 . \quad (7.5.43)$$

Section 3.5 shows how to use this also to calculate the rates of absorption and stimulated emission of radiation.

Calculations are made easier if we replace matrix elements of momentum vectors with matrix elements of position vectors. For this, we use the commutator

$$[\mathbf{X}_n, H] = \left[\mathbf{X}_n, \sum_{n'} \frac{1}{2m_{n'}}\mathbf{P}_{n'}^2\right] = \frac{i}{m_n}\mathbf{P}_n ,$$

so

$$(\Phi_{f,\mathbf{p}_f+\mathbf{q}} , \mathbf{P}_n\Phi_{i,\mathbf{p}_i}) = -i(E_i - E_f)m_n(\Phi_{f,\mathbf{p}_f+\mathbf{q}} , \mathbf{X}_n\Phi_{i,\mathbf{p}_i})$$
$$= -i|\mathbf{q}|m_n(\Phi_{f,\mathbf{p}_f+\mathbf{q}} , \mathbf{X}_n\Phi_{i,\mathbf{p}_i}) .$$

Therefore the decay rate (7.5.43) may be written

$$A_i^f = \frac{4|q|^3}{3}\left|\sum_n e_n(\mathbf{X}_n)_{fi}\right|^2 , \quad (7.5.44)$$

where

$$(\Phi_{f,\mathbf{p}_f+\mathbf{q}}, \mathbf{X}_n\Phi_{i,\mathbf{p}_i}) = \delta^3(\mathbf{p}_f - \mathbf{p}_i + \mathbf{q})(\mathbf{X}_n)_{fi} . \quad (7.5.45)$$

In cgs units, Eq. (7.5.44) takes the form

$$A_i^f = \frac{4|\omega|^3}{3\hbar c^3}\left|\sum_n e_n(\mathbf{X}_n)_{fi}\right|^2 ,$$

where $\omega = |\mathbf{q}|c/\hbar$ is the photon circular frequency. This formula was guessed by Heisenberg[18] in 1925 by setting the radiation power emitted in the transition $i \to f$ that had been calculated in classical electrodynamics[19] equal to $\hbar\omega A_i^f$. He used this formula as a starting point in his matrix mechanics approach to quantum mechanics. The quantum-mechanical derivation was first given by Dirac[20] in 1927.

Selection Rules

As we have already warned, the electric dipole approximation is not useful if selection rules give zero for the decay matrix element. We can derive the selection rules from either Eq. (7.5.43) or Eq. (7.5.44). First, as shown in Section 5.2, the components of the operator $\mathbf{X}$ can be assembled into the spherical harmonics for $\ell = 1$,

$$\mp\sqrt{\frac{3}{8\pi}}(X_1 \pm iX_2) = |\mathbf{X}|Y_1^{\pm 1}(\mathbf{X}/|\mathbf{X}|) , \quad \sqrt{\frac{3}{4\pi}}X_3 = |\mathbf{X}|Y_1^0(\mathbf{X}/|\mathbf{X}|) .$$

According to the rules for addition of angular momenta set out in Section 5.4, if the initial atom at rest has total angular momentum quantum number j_i, then the states $X_k\Phi_{i,\mathbf{p}_i}$ for $\mathbf{p}_i = 0$ can only have total angular momentum quantum number j_f equal to $j_i + 1$, j_i, or $j_i - 1$ and, furthermore, if $j_i = 0$ then only $j_f = 1$ is possible; $j_f = 0$ is only possible if $j_i = 1$. Hence radiative decay does not occur in the electric dipole approximation unless the initial and final atomic states satisfy the selection rule

$$|j_f - j_i| \leq 1 \leq j_i + j_f . \qquad (7.5.46)$$

There is a further selection rule that follows from space inversion symmetry. As we saw in Section 5.4, if we change the sign of each of the three Cartesian coordinates, any state vector Φ is changed to $\Pi\Phi$, where the operator Π is unitary in the sense that $\Pi^\dagger\Pi = 1$, and, since making two space inversions in succession changes nothing, also $\Pi^2 = 1$. Physical states therefore can be chosen as eigenstates of Π with eigenvalue, known as the *parity* of the state, equal to $+1$ or -1. The coordinate vector is obviously odd under space inversion, so

---

[18] W. Heisenberg, Z. Physik **33**, 879 (1925); reprinted in English in Van der Waerden, *Sources of Quantum Mechanics*, listed in the bibliography.

[19] J. Larmor, Phil. Mag. S.5 **44**, 503 (1897).

[20] P. A. M. Dirac, Proc. Roy. Soc. A **114**, 710 (1927).

$\Pi \mathbf{X}_n \Pi = -\mathbf{X}_n$. Hence if the initial and final atomic states have parity π_i and π_f, the transition rate will vanish in the electric dipole approximation unless the initial and final parities satisfy the selection rule

$$\pi_f = -\pi_i \,. \tag{7.5.47}$$

For instance, in hydrogen the transition $2p \to 1s$ (ignoring spin) has $j_i = 1$, $\pi_i = -$, $j_f = 0$, $\pi_f = +$, so it satisfies the selection rules (7.5.46) and (7.5.47) and is therefore predominantly an electric dipole transition. This is the Lyman alpha ultraviolet transition. On the other hand, in the electric dipole approximation the $2s$, $3s$, and $3d$ states are forbidden by both selection rules from decaying into the $1s$ ground state.

Of course the electric dipole approximation is just an approximation. Instead of simply replacing the coordinates $\mathbf{X}_n$ in the exponent in Eq. (7.5.32) with the center-of-mass coordinate vector $\overline{\mathbf{X}}$, we can expand the exponential in powers of the small quantity $\mathbf{q}\cdot[\mathbf{X}_n - \overline{\mathbf{X}}]$. With one factor of this quantity, the operator in the matrix element involves two factors of coordinates, which can be assembled into the spherical harmonics Y_2^m and Y_1^m, which are respectively known as *electric quadrupole* and *magnetic dipole* terms. With two factors of coordinates, these operators are even under space inversion, so these contributions to the matrix element vanish unless the initial and final states satisfy the selection rules

$$|j_i - j_f| \leq 2 \leq j_i + j_f \,, \qquad \pi_i = \pi_f \qquad \text{electric quadrupole} \tag{7.5.48}$$

$$|j_i - j_f| \leq 1 \leq j_i + j_f \,, \qquad \pi_i = \pi_f \qquad \text{magnetic dipole}\,. \tag{7.5.49}$$

For instance, in hydrogen the transition $3d \to 1s$ occurs as an electric quadrupole transition. The rates of both electric quadrupole and magnetic dipole transitions are suppressed relative to electric dipole transitions by factors of order $(qr/\hbar)^2$, where r is a characteristic atomic radius; for optical transitions, this is of order 10^{-7}.

We can go on, including higher and higher powers of $\mathbf{X}_n - \overline{\mathbf{X}}$ in the expansion of the exponential in Eq. (7.5.32), and also including effects of electron spin. But whatever effects we include, there is one kind of transition in which single-photon emission is completely forbidden: transitions between states that both have total angular momentum zero. This is a simple consequence of angular momentum conservation. As we have seen, a photon of helicity ± 1 has an angular momentum component in the direction of motion ± 1, and therefore cannot be emitted in a transition between states that have zero total angular momentum. For instance, none of the excited states of ^{12}C or ^{16}O with $j = 0$ can emit a single photon in gamma decay to the $j = 0$ ground state. Such transitions require the emission of pairs of photons, or if enough energy is available, of electron–positron pairs.

Gauge Invariance and Charge Conservation

It is essential both for the consistency of the Maxwell equations and for gauge invariance that the current four-vector $J^\mu(x)$ defined by Eq. (7.5.7) should be conserved. In modern theories the matter with which electromagnetic fields interact is described by a field theory, so we need to ask, in what sort of field theory for matter is this current conserved? There is a simple answer. If charge is to be conserved, then the net charge destroyed by the product of fields in any term in the Lagrangian density must vanish, so if each field φ_n destroys a charge e_n and creates a charge $-e_n$, then

$$\sum_n e_n \left[\frac{\partial \mathcal{L}_{\text{mat}}}{\partial \varphi_n} \varphi_n + \frac{\partial \mathcal{L}_{\text{mat}}}{\partial(\partial \varphi_n/\partial x^\mu)} \frac{\partial \varphi_n}{\partial x^\mu} \right] = 0 \, , \tag{7.5.50}$$

with summation of course understood over the repeated spacetime index μ. This is the same as saying that the Lagrangian density is invariant under the phase transformation

$$\varphi_n \to [1 + i\epsilon e_n]\varphi_n \, , \tag{7.5.51}$$

with ϵ an arbitrary infinitesimal. Using the Euler–Lagrange equation (7.1.8) allows us to write Eq. (7.5.50) as a conservation equation

$$\frac{\partial J^\mu}{\partial x^\mu} = 0 \, , \tag{7.5.52}$$

where

$$J^\mu = -i \sum_n e_n \frac{\partial \mathcal{L}_{\text{mat}}}{\partial(\partial \varphi_n/\partial x^\mu)} \varphi_n \, . \tag{7.5.53}$$

(This is an example of the relation between symmetry principles and conservation laws first expressed in the Noether theorem discussed in Section 5.7.) The factor $-i$ is inserted here so that

$$J^0 = -i \sum_n e_n \pi_n \varphi_n \, , \tag{7.5.54}$$

and therefore

$$\left[\int J^0 d^3x \, , \, \varphi_n \right] = -e_n \varphi_n \, , \tag{7.5.55}$$

which tells us that e_n is indeed the value of the charge $\int J^0 d^3x$ that is destroyed by the field φ_n.

For the vector potential to interact with this conserved current, in the sense of Eq. (7.5.7), it is sufficient to arrange that $\partial \varphi_n/\partial x^\mu$ and A_μ always occur in the matter Lagrangian density in the combination

$$D_\mu \varphi_n \equiv \left[\frac{\partial}{\partial x^\mu} - ie_n A_\mu\right]\varphi_n \tag{7.5.56}$$

so that

$$\frac{\partial \mathcal{L}_{\text{mat}}}{\partial A_\mu} = \sum_n -ie_n \frac{\partial \mathcal{L}_{\text{mat}}}{\partial(\partial\varphi_n/\partial x^\mu)} = J^\mu \,. \tag{7.5.57}$$

From this, Eq. (7.5.7) follows immediately.

For instance, we can use this prescription to include electromagnetic interactions in the Lagrangian density (7.4.5) for a complex scalar field φ that destroys charge e:

$$\mathcal{L}_{\text{mat}} = -\eta^{\mu\nu}\left[\frac{\partial\varphi}{\partial x^\mu} - ieA_\mu\varphi\right]^\dagger\left[\frac{\partial\varphi}{\partial x^\nu} - ieA_\nu\varphi\right] .$$

Also, we used this prescription in the previous section to include electromagnetic interactions in the Lagrangian density for Dirac fields.

There is a more general possibility, that in addition to depending on the φ_n and $D_\mu\varphi_n$, the matter Lagrangian density may also depend on the gauge-invariant field-strength tensor $F_{\mu\nu}$. In this case, there is an additional term in the current defined by Eq. (7.5.7):

$$J^\mu = -i\sum_n e_n \frac{\partial \mathcal{L}_{\text{mat}}}{\partial(\partial\varphi_n/\partial x^\mu)}\varphi_n + \frac{\partial}{\partial x^\nu}\frac{\partial \mathcal{L}_{\text{mat}}}{\partial F_{\mu\nu}} \,. \tag{7.5.58}$$

Because of the antisymmetry of $F_{\mu\nu}$, the new term in J^μ is separately conserved. The possibility of this new term alerts us that the general principles of electrodynamics do not in themselves fully dictate the parameters in the Lagrangian that characterize the interaction of matter and radiation, including the magnetic moments of various particles.

Local Phase and Matrix Transformations

These prescriptions can be framed as consequences of an extended version of gauge invariance. We have already noted that the condition (7.5.50) of charge conservation is equivalent to the invariance of the Lagrangian density under the phase transformation (7.5.51). But if $\partial\varphi_n/\partial x^\mu$ only appears in the Lagrangian in the form (7.5.56), then the Lagrangian is invariant under a *local* phase transformation, with ϵ an arbitrary infinitesimal function of spacetime coordinates,

$$\varphi_n(x) \to [1 + i\epsilon(x)e_n]\varphi_n(x) \,, \tag{7.5.59}$$

provided that the vector potential at the same time undergoes the gauge transformation

$$A_\mu(x) \to A_\mu(x) + \frac{\partial\epsilon(x)}{\partial x^\mu} \,. \tag{7.5.60}$$

Today this reasoning is often run in reverse. It is assumed that the Lagrangian density is invariant under the local phase transformation (7.5.59), with $\epsilon(x)$ an arbitrary infinitesimal function of spacetime coordinates, and from this the existence is deduced of a vector field $A_\mu(x)$ whose properties are governed by invariance under the gauge transformation (7.5.60).

Indeed, our Standard Model of elementary particles and forces is based on an assumed invariance under a larger group of local transformations, not just by x-dependent phases, as in Eq. (7.5.59), but transformations by x-dependent matrices similar to those for isotopic spin rotations. From this, one deduces the existence of a number of photon-like particles: some, the gluons with zero mass, whose strong interactions prevent them from being observed in isolation, and others that are observed, the $W^\pm$ and Z^0, that become massive as a result of a spontaneous breakdown of the local gauge symmetry. But these matters are beyond the scope of this book.

Assorted Problems

1. Suppose that in a diatomic gas such as H_2, the vibrational degrees of freedom are fully excited, along with the rotational and translational degrees of freedom. What is the ratio of the energy density of the gas to its pressure? What does this tell you about the speed of sound in the gas?

2. Suppose that Einstein in 1905 had assumed that, in the radiation at temperature T in a cubical enclosure, the number n of photons for each wave number and polarization is not any positive integer but can only be $n = 0$, $n = 1$, or $n = 2$. What would he have found for the energy density $\mathcal{E}(\nu, T)$ per unit frequency interval at frequency ν and temperature T?

3. Suppose that in the 1910 experiment that revealed the existence of the nucleus of the atom, the nucleus had been moving toward the radon alpha ray source with speed v_0. What could one conclude about the mass of the nucleus from the observation that alpha particles are sometimes scattered straight backwards from the atom?

4. Suppose that the potential energy of an electron in the field of a nucleus is not $-Ze/r$ but rather $V(r) = -gr^{-\eta}$, where g and η are positive-definite constants, but that Bohr's quantization condition $m_e v_n r_n = n\hbar$ is still valid, with $\hbar$ some constant and n running over all positive-definite integers.

   - What would Bohr in 1913 have found for the radii r_n, velocities v_n, and energies E_n ?
   - For what values of η do circular orbits exist that have $E_n < 0$?
   - What would be found for the relation between $\hbar$ and h if one imposed Bohr's correspondence principle on the orbits with $n \gg 1$?

5. How does the pressure in a non-relativistic ideal gas vary when the mass density varies adiabatically, assuming that the internal energy density is either

- much bigger than the pressure, or
- equal to the pressure, or
- one percent of the pressure?

6. It is summer in Texas. The temperature outside is 104 °F (40 °C). In order to keep the inside of your house at a comfortable 68 °F (20 °C) you need to take 10^4 joules per second of heat energy from inside to outside. For this purpose you use an inverse Carnot cycle. How much power will you need to run this?

7. In the electrolysis of water, how long does it take a 1 ampere current to produce 1 gram of oxygen gas?

8. Fifteen grams of element X combine with three grams of hydrogen to produce 18 grams of a compound Y of element X and hydrogen, with nothing left over. Also, as gases all at the same temperature and pressure, 2 liters of element X combine with 3 liters of hydrogen (H_2) to give 2 liters of compound Y. What is the chemical formula for compound Y, and what is the atomic weight of element X? (Take the atomic weight of hydrogen atoms as 1, and assume the validity of Avogadro's principle.)

9. Consider a particle of mass m and velocity $\mathbf{v}$, with $|\mathbf{v}| \ll c$. Find the term in the energy of this particle of order mv^4/c^2, and the term in the momentum of order mv^3/c^2.

10. Suppose an observer who uses coordinates x^μ sees a uniform magnetic field of magnitude B_1, pointing in the 1-direction, and zero electric field. A second observer uses coordinates $x'^\mu = \Lambda^\mu{}_\nu x^\nu$, where $\Lambda^\mu{}_\nu$ is the Lorentz transformation (4.2.6) that gives a body at rest a velocity with magnitude v in the 3-direction. What are the values of the components of the electric and magnetic fields seen by the second observer?

11. In a spacetime with two space dimensions and one time dimension, the electromagnetic field consists of a two-component electric field $\mathbf{E}$ and a one-component magnetic field B. They satisfy differential equations

$$4\pi J_1 = -\frac{\partial E_1}{\partial t} + c\frac{\partial B}{\partial x_2}\,, \qquad 4\pi J_2 = -\frac{\partial E_2}{\partial t} - c\frac{\partial B}{\partial x_1}\,,$$

$$4\pi\rho = \frac{\partial E_1}{\partial x_1} + \frac{\partial E_2}{\partial x_2}\,, \qquad 0 = \frac{\partial E_2}{\partial x_1} - \frac{\partial E_1}{\partial x_2} + \frac{1}{c}\frac{\partial B}{\partial t}\,.$$

Find what kind of transformation properties, under (2 + 1)-dimensional Lorentz transformations, we can give the field components E_1, E_2, and B and the densities J_1, J_2, and ρ so that the above equations are Lorentz invariant, in the sense that they are invariant under linear transformations on spacetime intervals that leave $(\Delta x^1)^2 + (\Delta x^2)^2 - c^2(\Delta t)^2$ invariant.

Show that with the fields and densities transforming in this way, these equations really are Lorentz invariant.

12. A particle known as a K meson, with mass 494 MeV/c^2, decays at rest into a muon, with mass 106 MeV/c^2, and a neutrino, with negligible mass. Use the conservation of energy and momentum to find the velocity of the muon.

13. What second-order partial differential equation (second-order in both time and space derivatives) is satisfied by the de Broglie wave function for a free particle when we do *not* assume that its velocity is much less than c?

14. When a beam of electrons of some definite energy is directed at a perfect crystal, it is found that the largest angle θ between the incident and reflected waves at which reflection is enhanced by constructive interference is $150°$. At what other value or values of θ is reflection enhanced by constructive interference?

15. Suppose we measure the position of the electron in the lowest-energy state of a hydrogen atom. What is the probability of finding that the electron is farther than 10^{-8} cm from the nucleus?

16. Consider an electron in a $d_{3/2}$ state with orbital angular momentum quantum number $\ell = 2$, total angular momentum quantum number $j = 3/2$, and total angular moment 3-component $J_3 = \hbar/2$. Suppose we measure the 3-component S_3 of the spin. What are the probabilities of getting the results $S_3 = \hbar/2$ and $S_3 = -\hbar/2$? (Calculate whatever Clebsch–Gordan coefficients you need – do not just look them up in a table.)

17. Suppose the electron has spin 3/2 rather than 1/2, but that all other properties of electrons and nuclei are as they are in the real world. What would you expect would be the atomic numbers Z of the two lightest halogen elements, that behave like fluorine and chlorine in our world?

18. When a free electron is placed in a uniform magnetic field $\mathbf{B}$ pointing in the 1-direction, the Hamiltonian becomes

$$H = \frac{\mathbf{p}^2}{2m_e} + \mu|\mathbf{B}|S_1$$

where $\mathbf{S}$ is the operator representing the electron spin vector and μ is a constant, related to the electron magnetic moment. Suppose that at $t = 0$ the expectation value of the spin vector has components

$$\langle S_1 \rangle = \langle S_2 \rangle = 0\,, \quad \langle S_3 \rangle = \hbar/2\,.$$

What are the expectation values of the spin vector components at any later time?

19. Suppose that the interaction of the electron in a hydrogen atom with some sort of external field produces a term in the potential

$$\Delta V(r) = gr ,$$

where g is a small constant. Calculate the terms in the resulting shift in the energy of the $1s$ state that are of first and second order in g.

20. Suppose that the spin–orbit coupling of the electron in hydrogen produces a term in the Hamiltonian

$$\Delta H = \xi \mathbf{L} \cdot \mathbf{S}$$

where ξ is a constant, and $\mathbf{L}$ and $\mathbf{S}$ are the orbital angular momentum and spin angular momentum of the electron. What does this term contribute to the fine-structure splitting between the $2p_{1/2}$ and $2p_{3/2}$ states of hydrogen?

21. Consider the scattering of a spinless particle of mass m and momentum p by a central potential

$$V(r) = V_0 \exp\left(-r^3/R^3\right)$$

where V_0 and R are constants. Use the Born approximation to give a formula for the scattering amplitude in the limit $pR \ll \hbar$.

22. Consider a one-particle system with a Lagrangian

$$L = \frac{m}{2}\left(\frac{d\mathbf{X}}{dt}\right)^2 + \left(\frac{d\mathbf{X}}{dt}\right) \cdot \mathbf{V}(\mathbf{X}) ,$$

where $\mathbf{V}$ is some vector function of position $\mathbf{X}$.

- What equation of motion is satisfied by $\mathbf{X}$?
- Find the Hamiltonian of this theory.
- What is the differential equation satisfied by the wave function $\psi(\mathbf{x})$ of a state with a definite energy E?

23. Consider a particle of charge e and mass m in classical electromagnetic potentials that depend on time as well as position, with Hamiltonian

$$H(\mathbf{X}, \mathbf{P}) = \frac{1}{2m}\left[\mathbf{P} - \frac{e}{c}\mathbf{A}(\mathbf{X}, t)\right]^2 - e\phi(\mathbf{X}, t) .$$

Suppose you perform a time-dependent and position-dependent gauge transformation, to new potentials

$$\mathbf{A}^{\#} = \mathbf{A} + \nabla\xi , \quad \phi^{\#} = \phi - \frac{1}{c}\frac{\partial \xi}{\partial t} ,$$

where ξ is an arbitrary real function of position and time. What is the relation between the wave function $\psi^{\#}(\mathbf{x}, t)$ that satisfies the time-dependent

Schrödinger equation for the new potentials and the wave function $\psi(\mathbf{x}, t)$ that satisfies the time-dependent Schrödinger equation for the original potentials?

24. Find the coordinate-space wave function of the one-particle state with angular momentum $\ell = 0$ and energy $V_0 + 2\hbar\omega$ in the harmonic oscillator potential (6.3.1).

25. Suppose that the potential felt by an alpha particle for radius r outside the nuclear radius R is not the Coulomb potential, but instead $V(r) = g/r^2$, where g is some positive constant. Calculate the exponential suppression factor in the rate of decay of an unstable alpha particle state with energy $E \ll g/R^2$.

26. Consider the theory of a neutral spinless particle A and a non-neutral spinless particle B, with Lagrangian density

$$\mathcal{L} = -\frac{1}{2}\eta_{\mu\nu}\frac{\partial\varphi_A}{\partial x^\mu}\frac{\partial\varphi_A}{\partial x^\nu} - \frac{m_A^2}{2}\varphi_A^2 - \eta_{\mu\nu}\frac{\partial\varphi_B^\dagger}{\partial x^\mu}\frac{\partial\varphi_B}{\partial x^\nu} - m_B^2\varphi_B^\dagger\varphi_B$$
$$- g\varphi_A\varphi_B^\dagger\varphi_B\,.$$

Calculate the S-matrix elements for the processes $A + B \to A + B$ and $B + \overline{B} \to A + A$ to lowest order in g, where $\overline{B}$ is the antiparticle of B.

27. Calculate the rate for emission of a photon in the transition $2p \to 1s$ in hydrogen. Derive formulas and use them to find numerical values. You can use the facts that the proton is much heavier than the electron, and that the wavelength of the photon emitted in this process is much larger than the atomic size, and you can neglect electron spin.

28. What powers of the photon wave number appear in the rates for single-photon emission in the decays of the $4f$ state of hydrogen into the $2s$ and $2p$ states?

Bibliography

R. T. Beyer, *Foundations of Nuclear Physics* (Dover Publications, New York, 1949) [Chapters 3, 6]. Facsimiles of 19 early papers on nuclear physics, including papers of Rutherford from 1911 and 1919.

Max Born, *Atomic Physics* (Blackie & Sons, London, 1937; 6th edn. Hafner Publishing, New York, 1956) [Chapters 2, 3, 5]. An excellent survey of quantum theory and its applications, from a founder of the theory, with 39 useful mathematical appendices.

Stephen G. Brush, *The Kinetic Theory of Gases – An Anthology of Classic Papers with Historical Commentary* (Imperial College Press, London, 2003) [Chapters 1 and 2]. This is an invaluable collection of original papers, including work by Boyle, Newton, Bernoulli, Joule, Clausius, Maxwell, and Boltzmann cited in the text, along with reprints of interesting historical discussions by Brush.

J. Chadwick, ed., *The Collected Papers of Lord Rutherford of Nelson* (Interscience, 1963) [Chapters 3, 6].

P. A. M. Dirac, *Principles of Quantum Mechanics* (Clarendon Press, Oxford, 1930; 4th edn. 1958) [Chapter 5]. Long the leading treatise on quantum mechanics.

Enrico Fermi, *Thermodynamics* (Prentice Hall Co., New York, 1937; reprinted by Dover Press in 1956) [Chapter 2]. This is a masterpiece of scientific exposition, based on lectures that Fermi gave at Columbia University in 1936.

G. Holton, Am. J. Phys. **28**, 627 (1960) [Chapter 4]. Insightful assessment of contributions of Einstein, Lorentz, and Poincaré to special relativity theory.

A. J. Ihde, *The Development of Modern Chemistry* (Harper & Row, 1964) [Chapter 1].

Martin J. Klein, ed., *Letters on Wave Mechanics* (Philosophical Library, New York, 1967) [Chapter 5]. Correspondence among Einstein, Lorentz, Planck, and Schrödinger, translated into English with useful commentary by Klein.

Thomas S. Kuhn, *Black-Body Theory and the Quantum Discontinuity, 1894–1912* (Oxford University Press, New York, 1978) [Chapter 3]. This is a detailed analysis of the work of Planck and Einstein on black-body radiation.

L. D. Landau and E. M. Lifshitz, *Fluid Mechanics* (Pergamon Press, London, 1959) [Chapter 2]. This is the classic text on many aspects of fluid mechanics, including the hydrodynamics of viscous fluids. It is translated from the Russian by J. B. Sykes and W. H. Reid.

Arthur I. Miller, *Albert Einstein's Special Theory of Relativity: Emergence (1905) and Early Interpretation (1905–1911)* (Addison-Wesley, Reading, MA, 1981) [Chapter 4]. Detailed analysis of the founding of special relativity.

M. J. Nye, *The Question of the Atom – From the First Karlsruhe Conference to the First Solvay Conference* (Tomash Publishers, Los Angeles, 1984) [Chapters 2 and 3]. English language version of research reports from 1860 to 1911, including papers by Boltzmann, Einstein, Mendeleev, Perrin, Rutherford, J. J. Thomson.

Jean Perrin, *Brownian Motion and Molecular Reality* (Taylor and Francis, London, 1910) [Chapter 2]. This is the translation from the French by F. Soddy of Perrin's review of his experiments on diffusion, published in September 1909 in the Annales de Chemie et de Physique, 8th Series.

Wayne Saslaw, "A History of Thermodynamics: The Missing Manual," Entropy **22**, 27 (2020).

J. F. Shearer and W. M. Deans, *Collected Papers on Wave Mechanics* (Blackie and Son, London, 1928) [Chapter 5]. Papers of Schrödinger and others, translated into English.

John Stachel, ed., *Einstein's Miraculous Year* (Princeton University Press, Princeton, NJ, 1998) [Chapters 2, 3, 4]. This is an invaluable collection of Einstein's papers from about 1905 on Brownian motion, special relativity, and the photon.

B. L. Van der Waerden, *Sources of Quantum Mechanics* (North-Holland Publishing, Amsterdam, 1967; reprinted by Dover Publications, New York, 1968) [Chapters 3, 5]. Papers on early quantum theory and matrix mechanics by Bohr, Dirac, Einstein, Heisenberg, Pauli, and others, all in English or English translation.

Steven Weinberg, *The Discovery of Subatomic Particles* (Scientific American Library, 1983; revised edn. Cambridge University Press, Cambridge, UK, 2003) [Chapters 1, 3, 6]. This is a non-mathematical historical account of the discoveries of the electron, proton, neutron, photon, etc., going back to the beginnings of chemistry, with algebra-based technical appendices.

Steven Weinberg, *Lectures on Quantum Mechanics* (Cambridge University Press, Cambridge, UK, 2012; 2nd edn. 2015) [Chapters 3, 5, 6, and 7]. This is a graduate-level introduction to quantum mechanics, with some historical discussion of the early quantum theory.

Steven Weinberg, "Half a Century of the Standard Model," Phys. Rev. Lett. **121**, 220001 (2018) [Chapters 6, 7]. A historical review of the Standard Model.

L. Pearce Williams, *Relativity Theory: Its Origins and Impact on Modern Thought* (John Wiley and Sons, New York, 1968) [Chapter 4]. Contains an 1887 article by Michelson and Morley and text of a 1904 talk by Poincaré.

Author Index

Subject Index